Power Systems

For further volumes:
http://www.springer.com/series/4622

Siddhartha Kumar Khaitan and Anshul Gupta (Eds.)

High Performance Computing in Power and Energy Systems

Editors
Dr. Siddhartha Kumar Khaitan
Iowa State University
Ames, IA
USA

Dr. Anshul Gupta
IBM T.J. Watson Research Center
Yorktown Heights
NY
USA

ISSN 1612-1287 e-ISSN 1860-4676
ISBN 978-3-642-32682-0 e-ISBN 978-3-642-32683-7
DOI 10.1007/978-3-642-32683-7
Springer Heidelberg New York Dordrecht London

Library of Congress Control Number: 2012944446

© Springer-Verlag Berlin Heidelberg 2013

This work is subject to copyright. All rights are reserved by the Publisher, whether the whole or part of the material is concerned, specifically the rights of translation, reprinting, reuse of illustrations, recitation, broadcasting, reproduction on microfilms or in any other physical way, and transmission or information storage and retrieval, electronic adaptation, computer software, or by similar or dissimilar methodology now known or hereafter developed. Exempted from this legal reservation are brief excerpts in connection with reviews or scholarly analysis or material supplied specifically for the purpose of being entered and executed on a computer system, for exclusive use by the purchaser of the work. Duplication of this publication or parts thereof is permitted only under the provisions of the Copyright Law of the Publisher's location, in its current version, and permission for use must always be obtained from Springer. Permissions for use may be obtained through RightsLink at the Copyright Clearance Center. Violations are liable to prosecution under the respective Copyright Law.

The use of general descriptive names, registered names, trademarks, service marks, etc. in this publication does not imply, even in the absence of a specific statement, that such names are exempt from the relevant protective laws and regulations and therefore free for general use.

While the advice and information in this book are believed to be true and accurate at the date of publication, neither the authors nor the editors nor the publisher can accept any legal responsibility for any errors or omissions that may be made. The publisher makes no warranty, express or implied, with respect to the material contained herein.

Printed on acid-free paper

Springer is part of Springer Science+Business Media (www.springer.com)

Preface

High performance computing (HPC) offers significant potential in critical infrastructure applications areas like power and energy systems. Recent years have seen a dramatic increase in the performance, capabilities, and usability of HPC platforms, along with the introduction of several supporting tools, libraries, compilers, and language extensions, etc. HPC, through the use of parallelism, can deliver orders of magnitude performance gains over traditional sequential approaches, making many of the applications computationally amenable that were previously considered infeasible due to their complexity, computational intensive nature, and large execution times. HPC is already seeing widespread use in applications such as energy systems, image processing, weather prediction, fluid dynamics, power systems, bioinformatics, aerospace simulation etc. This book serves to bring together recent advances in the application of HPC in accelerating computations in complex power and energy systems. The aim of the book is to equip the engineers, designers and architects with the understanding of complex power and energy systems, and to help them leverage the potential of HPC for solving the problems arising in these areas.

Following is a brief introduction to various chapters in the book.

The chapter entitled "High Performance Computing in Electrical Energy Systems Applications" presents a brief survey of areas of HPC application in electrical energy systems. Recent results for HPC applications in simulation of electromechanical and electromagnetic transients, composite reliability evaluation, tuning of multiple power system stabilizers and optimization of hydrothermal systems operation are also presented.

The chapter entitled "High Performance Computing for Power System Dynamic Simulation" presents the design of the next-generation high-speed extended term (HSET) time domain simulator as a decision support tool for power systems operators. It motivates the application of HPC through detailed power system component modeling and leverages efficient numerical algorithms, especially direct sparse solvers, for enabling real-time simulations. The developed simulator is validated against commercial simulators and performance results are presented.

The chapter entitled "Distributed Parallel Power System Simulation" discusses key concepts used in designing and building efficient distributed and parallel application software. It also discusses the challenges faced by traditional power system simulation software packages when being adapted to a modern networked enterprise computing environment.

The chapter entitled "A MAS-Based Cluster Computing Platform for Modern EMS" presents a multi-agent based cluster computing platform for implementing advanced application software of EMS. It discusses the implementation of a middleware named "WLGrid" for integrating distributed computers in Electric Power Control Center (EPCC) into a super computer. Several high performance techniques, such as task partitioning, concurrent dispatching, dynamic load-balancing and priority-based resource allocation are presented to accelerate the computation and to improve the real-time property of the whole system.

The chapter entitled "High-Performance Computing for Real-Time Grid Analysis and Operation" discusses the use of HPC in power grid applications such as state estimation, dynamic simulation, and massive contingency analysis. It compares static and dynamic load balancing techniques and also discusses look-ahead dynamic simulation for enabling faster than real-time simulation of power systems.

The chapter entitled "Dynamic Load Balancing and Scheduling for Parallel Power System Dynamic Contingency Analysis" proposes work-stealing based dynamic load-balancing technique for massive contingency analysis and compares it to traditional load balancing techniques such as master-slave load balancing techniques. It also presents computational gains achieved by using work-stealing for simulations of a large real system as compared to traditional methods.

The chapter entitled "Reconfigurable Hardware Accelerators for Power Transmission System Computation" presents design and prototype of reconfigurable hardware implemented on a Field Programmable Gate Array (FPGA) to speedup linear algebra subroutines used in system security analysis. It demonstrates the use of a specialized sparse LU decomposition hardware for gaining magnitude order speedup for power system Jacobian matrix sparse factorization.

The chapter entitled "Polynomial Preconditioning of Power System Matrices with Graphics Processing Units" discusses a GPU-based preconditioner designed to specifically handle the large sparse matrices encountered in power system simulations. It also discusses the preconditioning technique based on Chebyshev polynomial, and compares the GPU implementation with CPU implementation to show the computational gains by using GPU.

The chapter entitled "Reference Network Models: A Computational Tool for Planning and Designing Large-Scale Smart Electricity Distribution Grids" discusses reference network models (RNMs) which are large-scale distribution planning tools that can be used by policy makers and regulators to estimate distribution costs. This enables the distribution companies to meet the regulations and move towards the era of a low-carbon, sustainable power and energy sector.

The chapter entitled "Electrical Load Modeling and Simulation" discusses the challenges of load modeling and its impact on the performance of computational models of the power system. It also discusses the natural behavior of electrical loads and their interactions with various load control and demand response strategies.

Preface   VII

    The chapter entitled "On-Line Transient Stability Screening of a Practical 14,500-bus Power System: Methodology and Evaluations" presents a methodology for on-line screening and ranking of a large number of contingencies. It also discusses the computational gains achieved from the proposed methodology and its suitability for real-world deployment.

    The chapter entitled "Application of HPC for Power System Operation and Electricity Market Tools for Power Networks of the Future" presents an approach for using HPC to solve core optimization problems in grid operations and planning. It also highlights the benefits obtained from using HPC in increasing the operational standards and the level of renewable energy integrated into power system.

    Researchers, educators, practitioners and students interested in the study of high performance computing in power and energy systems should find this collection very useful. This book should also serve as an excellent state-of-the-art reference material for graduate and postgraduate students with an interest in future efficient and reliable power grid with large renewable energy penetration.

Contents

High Performance Computing in Electrical Energy Systems Applications 1
Djalma M. Falcao, Carmen L.T. Borges, Glauco N. Taranto

High Performance Computing for Power System Dynamic Simulation 43
Siddhartha Kumar Khaitan, James D. McCalley

Distributed Parallel Power System Simulation 71
Mike Zhou

A MAS-Based Cluster Computing Platform for Modern EMS 101
Boming Zhang, Wenchuan Wu, Chuanlin Zhao

High-Performance Computing for Real-Time Grid Analysis and Operation 151
Zhenyu Huang, Yousu Chen, Daniel Chavarría-Miranda

Dynamic Load Balancing and Scheduling for Parallel Power System Dynamic Contingency Analysis 189
Siddhartha Kumar Khaitan, James D. McCalley

Reconfigurable Hardware Accelerators for Power Transmission System Computation 211
Prawat Nagvajara, Chika Nwankpa, Jeremy Johnson

Polynomial Preconditioning of Power System Matrices with Graphics Processing Units 229
Amirhassan Asgari Kamiabad, Joseph Euzebe Tate

Reference Network Models: A Computational Tool for Planning and Designing Large-Scale Smart Electricity Distribution Grids 247
Tomás Gómez, Carlos Mateo, Álvaro Sánchez, Pablo Frías, Rafael Cossent

Electrical Load Modeling and Simulation 281
David P. Chassin

On-Line Transient Stability Screening of a Practical 14,500-Bus Power System: Methodology and Evaluations 335
Hsiao-Dong Chiang, Hua Li, Jianzhong Tong, Yasuyuki Tada

Application of HPC for Power System Operation and Electricity Market Tools for Power Networks of the Future 359
Ivana Kockar

Author Index 379

List of Contributors

Borges, Carmen L.T.
Federal University of Rio de Janeiro
Brazil

Chassin, David P.
Pacific Northwest National Laboratory
Richland, Washington, USA

Cossent, Rafael
Institute for Research in Technology (IIT)
ICAI School of Engineering
Comillas Pontifical University
Spain

Chiang, Hsiao-Dong
Cornell University
Ithaca, NY
USA

Chen, Yousu
Pacific Northwest National Laboratory
USA

Chavarría-Miranda, Daniel
Pacific Northwest National Laboratory
USA

Falcao, Djalma M.
Federal University of Rio de Janeiro
Brazil

Frías, Pablo
Institute for Research in Technology (IIT)
ICAI School of Engineering
Comillas Pontifical University
Spain

Gómez, Tomás
Institute for Research in Technology (IIT)
ICAI School of Engineering
Comillas Pontifical University
Spain

Huang, Zhenyu
Pacific Northwest National Laboratory
USA

Johnson, Jeremy
Drexel University
Philadelphia
USA

Khaitan, Siddhartha Kumar
Iowa State University
USA

Kamiabad, Amirhassan Asgari
University of Toronto
Canada

Kockar, Ivana
Institute for Energy and Environment
Department of Electronic and Electrical Engineering
University of Strathclyde
Glasgow, Scotland, UK

Li, Hua
Bigwood Systems, Inc.
Ithaca, NY
USA

McCalley, James D.
Iowa State University
USA

Mateo, Carlos
Institute for Research in Technology (IIT)
ICAI School of Engineering
Comillas Pontifical University
Spain

Nagvajara, Prawat
Drexel University
Philadelphia, USA

Nwankpa, Chika
Drexel University
Philadelphia, USA

Sánchez, Álvaro
Institute for Research in Technology (IIT)
ICAI School of Engineering
Comillas Pontifical University
Spain

Tate, Joseph Euzebe
University of Toronto
Canada

Taranto, Glauco N.
Federal University of Rio de Janeiro
Brazil

Tong, Jianzhong
PJM Interconnection
LLC, Norristown, PA
USA

Tada, Yasuyuki
Tokyo Electric Power Company
R&D center, Yokohama, Tokyo
Japan

Wu, Wenchuan
Tsinghua University, Beijing
China

Zhou, Mike
Chief Architect
InterPSS Systems, Canton, MI
USA

Zhang, Boming
Tsinghua University, Beijing
China

Zhao, Chuanlin
Washington State University
Washington
USA

List of Reviewers

Davis, Matt	PowerWorld, USA
Dinavahi, Venkata	University of Alberta, Canada
Feldmann, Peter	IBM T.J. Watson Research Center, USA
Krishnan, Venkat	Iowa State University, USA
Mittal, Sparsh	Iowa State University, USA
Nakka, Nithin	Teradyne Inc., USA
Paramasivam, Magesh	Iowa State University, USA
Pinar, Ali	Sandia National Lab., USA
Raju, Mandhpati	Convergent Science, USA
Santhanam, Ganesh Ram	Iowa State University, USA
Schenk, Olaf	University of Basel, Switzerland

Many reviewers preferred to be anonymous.

High Performance Computing in Electrical Energy Systems Applications

Djalma M. Falcao, Carmen L.T. Borges, and Glauco N. Taranto

Department of Electrical Engineering,
Federal University of Rio de Janeiro, Rio de Janeiro, Brazil
falcao@nacad.ufrj.br, carmen@nacad.ufrj.br,
tarang@coep.ufrj.br

Abstract. This chapter presents a review of the main application areas of high performance computing in electrical energy systems and introduces some results obtained with the implementation of parallel computing to some relevant problems encountered in the planning, operation and control of such systems. The research topics described in the paper are optimization of hydrothermal systems operation, tunning of multiple power system stabilizers, security constrained optimal power flow, composite reliability evaluation, dynamic security assessment, and distribution network reconfiguration. Most of the work described includes results obtained with tests conducted using actual models of electrical energy systems.

1 Introduction

Electrical energy systems comprise the aggregation of primary energy sources and loads through the electrical grid. Simulation, optimization, and control of such systems can be included in the category of highly computer intensive problems found in practical engineering applications. Modern electrical energy system studies require more complex mathematical models owing to the use of power electronic based control devices, the implementation of deregulation policies, and the introduction of smart grid technologies. New computing techniques such as those based on computational intelligence and evolutionary principles are also being introduced in these studies. All these facts are increasing even further the computer requirements of electrical energy system applications. High Performance Computing (HPC), encompassing from small desktop multi-core systems to cluster-based supercomputers, has achieved a stage of industrial development which allows economical use in this type of application.

The research work in the application of HPC to electrical energy systems has started around de 1080s mainly in universities and research centers. These early attempts focused in two main problems categories: naturally parallel applications, like Monte Carlo simulation and contingency analysis, and coupled problems, an example of which is the simulation of electromechanical transients. Most of this work was performed in experimental parallel machines using proprietary software tools. Some of the results obtained in these research activities were reported by Tylavsky et al [1992] and Falcao [1997].

The availability of affordable medium size computer clusters and multicore multiprocessors, together with the increased complexity of electrical energy systems problems, has induced a revival of interest un the application of HPC in electrical energy systems [Zhang et al 2010] and [Ott 2010].

2 Overview of HPC in Electrical Energy System Applications

This section presents an overview of the potential areas of application of high performance computing in electrical energy systems.

2.1 Electromechanical and Electromagnetic Dynamic Simulation

Electrical energy systems are very complex physical systems including electrical, electromechanical, mechanical, electrochemical, etc., components. Simulation studies are largely used in planning, operational planning, and real-time control. Also, real-time simulation, i.e., simulation in the same time frame as the physical phenomena happens, is used for testing protection and control devices and schemes. The dynamic behavior of such systems can be studied in at least three different time frames:

- *Quasi steady-state*: associated with the slow variation of loads and generation. Usually simulated using a steady-state model of the power network (power flow model) and ad hoc choice of load levels. Time constants assumed are in the order of several minutes.
- *Slow transients*: associated with the electromechanical behavior of turbine-generators as well as electromechanical loads. The simulation model comprises a set of nonlinear ordinary differential equations representing the electromechanical devices and a set of algebraic equations representing the electrical network. Time constants associated with this phenomenon vary from a few seconds to a few minutes.
- *Fast transients*: related transients caused by lightning, switching operations, etc., that propagate alongside the electrical grid. The simulation model is composed by partial differential equations (distributed circuit elements) and ordinary differential equations (lumped circuit elements). Time constants are in the order of milliseconds.

Electromechanical dynamic simulation software used in the electrical energy industry follow two general simulation schemes:

- Alternate scheme: the ordinary differential equations are converted to algebraic equations using an integration rule and solved alternatively with the algebraic equations modeling the network;
- Simultaneous scheme: both sets of algebraic equations are lumped together and solved by a Newton-based method.

The difficulties for the parallelization of the dynamic simulation problem in the alternating scheme are concentrated on the network solution. The differential equations associated with the synchronous machines and their controllers are naturally decoupled and easy to parallelize. On the other hand, the network equations constitute a tightly coupled problem requiring ingenious decomposition schemes and solution methods suitable for parallel applications. The simultaneous scheme also requires the parallel solution of linear algebraic equations sets in every integration step, with difficulties similar to the ones described for the alternating scheme.

Intense research work has been developed in the eighties and nineties aiming to the development of parallel implementation of simulation schemes reported above. This effort can be roughly divided into two approaches:

- Paralellization of the linear algebraic equations by a direct method with the scheduling of elementary operations based on precedence relationships defined by the system's topology and optimal node ordering [Chai and Bose 1993];
- Decomposition of the electrical system into smaller subsystem and the solution of the equations by block direct or iterative methods [Decker et al, 1996, Tomim et al 2009].

The results obtained with the research work cited above were less encouraging than expected. Efficiencies superior to fifty percent were seldom achieved in the tests. These poor results can be attributed to the inadequacy of the parallel computer systems used, most of them of the distributed memory type, and the relatively small dimension of the models used. The advent of multi-core processors, with several CPUs accessing a shared memory, opens up new opportunities for efficient parallel implementation of this kind of application.

In the usual model of the power network for electromagnetic transient simulation, all network components, except transmission lines, are modeled by lumped parameter equivalent circuits composed of voltage and current sources, linear and nonlinear resistors, inductors, capacitors, ideal switches, etc. These elements are described in the mathematical model by ordinary differential equations which are solved by step-by-step numerical integration, often using the trapezoidal rule, leading to equivalent circuits consisting of resistors and current sources. Transmission lines can be adequately represented by a traveling wave model consisting of two disjoint equivalent circuits containing a current source in parallel with an impedance in both ends of the line. The value of the current sources are determined by circuit variables computed in past integration steps known as history terms.

The mathematical model based on the above line model is nicely structured for parallel processing. Subnetworks of lumped parameter circuit elements representing a group of devices in a substation, connected by transmission lines, can be represented by sets of nodal equations that interface with other groups of equations by the variables required to calculate the current sources in the transmission line equivalent circuits in past integration steps. The exploitation of this characteristic of the network model in the partitioning of the set of equations for parallel processing conduces to naturally decomposed models which, as long as the computer load balancing is efficiently performed, may lead to very efficient parallel implementation [Falcao et al 1993].

2.2 Real-Time Control

Modern energy system are supervised and controlled from several control centers which process information gathered alongside the electrical grid. Several functions, related to economical and security objectives, are carried out continuously in these centers. The most computationally demanded of these functions are the ones concerned with the safeguard of the system integrity against contingencies (failures) that can torn apart the system and provoke interruption in the electrical energy supply to business and homes.

The security functions are implemented in two steps:

- Assessment of the effects of a set of credible contingencies in the system behavior in order to quantify the potential of each of these contingencies to cause a partial or total system blackout;
- For contingencies considered dangerous to system security, how to modify (reschedule) the system present operational condition in order to prevent their dangerous consequences.

The first step above is usually performed through steady-state and electromechanical dynamic simulation software which reproduces the behavior of the electrical energy system in the events of each contingency. The second step, is implemented trough an algorithm known as Optimal Power Flow which performs the operational point rescheduling with a secure criterion and taking into consideration all operational constraints.

The parallel implementation of the first step is straightforward consisting in the distribution of the simulation procedures corresponding to each contingency to different processors. Relatively simple scheduling algorithms may lead to high efficient implementations. The parallelization of the OPF algorithms, however, is a much more complicated matter. In section 5 of this chapter, a possible solution to this problem is described. The parallel implementation of a real-time dynamic security assessment method is introduced in section 6.

2.3 Operational Planning and Optimization

The operation of large electrical energy system requires a carefully planning of the operational strategy in order to optimize the energy resources and guarantee the system integrity. Uncertainties regarding the demand, interruptible power supply like wind and solar sources, availability of primary resources like fuel for the thermal plants and enough water inflows for the hydro plants, etc. Moreover, the liberalization of electricity markets has introduced more uncertainties in the planning studies as extreme operating conditions, forced by commercial transactions, may occur.

Two large scale optimization problems have been thoroughly researched associated with the operational problems of different types of electrical energy systems:

- *Security Constrained Unit Commitment* (SCUC) is an extension to the conventional unit commitment problem, very relevant in predominately thermal generation energy systems, in which are determined the time instant generators are committed (turned on or off) and dispatched (power output level), including the representation of pre- and post-contingent transmission constraints. The SCUC is a procedure fundamental to drive the day-ahead operation of most modern electric power markets. Due to the tendency of representing very large system models, the SCUC presents itself as very large and complex mixed integer nonlinear optimization problem. The SCUC problem has been tackled by several methods, including Lagrangian Relaxation and Mixed-Integer Programming with special treatment of the transmission constraints. Both approaches can be optimized for parallel execution [Li and Shahidehpour 2005].
- *Hydro-thermal System Optimization* is an important aspect in the operation of hydro dominated electrical energy systems. In such systems, thermal generation is used as a complementary source of energy as the cost of hydro power is generally much lower. However, hydro power is limited and uncertain due to the stochastic nature of the water inflows. The optimization of a hydro-thermal system involves obtaining an operation strategy which, for each stage of the planning period, produces generation targets for each power plant, minimizing the total operation const and the risk of energy deficit, while satisfying hydraulic an electrical constraints. For large energy system and long periods of planning, this problem is a huge dynamic optimization problem. The usual approach to solve this problem is to decompose it in a chain of optimization problem with decreasing time horizon and increased model details [Maceira et al 2002]. The approach followed to solve some of the components of the optimization chain referred to above use dual stochastic dynamic program (Pereira and Pinto, 1991). In Section 3 of this chapter it is presented an approach for the solution of this problem in clusters of computers.

2.4 Generation and Transmission Expansion Planning

Generation and transmission expansion planning studies aiming to the determination of when, where and what type of new generation and transmission equipment should be included in the system in order to prepare the system to meet a forecasted demand. Taking into consideration the actual system dimensions, the large expansion options and the time period of several years considered in practice, this problem presents itself as a huge optimization problem with all the unfriendly characteristics of such problems: nonlinearities, nonconvexity, time dependence, integer and discrete variables, etc.

In an ideal situation, both generation and transmission expansion should be modeled simultaneously. However, to make the problem more treatable, a decomposition approach is usually adopted: the generations and expansion studies are solved individually and iteratively. Moreover, the time dependence inherent to the expansion problems is usually relaxed in several ways to reduce the computational requirements. Approximations in the models of generation and transmission systems are also usually introduced.

Several optimization approaches have been used to solve the expansions problems including mixed integer nonlinear, dynamic programming, and more recently, several metaheuristic approaches [Zhu and Chow 1997, Latorre et al, 2003]. In all approaches, most of the computational burden is associated with the performances assessment of the expanded system regarding different operational conditions. This task involves obtaining the solution of analysis or optimization of independent problems which can be solved in parallel. This fact has been explored extensively in different research works [Faria Jr. et al 2005, Rodriguez et al 2009].

2.5 Probabilistic Assessment

Probabilistic assessment of electric energy systems has been increasing its importance in expansion and operational planning of such systems. Compared with the traditional worst case deterministic approaches, it conduces to more economical solutions and more effective basis for the comparison of potential solutions. More recently, the introduction of uncertain generation technologies, like wind and solar generation, has increased even more the need of probabilistic assessment. Reliability assessment has been the main area of development but stochastic optimization has also played important role [Allan and Billinton 2000]. In Section 3 of this chapter, the hydrothermal operation problem is approached as a stochastic optimization problem while in Section 6 the reliability assessment of combined generation and transmission system is analyzed.

2.6 Smart Grid Potential Applications

The term Smart Grid should be understood more as a concept than a specific technology or device. It is based on intensive use of automation technology, computing and communications to monitor and control the power grid, which will allow the implementation of strategies to control and optimize the network much more efficient than those currently in use.

The introduction of the concept of Smart Grid will produce a marked convergence between the infrastructure for generation, transmission and distribution infrastructure and digital communications and data processing. The latter will operate as an Internet of Devices, linking the so-called IEDs (Intelligent Electronic Devices) and exchanging information and control activities among the various segments of the grid. This convergence of technologies requires the development of new methods of control, automation and optimization of the electrical system, with a strong tendency to use techniques of distributed problem solving [Ipackchi 2009].

In the electrical distribution networks [Fan 2009], the combination of millions of smart meters installed in residential and business consumer premises, together with smart sensors and instrumentations, and the introduction of electric vehicle technology, will produce a vast amount of data that will require large scale processing system combining technique of feature extraction, data mining and intelligent monitoring. In the electrical transmission system (Stanley 2010), the introduction of Phasor Measurement Units (PMUs), which allows a very precise and fast measurement of electrical quantities time synchronized using the General Positioning System (GPS), has open up the possibilities of wide area monitoring and control of electrical energy system. For this kind of control to be implemented efficiently, high performance computing facilities must be used to meet the time requirements for decision making. Also, situational awareness techniques, processing information of the electrical energy system together with meteorological data, events information (TV attractions, traffic, etc.) will also introduce even more stringent computational performance.

3 Optimization of Hydrothermal Systems Operation

The basic objective of the operational planning of hydrothermal energy systems is to determine, for each stage (hours to months) of the planning period (months or years), the generation goals for each power plant which meet the demand and minimize the expected value of the generation costs in the period. The Brazilian system [ONS] is predominately based on hydroelectric generation (more than 93% of generated power) with a transmission network whose length is higher than 85.000 km. Due to the system characteristics such as the stochastic behavior of the water inflows, the existence of large reservoirs, the location of the plants far away from the load centers and the systems interconnections, the operation planning problem is represented as a huge stochastic optimization problem. To avoid this very computationally intensive problem, the operational planning is divided in three steps with different horizons/discretization periods [Maceira 2002] medium-term (5 years/monthly), short-term (months/weeks), and very-short-tem (1 week/hours).

In this section, the medium-term problem is addressed. In this problem, it is adopted the use of equivalent reservoirs to represent groups of hydro plants and their individual reservoirs, considering that the individual representation would be offset by the uncertainties in the inflows. The problem has the following characteristics:

- Time coupling, i.e., it is necessary to evaluate the future consequences of a decision taken in the present. The optimal solution is the equilibrium between the benefits of the present water usage and the future benefits of its storage, measured by the expected economy in the fuel consumption in the thermal power plants.
- Stochastic nature due to the uncertainties in the water inflows and the power demand.
- Space coupling owing to the dependence among the operation of cascaded hydro plants.
- The value of the energy produced by hydro plants cannot be obtained directly but as the economy resulting from the reduction of thermal plant costs and the avoided interruption costs.
- There is a conflict between the objectives of economic operation and reliability of supply. The maximization of the hydraulic energy minimizes costs but increases the risk of energy deficit. On the other hand, the maximization of the reliability of supply implies in to higher operational costs.

3.1 Problem Formulation

The optimization problem can be formulated as [Pinto, Borges, and Maceira 2009]:

> *Minimize Generation Cost + Deficit Cost + Future Cost*
> *Subject to*
> > *Water balance equations*
> > *Demand balance equations*
> > *Future cost function equations*
> > *Hydro generation equations*
> > *Hydro network balance equations*
>
> *and upper and lower limits on*
> > *Subsystems interchanges*
> > *Power output of thermal plants*
> > *Energy stored in the equivalent reservoirs at the end of each stage.*

The detailed expressions used for the objectives and constraints in the optimization problem formulated above can be found in [Pinto, Borges, and Maceira 2009].

3.2 Solution Approach

Pereira and Pinto [1991] introduced an approach for the solutions of the optimization problem reported in the last section referred to as Dual Stochastic Dynamic Programming (DSDP) which is an evolution of the Stochastic Dynamic Programming. This solution approach involves determining an operation strategy such that at each stage are defined generation goals for each subsystem that minimizes de expected value of the total operation cost in the whole planning horizon. This strategy, represented by the future cost function (FCF), is approximated by a piece-wise linear function, built iteratively trough Benders cuts.

The method is an iterative process for successive approximation of the optimal solution strategy in which the stochastic nature of the problem is represented by several scenarios of water inflows to equivalent reservoirs. At each stage and for each system state (storage level and water inflows in the last months), the hydro-thermal operational problem is modeled as a linear programming (LP) problem and the dual variables associated to the solution of these problems are used to built the Benders cuts. This recursive process is named *backward process*. To obtain an estimate of the expected value of the operational cost in the planning horizon, considering the FCF built in the backward process, a simulation of the system operation is performed for the various water inflows scenarios. This procedure is named the *forward process*. In a given stage, both in the backward and in the forward processes, as many LP problems are solved as there are water inflow scenarios.

The solution of a problem corresponding to one scenario is completely independent from the others.

3.3 Parallel Implementation

Pinto et al [2007] and Pinto, Borges, and Maceira [2009] reported the parallel implementation of the approach reported in the last section. In the backward process, in each stages of the planning horizon, each processor receives a set of LP problems corresponding to the economic dispatch associated with the different water inflow scenarios. Once solved each of these problems, a Benders cut is generated in each processor and sent to master processor. In the master processor, a FCF function is built using all the Benders cut which will be used in stage t-1 and transmitted to all processors again. This procedure is illustrated in Figure 1.

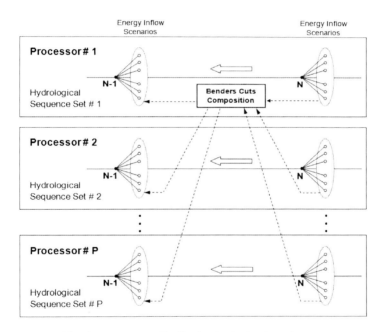

Fig. 1. Benders cut distribution in the *backward process*

In the forward process, the processors receive the set of LP problems (each one corresponding to a hydrothermal operational problem under a specific water inflow scenario) in each stage whose solution are sent to master processor as illustrated in Figure 2. In the case that the convergence of the process has not been achieved, a new iteration (backward and forward processes) is repeated.

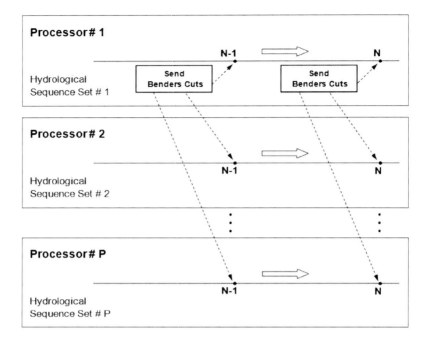

Fig. 2. Benders cut distribution in the *forward process*

3.4 Results

The computational environment used was an IBM cluster composed of a front-end server model x3550 with a quad-core Intel Xeon processor, 2GHZ, cache L2 12Mbyte, and 2Gbytes RAM. The processor nodes are a set of 42 blade boards, each one equipped with 2 quad-core Intel Xeon, 2.66 MHz, cache L2 with 8Mbytes, and two communication interfaces 10/100/1000Mbps. The message passing environment used is MPICH2, version 1.08.

The test case analyzed is the actual data set for the Monthly Operational Program used by the Brazilian Independent System Operator for March 2009. This data has the following dimensions:

- Study period of five years and post-study period of five years, implying in a horizon study of 120 months;
- Number of water inflow scenarios: 200;
- Number of openings in the backward cycle: 20;
- Number of water inflow scenarios in the final simulation: 2000.

This test case was executed three times with 1 to 48 cores. The results obtained are shown in Table 1.

Table 1. CPU time of the tested cases (hours:minutes:seconds)

Runs	Number of Processing Cores								
	1	2	4	8	16	24	32	40	48
1	11:53:38	06:07:17	03:10:08	01:44:08	01:02:14	00:55:22	00:48:30	00:41:08	00:39:29
2	11:53:57	06:07:12	03:10:13	01:44:06	01:02:05	00:54:58	00:48:34	00:40:43	00:39:18
3	11:53:33	06:07:30	03:10:18	01:44:09	01:01:55	00:55:45	00:48:16	00:40:43	00:39:07
Average	11:53:43	6:07:20	3:10:13	1:44:08	1:02:05	0:55:22	0:48:27	0:40:51	0:39:18
St. Dev.	0:00:13	0:00:09	0:00:05	0:00:02	0:00:10	0:00:24	0:00:09	0:00:14	0:00:111
Blades	1	1	1	1	2	3	4	5	6

The speedups and efficiencies achieved in this study are shown in Figures 3 and 4, respectively.

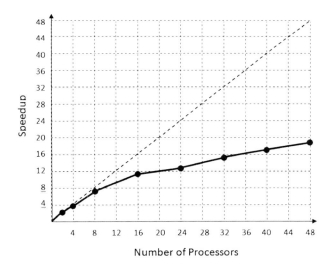

Fig. 3. Average speedup for the test case

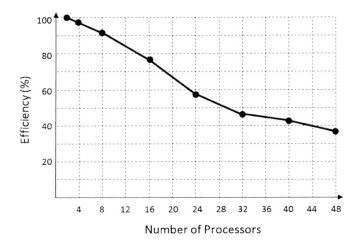

Fig. 4. Average efficiency for the test case

4 Tunning of Multiple Power System Stabilizers

This section presents a Power System Stabilizer (PSS) tuning procedure based on a Parallel Genetic Algorithm (PGA). The procedure is applied to the simultaneous tuning of 61 PSSs, using their actual models in the entire Brazilian Interconnected Power System (BIPS) modeled with 3849 buses. Heavy, median and light loading conditions were considered during the tuning process, to ensure control robustness. A PGA with small populations distributed in various CPU is utilized to accelerate the searching for the solution. The PGA utilizes the genetic operator called migration. Real number chromosome representation, arithmetic crossover, variable exponential mutation and variable search space are some techniques exploited.

4.1 Problem Description and Formulation

The use of high performance computing in the problem of coordinated tuning of power system damping controllers started in our group more than ten years ago when we published [Bomfim et al 2000]. In [Bomfim et al 2000] the better computational power we had at that time was an IBM RS/6000 SP with 4 CPU. Modeling simplifications for the BIPS in terms of state-space reduction had to be done at that time, in order to have reasonable solution time for the tuning of 22 PSS. This section summarizes the newest results we have obtained for the same problem [Bomfim et al 2010]. The results we present were obtained for the entire BIPS with very few modeling simplifications. The problem involves in the coordinated tuning of 61 PSS using their actual modeling structure. The computer used to perform the PGA is the ALTIX ICE having 64 CPU Quad Core Intel Xeon 2.66 GHz 256 cores from the Center for Service in High Performance Computing [Nacad].

The implemented PGA is customized to consider a searching space that varies its bounds along the process, migration and variable mutation rate. The bound changes in the searching space are performed after a specified number of generations, becoming larger or smaller according to some defined criteria. It improves PGA search to reach the best solution. Initial mutation rate is reset at each change of the searching space. Migration operator is performed after a new space is created.

The method is based on the optimization of a function related to the global minimum damping ratio and the minimum interarea damping ratio, related to the closed-loop spectrum, constrained in the controller parameter space. The robustness of the controllers is taking into account during the tuning process, simply by considering a pre-specified set of system operating conditions into the objective function. A decentralized coordinate design is performed as the controller channels are closed simultaneously with all the cross-coupling signals amongst controllers not taken into consideration.

Subtransient rotor effects are modeled for 92 generators. The open-loop linearized dynamic representation of the system ended up with a system matrix with 1516 state variables.

Power system damping controllers are usually designed to operate in a decentralized way. Input signals from remote sites are not considered timely and reliably enough with today's technology. The advent of Phasor Measurement Units (PMU) may change this scenario in the near future. The robustness of the controllers is ensured by considering the performance of the control system for several different operating conditions. The choice of these conditions is based on experience and simulation studies.

Tuning of power system damping controllers uses a small-signal model represented by the well-known state space equations (4.1).

$$\dot{x}(t) = Ax(t) + Bu(t)$$
$$y(t) = Cx(t) + Du(t) \qquad (4.1)$$

Where x is the vector of the state variables, such as the machine speeds, machine angles, and flux linkages; u is the vector of the input variables, such as the control signals, y is the vector of the measured variables, such as bus voltages and machine speeds; A is the power system state matrix; B is the input matrix; C is the output matrix; and D is the feedforward matrix.

Stability of (4.1) is determined by the eigenvalues of matrix A.

Figure 5 shows the damping control loop via the excitation system. AEG(s) represents the combined transfer function for the automatic voltage regulator (AVR), the exciter and the generator, and $PSS_{SC}(s)$ and $PSS_{GA}(s)$ when combined represent the PSS(s) as given by (4.2).

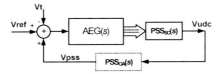

Fig. 5. Close-loop setup

$$PSS(s) = PSS_{GA}(s) \cdot PSS_{SC}(s) \tag{4.2}$$

One should note that the decomposition of the PSS structure in two transfer functions, is made to preserve the actual structure of the PSSs modeled in the official database [ONS] utilized in the studies of the BIPS. $PSS_{SC}(s)$ involves all signal conditioning filters, such as ramp-tracking filter, washout, reset functions, etc, and it can receive multiple input signals, as it is the case in many existing PSSs in the BIPS.

$PSS_{GA}(s)$ has a fixed structure shown in Figure 6, and represents the part of the PSS tuned by the PGA.

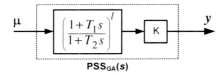

Fig. 6. Block diagram of PSS to be tuned by the PGA

The expression that defines the $PSS_{GA}(s)$ to be tuned is shown in (4.3).

$$PSS_{GA_i}(s) = K_i \frac{\left(1 + \frac{\sqrt{\alpha_i}}{\omega_i} s\right)^l}{\left(1 + \frac{1}{\omega_i \sqrt{\alpha_i}} s\right)^l}, \quad i = 1, 2, \ldots, p \tag{4.3}$$

Where l is the number of lead-lag blocks in the model, and p is the number of PSSs to be tuned. The parameters of each controller to be determined are K_i, α_i and ω_i.

To form the open-loop state-space matrices we consider the existing gain equal to zero and the lead-lag blocks, for phase compensation, disconnected from the controller. This resource is readily available in PacDyn program [PacDyn], where the transfer function considered to creating the state matrices is given by V_{ref}/V_{udc}.

The decentralized control design requires a control law such that the closed-loop system is stable and has no eigenvalues with less damping than minimum specified damping ratio (ς_{min}) in all nc operating conditions.

Formulation as an Optimization Problem

Experience showed us that the initial objective function of maximizing the minimum eigenvalue damp [Bomfim et al 2000], was always constrained by local electromechanical modes, impairing improvements to the damping of interarea modes. To mitigate this problem, the objective function was stratified according to the frequency of the electromechanical modes. A double weight is given to the interarea modes, and a weighted sum of minimum damping factors of interarea and local modes became a new objective function (4.4).

$$Max\ F = Max \left[Min \left(0.33 damp_{min} + 0.67 damp_{interarea} \right)_{nc} \right] \quad (4.4)$$

Where nc is the number of operating condition; $damp_{min}$ is the minimum damping ratio of the local electromechanical modes, and $damp_{interarea}$ is the minimum damping ratio of the interarea modes. The considered threshold frequency bounds separating the interarea set of modes has a lower bound of 0.25 rad/s and an upper bound of 5 rad/s. The lower bound was considered to exclude modes related to the governors.

Subject to (4.5),

$$\begin{aligned} K_{i\ min} &\leq K_i \leq K_{i\ max} \\ \alpha_{i\ min} &\leq \alpha_i \leq \alpha_{i\ max} \\ \omega_{i\ min} &\leq \omega_i \leq \omega_{i\ max} \end{aligned} \quad (4.5)$$

This modification yielded much better solutions for the BIPS, in which three important interarea modes are present, namely South/Southeastern mode, North/Northeastern mode and North/Southern mode.

4.2 Parallel Implementation

The solution to the optimization problem defined in (4.4) and (4.5) can be obtained using a PGA [Cantú-Paz 1999].

PGA Elements

Fitness Function: The fitness function *ff* used by PGA is defined in (4.4), where *ff*=*max F*.

Parameter encoding and limits: The controller parameters are encoded as real number string using concatenated multiparameter representation. The PGA is customized to consider variable searching space. The upper and lower limits change along the searching process, and they depend on the size of the searching region. After 10 generations the limits change to reduce the region, and after 5 generations the limits change to expand the region [Gen 1997].

The variable searching space is performed every 5 or 10 generations, becoming smaller or larger according to some defined criteria. The enlargement of the searching space is performed every 10 generations and it helps to introduce new values to the searching-space bounds. The reduction of the searching space is performed every 5 generations and it helps to fine tune the best solutions found during the enlarged-space search.

Genetic Operators: An elitism strategy was used, i.e., the best individual of current population is guaranteed to be present in the next population by replacement of the worst individual of next generation by the best individual of the previous one. Arithmetic uniform crossover operator, as described in [Gen 1997] and [Kim 2002], was used with 0.85 for crossover rate, 0.5 for mask rate based on extensive tests. The factor to change the value position is randomly chosen in the interval [0.0, 1.0]. Mutation was implemented using an exponential decreasing mutation rate, which starts at a specified maximum value and decreases exponentially until it reaches a given minimum value, and return to initial value after each change in the searching space. In this genetic operator the randomly selected gene is replaced by another real number randomly selected within its bounds. The exponential mutation rate used was 0.02 at the beginning and truncated at 0.004.

Depending on the computer power in hand, the number of CPUs may be greater than the number of individuals in the population (Not unusual nowadays!). In such situations the population in each CPU would be the same. To avoid this problem, a smaller subset from the best chromosomes needs to be chosen according to some ranking criterion. In this work, we have chosen only the 4 best individuals among the best 64. This strategy was named *elitism of migration*. Migration operator is performed after the search space is changed to avoid creating the same region in each CPU. PGA allows small populations in each CPU, but the total individuals that PGA will combine, is the sum of individuals in all CPUs, as shown in (4.6).

$$NIPGA = \sum_{1}^{NCPU} NICPU \qquad (4.6)$$

Where *NIPGA* is the number of all individuals in the PGA; *NCPU* is the number of CPUs used; and *NICPU* is the number of individuals in each CPU. In this work it was used *NCPU*=64 CPUs and *NICPU*=20 individuals, yielding *NIPGA*=1280 individuals.

The stopping rule is the maximum number of generations. The number of eigenvalue calculations (NQR) performed until the end of the search is given by (4.7).

$$NQR = NIPGA * NGER \qquad (4.7)$$

The maximum number of generations was 160, then NQR = 204,800 eigenvalue calculations performed by the PGA to reach the best solution. Each processor performs 3,200 eigenvalue calculations. The time to perform each one is 83 seconds.

4.3 Results

To verify the performance of the proposed methodology the BIPS was used. It is comprised of 3849 buses, 5519 transmission lines and transformers and 202 generators. Heavy, median and light loading operating conditions from September 2008 were considered.

Figure 7 depicts a pictorial view of the Brazilian Interconnections regions, highlighting the major interarea modes: the South/Southeastern (S/SE) mode, the North/ Northeastern (N/NE) mode and the North/Southern (N+NE/S+SE) mode.

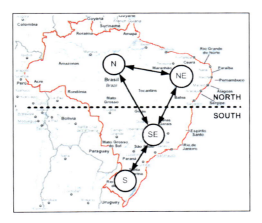

Fig. 7. Interconnection sub-systems of BIPS

The state-space representation of the open-loop system has 1516 state variables and the close-loop has 1638 state variables. The objective is the tuning of 61 PSSs from the larger units of the BIPS. No simplifying assumptions regarding the actual PSS structures were made. The PSSs that remained out of the tuning process were left unchanged with respect to their current parameters.

The open-loop eigenvalues are shown in Figure 8. Note that the BIPS is open-loop unstable.

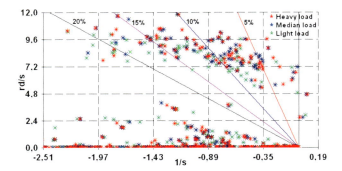

Fig. 8. Open-loop eigenvalues for the BIPS

Figure 9 shows the close-loop eigenvalues after tuning the 61 PSSs. Note that the system becomes stable with minimum damping ratio greater than 10%, achieved in a local mode in the light loading condition.

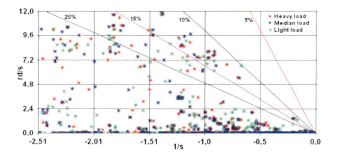

Fig. 9. Close-loop system poles with 61 PSS's tuning by PGA

The N/S interarea mode is the eigenvalue $-0.571+j2.051$ with 27.1% of damping. As shown in Figure 9 all intearea modes have a damping ratio greater than 20% in all loading conditions.

The best PSS tuning solutions were also validated in nonlinear simulation tests with the BIPS facing large perturbations as single-phase fault applied on Ibiuna 500 kV AC-Bus. The fault is cleared after 80 ms by opening two Ibiuna–Bateias 500 kV transmission lines that connect South to Southeast regions. Results presented in [Bomfim 2010] show the good damping performance of the tuned PSS for large perturbations.

5 Security Constrained Optimal Power Flow

This section reports the work developed by Alves, Borges, and Filho [2007] in the implementation of a parallel version of a Security Constrained Optimal Power Flow approach. An actual size model of the Brazilian interconnected electrical energy system was used in the performed tests.

5.1 Problem Description

The Security Constrained Optimal Power Flow (SCOPF) has the objective to determine a feasible point of operation that minimizes an objective function, guaranteeing that even if any of the contingencies obtained from a list occurs, the post-contingency state will also be feasible, i.e., without limits violations [Monticelli, Pereira, and Granville, 1987]. From a given list of N possible contingencies, the SCOPF problem can be represented mathematically as:

$$\min f(z_o)$$
$$s.t.$$
$$a_0(z_0) \leq b_0$$
$$a_i(z_i) \leq b_i \quad (5.1)$$
$$for \quad i = 1,2,...,N$$

where:

$f(.)$ is the objective function;

$a(.)$ represents the non-linear balance equations of the electric network together with the operative constraints;

z represents the variables that will be optimized in the solution of the problem (state and control variables).

Each set of constraints $a_i(z_i) \leq b_i$, for $i = 1, 2, ..., N$, is related with the configuration of the network under contingency and must respect the operations constrains in this condition.

The objective function to be minimized in the problem depends on the purpose of the tool utilization. For the use in control centers as a security control tool, the common objective functions are minimum loss, minimum deviation of the programmed operation point and minimum deviation of the scheduled area interchange. Other objective functions are also used in SCOPF problems, such as minimum reactive allocation, minimum load shed, minimum generation cost, etc. The state variables are, usually, the bus voltages and angles. The control variables, which are modified in order to obtain the optimal operation point, are the generators active injection, terminal voltages and reactive injection, transformers tap position, areas interchange, etc.

The SCOPF can be interpreted as a two-stage decision process [Granville and Lima 1994]:

- In the first stage, find an operation point z_o for the base case (present network configuration and operating conditions) problem $a_o(z_o) \leq b_o$;
- In the second stage, given the operating point z_o, find new operating points z_i that meet the constraints $a_i(z_i) \leq b_i$, for each contingency configuration.

The solution method used in this work is based on Benders Decomposition, where the base-case problem represents the current operation point and each sub-problem represents one of the N contingencies. Figure 10 shows the flowchart of the SCOPF solution algorithm based on Benders Decomposition.

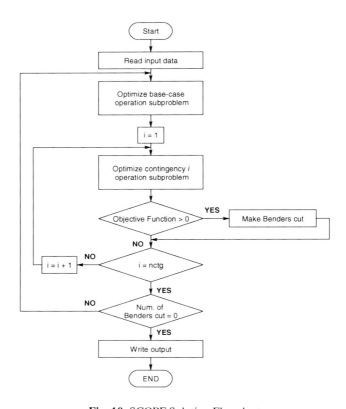

Fig. 10. SCOPF Solution Flowchart

The SCOPF solution algorithm consists, then, in solving the base-case optimization problem and then, from the operation point obtained in the base-case solution, to solve each of the contingency sub-problems. For each unfeasible contingency, a Benders Cut is generated. At the end of all contingencies solution, the generated Benders Cuts are introduced in the new solution of the base-case. The convergence is achieved when no contingency sub-problem generates Benders Cut. The contingencies sub-problems correspond to conventional OPF problems representing the configurations of the network under contingency. The base-case sub-problem is formulated as an OPF augmented by the constraints relative to the unfeasible contingencies (Benders Cuts). Each OPF problem is solved by the Interior Points Method and the network equations are formulated by the non linear model.

5.2 Parallel Implementation

It can be noticed that the N contingencies sub-problems can be solved independently, once they only depend on the incoming operation point of the base-case, z_o. In that sense, the solution of the SCOPF can be directly benefited from the use of distributed processing, due to the natural parallelism that exists in the problem.

The parallelization strategy developed is based on the master-slaves computation topology. When the parallel processing begins, each processor receives an identification number, the master being processor number zero and the slaves, processors number 1 to (*nprocs*-1), where *nprocs* is the total number of processors available. Figure 11 shows an example of contingencies allocation for a list of 10 contingencies distributed among 3 processors (1 master and 2 slaves).

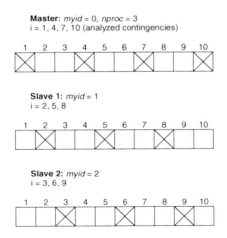

Fig. 11. Example of Contingencies Distribution among the Processors

The allocation of the contingencies sub-problems to the processors is performed by an asynchronous algorithm, based on the identification number of the processors (*myid*). In this way, the list of contingencies to be analyzed is distributed evenly among the participating processors in accordance with the value of *myid*, each one being responsible for the analysis of the contingencies of numbers (*myid* + 1 + *i. nprocs*), $i = 0, 1, 2, 3,...$ until the end of the list. It is important to emphasize that the master processor also participates in the contingencies analysis task, guaranteeing a better efficiency for the whole process, once the processor is not idle while the slaves work.

Depending on the number of contingencies in the list and the number of processors, it can happen that some processors receive one contingency more than the others. The processing time for each contingency can also vary, since the number of iterations required for the OPF of the contingencies to converge varies from case to case. However, in the contingencies distribution strategy adopted, each processor, after finishing the analysis of a contingency, immediately begins the analysis of another without needing the intervention of a control process. In that way, the computational load of each processor is, on average, approximately the same for large contingencies lists, ensuring an almost optimal load balancing.

The algorithm of the distributed application based on message passing can be summarized in the following steps:

Step 1: All processors read the input data.

Step 2: All processors optimize the base-case sub-problem.

Step 3: Parallel contingency sub-problems optimization by all processors.

Step 4: Master processor collects from all processors the partial Benders Cuts data structure (Synchronization Point).

Step 5: Master processor groups and reorganizes the complete Benders Cuts data structure.

Step 6: Master processor sends the complete Benders Cuts data structure to all processors (Synchronization Point).

Step 7: Verify if any of the N contingencies sub-problems is unfeasible. In the positive case, return to step 2.

Step 8: Master processor generates output reports.

All processors read the input data simultaneously, since the input files can be accessed by all via a shared file system, what eliminates the need to send the data read by just one processor to the others. The solution of the base-case sub-problem is also done simultaneously by all processors in order to avoid the need to send the results calculated by just one processor to the others. The reorganization of the Benders Cut data structure is a task introduced due to the distributed processing. After the analysis of their lists, each slave processor has its own partial Benders Cut data structures, which are sent to the master processor to be grouped and reorganized and later sent back again to all slave processors.

The flowchart of the developed parallel algorithm is shown in Figure 12.

High Performance Computing in Electrical Energy Systems Applications

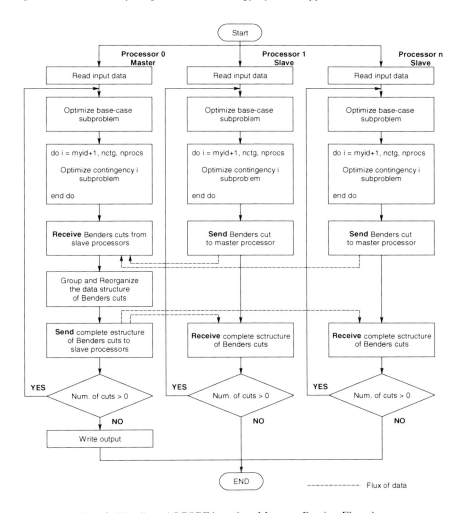

Fig. 12. Distributed SCOPF based on Message Passing Flowchart

The two synchronization points of the algorithm, associated with the collection of the partial Benders Cuts structures and the distribution of the updated complete structure to all processors, are the only communication points of the algorithm, and for that reason, a high efficiency is expected from the distributed implementation. However, the efficiency will also depend on other factors, such as the communication technology used and the number of base-contingencies interactions (base-case plus contingencies sub-problems) necessary for convergence, since more interactions cause more communication among processors.

5.3 Results

The computational environment used for the tests of the distributed implementation was a Cluster composed of 16 dual-processed 1.0GHz Intel Pentium III with 512MB of RAM and 256 KB of cache per node, Linux RedHat 7.3 operating system and dedicated Fast Ethernet network.

The test system used is an equivalent of the actual Brazilian Interconnected System for the December 2003 heavy load configuration. The studied system is composed of 3073 buses, 4547 branches, 314 generators and 595 shunt reactors or capacitors. The total load of the system is 57,947 MW and 16,007 Mvar, the total generation is 60,698 MW and 19,112 Mvar and the total losses are 2,751 MW and 3,105 Mvar.

For the security control, the objective function used was minimum losses together with minimum number of modified controls. The controls that could be modified for optimization of the base-case were: generated reactivate power, generator terminal voltage and transformer tap. The number of available controls is 963 controls, where 515 are taps, 224 are reactivate power generation and 224 are terminal voltages. The list of analyzed contingencies is formed by 700 lines or transformers disconnection. The contingency list was formulated in a way to obtain a good condition for tests, that is, the size of the list can be considered between medium and large, and some contingencies generate Benders Cuts during the optimization process.

The SCOPF solution process converged in 3 base-contingencies iterations. In the first base-contingencies iteration, 12 contingencies generated Benders Cuts, in the second iteration, only 1 contingency generated Benders Cuts and, finally, in the third iteration, no contingency generated cut. The total losses after the optimization were 2,706 MW and 2,935 Mvar, what represents a reduction of about 2% in the active losses of the system.

The number of controls modified to lead the system to an optimal secure operation point was small, only 7 modifications of generator terminal voltages. That is due to the use of the objective function of minimum number of modified controls together with the minimum losses. This is an important issue for the use of SCOPF in the real time system operation. If the list of control actions is very long, it becomes unfeasible for the operator to perform them in time to turn the system secure.

The performance of the distributed implementation has been evaluated using from 1 to 12 nodes of the Cluster, obtaining exactly the same results as the sequential program. Table 2 shows the execution time, while Table 3 shows the speedup and the efficiency, for different numbers of processors.

It can be observed that the distributed implementation presents an excellent efficiency, superior to 92% using 12 processors of the distributed platform. The processing time is significantly reduced, changing from 8 minutes 25 seconds of the sequential processing to 45.43 seconds in parallel using 12 processors.

Table 2. Execution Time

No. Processors	Time
1	8 min 25 s
2	4 min 16 s
4	2 min 10 s
6	1 min 27 s
8	1 min 7 s
10	54.26 s
12	45.43 s

Table 3. Speedup and Efficiency

No. Processors	Speedup	Efficiency (%)
1	1	-
2	1.97	98.56
4	3.88	97.10
6	5.79	96.49
8	7.50	93.70
10	9.31	93.06
12	11.11	92.62

Figures 13 and 14 show the Speedup evolution and processing time reduction with the number of nodes used for the distributed solution, respectively.

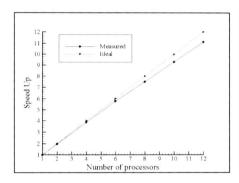

Fig. 13. Speedup Curve

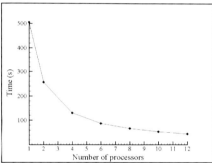

Fig. 14. Processing Time

It can be observed that the Speedup curve is almost ideal (linear). The Efficiency is only slightly reduced as the number of processors increases, what indicates that the parallel algorithm is scalable. From the good performance obtained, it can be expected that, if it is necessary to obtain a smaller response time for the real time application, this objective can be reached using a larger number of processors.

Therefore, the Security Control based on SCOPF is made possible in real time application using low cost and easily scalable platforms as a cluster of computers. Although there is no consensus about which execution time is acceptable for the use of this type of tool in the real time operation, the execution time obtained with 12 processors is perfectly compatible with the real time requirements of the Brazilian ISO nowadays, where a time slice of two minutes is available for the security control task.

6 Composite Reliability Evaluation

This section describes the work developed by Borges et al [2001a, 2001b] in which the implementation of different approaches for the composite reliability evaluation of electrical energy systems in parallel computer system was studied. The work used actual models of the Brazilian interconnected power systems.

6.1 Problem Description

Power system composite reliability evaluation consists of the calculation of several indices that are indicators of the system adequacy to power demand. The adequacy analysis is done taking into consideration the possibility of components failure at both the generation and the transmission sub-systems. A powerful technique used for power system composite reliability evaluation is Monte Carlo Simulation (MCS). MCS allows accurate modeling of power system components and operating conditions, provides the probability distributions of variables of interest, and is able to handle complex phenomena and a large number of severe events.

There are two different approaches for Monte Carlo simulation when used for composite system reliability evaluation: non-sequential MCS and sequential MCS. In non-sequential MCS, the state space is randomly sampled without concerning the system operation process chronology. In sequential MCS the system states are sequentially sampled for several periods, usually years, simulating a realization of the stochastic process of system operation. The expected values of the main reliability indices can be calculated by both approaches. However, estimates of specific energy supply interruption duration and the probability distribution of duration related indices can only be obtained by sequential MCS. In applications related to production cost evaluation, only the sequential MCS approach can be used. On the other hand, sequential MCS demands higher computational effort than non-sequential MCS. Depending on the system size and modeling level, sequential MCS computation requirements may become unacceptable at conventional computer platforms.

In both MCS approaches, reliability evaluation demands the adequacy analysis of a very large number of system operating states, with different topological configurations and load levels, in order to determine if the energy demand can be attended without operating restrictions and security violations. This requires the solution of a contingency analysis problem and, in some cases, of an optimal load flow problem for each sampled state.

In general terms, reliability indices calculation using MCS may be represented by the evaluation of expression $\bar{E}(F) = 1/N \sum F(.)$, where N is the number of samples. The reliability indices $\bar{E}(F)$ correspond to estimates of the expected value of different adequacy evaluation functions F. In non-sequential MCS, $F(.)$ is calculated for each individual state in a sample composed of N states. In sequential MCS, $F(.)$ is calculated over all system states that compound each yearly transition sequence for a sample composed of N yearly sequences.

The main differences between the two approaches are related to the way the system states are sampled (randomly in non-sequential MCS and sequentially in time in sequential MCS) and, consecutively, with the way the indices are calculated. In the non-sequential MCS approach, the adequacy analysis may be performed independently for each sampled state, what suggests the application of parallel processing for the reduction of computation time. The chronological dependency that exists in the sequential MCS approach, however, introduces much more complexity in developing a parallel algorithm. Although MCS techniques are used to sample the system state configuration, each of them is now coupled in time with the next one, and this introduces significant sequential constraints in the parallelization process.

A conceptual algorithm for composite reliability evaluation using non-sequential MCS may be described as:

(1) *Select an operating scenario $\underline{x}$ corresponding to a load level, components availability, operation conditions, etc.*
(2) *Calculate the value of an evaluation function $F(\underline{x})$ which quantifies the effect of violations in the operating limits in this specific scenario. Corrective actions such as generation re-scheduling, load shedding minimization, etc., can be included in this evaluation.*
(3) *Update the expected value of the reliability indices based on the result obtained in step (2)*
(4) *If the accuracy of the estimates is acceptable, terminate the process. Otherwise, return to step (1).*

The accuracy of MCS may be expressed by the coefficient of variation α, which is a measure of the uncertainty around the estimates.

A conceptual algorithm for composite reliability evaluation using sequential MCS may be described as:

(1) *Generate a yearly synthetic sequence of system states y_k;*
(2) *Chronologically evaluate the adequacy of all system states within the sequence y_k and accumulate these results;*
(3) *Calculate the yearly reliability indices $E[F(y_k)]$ based on the values calculated in step (2) and update the expected values of the process reliability indices;*
(4) *If the accuracy of the estimates of the process indices is acceptable, terminate the process. Otherwise, return to step (1).*

The yearly synthetic sequence is generated by combining the components states transition processes and the chronological load model variation in the same time basis. The component states transition process is obtained by sequentially sampling the probability distribution of the components states duration, which may follow an exponential or any other distribution.

6.2 Parallel Implementation

6.2.1 Parallel Non-sequential Simulation

An asynchronous parallel methodology for composite reliability evaluation using the state sampling approach was developed, in which the adequacy analysis of the sampled system operating states are performed in parallel on different processors. In this way, a high degree of task decoupling is achieved.

The master-slaves paradigm is adopted, in which the master is responsible for scattering and gathering data, for controlling the convergence of the parallel process and calculating the reliability indices. The slaves, and the master too, are responsible for analyzing the adequacy of the system states allocated to them.

The problem initialization is executed by the master processor, followed by a broadcast of the data to all slaves. All processors, including the master, pass to the phase of states adequacy analysis, each one analyzing different states. After a time interval, each slave sends to the master the data relative to its own local convergence, independently of how many states it has analyzed so far, and then continues to analyze other states. This process is periodically repeated.

When the master receives a message from a slave, it verifies the status of the global parallel convergence. If it has not been achieved, the master goes back to the system states adequacy analysis task until a new message arrives. When the parallel convergence is detected or the maximum number of state samples is reached, the master broadcasts a message telling the slaves to stop simulating, upon what the slaves send back their partial results to the master. The master then calculates the reliability indices, generate reports and terminate. In this parallelization strategy, there is no kind of synchronization during the simulation process and the load balancing is established by the processors capacities and the system states analysis complexity.

The strategy adopted for distribution of the system states over the processors is to generate the different system states directly at the processors in which they will be analyzed. For this purpose, all processors receive the same seed and execute the same random numbers sampling, generating the same system states. Each processor, however, starts to analyze the state with a number equal to its rank in the parallel computation and jump to analyze the next states using as step the number of processors involved in the computation. Supposing that the number of available processors is 4, then processor 1 analyzes states numbered 1,5,9..., processor 2 analyzes states numbered 2,6,10..., and so on. This strategy avoids the problem of generating correlated random numbers sequences on different processors that may occur if different seeds are used by the processors. Moreover, it mimics the sequential execution, what tend to produce the same results and reduce the difficulties of debugging the parallel code.

A consideration introduced by the asynchronous parallel methodology is the extra simulation. During the last stage of the simulation process, the slaves execute some analyses that are beyond the minimum necessary to reach convergence. The convergence is detected based on the last message sent by the slaves and, between the shipping of the last message and the reception of the stop message, the slaves keep analyzing extra states. This, however, does not imply in loss of time for the computation as a whole, since no processor gets idle any time. The extra simulation is used for the final calculation of the reliability indices, generating indices still more accurate than the sequential solution, since in Monte Carlo methods the uncertainty of the estimate is inversely proportional to the number of analyzed state samples.

6.2.2 Parallel Sequential Simulation

The first approach to parallelize the sequential MCS, here called Methodology I, is to analyze the many simulation years necessary for convergence on different processors. Each processor analyzes the adequacy of all systems states within a year and, at the end of the analysis of the year, the slaves send the results to the master. The master, upon reception of any message, accumulates the yearly results and verifies the process convergence. When convergence is achieved, the master sends a message to each slave telling them to stop the simulation, receives from them the most updated results they have produced, calculates the process reliability indices, generates reports and terminates execution.

In order to allocate the yearly synthetic sequences to the processors, all processors receive the same seed and generate the same random numbers sequence, identical to the mono-processor simulation. This leads to the same yearly synthetic sequence generated on all processors. Each processor, however, starts to analyze the year numbered equal to its rank in the parallel computation and jumps to analyze other years using as step the number of scheduled processors. The approach is illustrated in Figure 15 for the case of four processors.

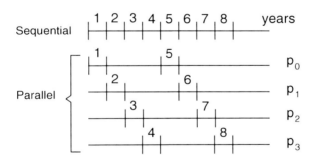

Fig. 15. Years Distribution Approach

The parallel methodology I is asynchronous. There is no synchronization after the tasks and no processor needs to wait for the others to complete their tasks. This, however, may result in some extra simulation performed by the slave processors at the end of the process. Since the algorithm is asynchronous, the slaves may analyze some years that are beyond those necessary to achieve convergence, due to the communication time and convergence check lag between the forwarding of the last message and the reception of the stop message by a slave. The extra simulation, however, is used in the reliability indices calculations, since the slaves return back to the master the results calculated including the extra years.

Because a year may be quite a large processing grain, the master does not wait until the end of the year it is analyzing to check for the process convergence. The convergence is checked as soon as a message arrives and when the process converges, the statistics of the year under analysis are disregarded. For the same reason, the slave does not finish to analyze the current year when the stop message from the master arrives. It ignores the results of this year and sends back to the master the results accumulated until the previous analyzed year.

Another approach to parallelize the sequential MCS, here called Parallel Methodology II, is to analyze each simulation year in parallel by allocating parts of one year simulation to different processors. Since within a yearly synthetic sequence the adequacy analysis of the system operating states should be chronologically performed, this parallelization strategy requires a more careful analysis of the problem and complex solution.

The number of system states contained in a yearly synthetic sequence is equally divided by the processors scheduled, the last processor getting the remainder if the division is not exact. For this purpose, every processor generates the same synthetic sequence by receiving the same seed and generating the same sequence of random numbers, as described in methodology I. Each processor, however, is responsible for analyzing a sub-sequence of this yearly sequence which is first determined by its rank in the parallel computation. At the end of each sub-sequence analysis, the slaves send to the master their partial results and start the simulation of another sub-sequence in the next year. The master is responsible for combining all these sub-sequence results and compounding a complete year simulation.

To balance the configuration evolution process between the processors, a cyclic sub-sequence allocation strategy has been developed. From the second year on, each slave analyzes the sub-sequence next to the one analyzed in the year before, until it reaches the last sub-sequence of a year, from where it returns to the second sub-sequence. The first sub-sequence is always analyzed by the master. This can be illustrated in Fig. 16, where it is supposed that four processors are scheduled.

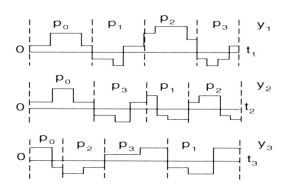

Fig. 16. Cyclic Sub-sequence Allocation Strategy

Parallel methodology II is also asynchronous. No processor needs to wait for the others to proceed analyzing a sub-sequence in the next year. The master can receive sub-sequences results of years posterior to the one it is currently analyzing but it can only verify the convergence when all sub-sequences of a particular year have been analyzed. The load balancing is then determined by the processing capacities of the processors and the analysis complexity of the system states within different sub-sequences.

6.3 Results

The parallel platform where the parallel methodologies have been implemented is a scalable distributed memory parallel computer IBM RS/6000 SP with 10 POWER2 processors interconnected by a high performance switch of 40 MB/sec full-duplex peak bandwidth. The message passing system used is MPI.

Three different power system models have been used to test the parallel methodologies: IEEE-RTS, composed of 24 buses, 38 circuits and 32 generators, and two representations of the Brazilian power system: BSO, composed of 660 buses, 1072 circuits and 78 generators, and BSE, composed of 1389 buses, 2295 circuits and 259 generators.

A tolerance of 5% in the coefficient of variation of the indices was used as convergence criterion. The load level is considered constant throughout the entire simulation period in the studied cases.

In the non-sequential MCS approach, an AC power flow and Optimal Power Flow models are used in the adequacy analysis of the system states, the last being solved by a successive linearizations technique. The efficiencies obtained by the parallel methodology on 4, 8 and 10 processors for the test systems, together with the CPU time of the sequential execution, are shown on Table 4.

Table 4. Non-sequential MCS - Parallel Results

System	CPU time	Efficiency (%)		
	p=1	p=4	p=8	p=10
RTS	35.03 sec	88.07	47.19	37.28
BSO	5.00 min	96.28	93.49	93.00
BSE	8.52 hour	97.04	94.91	93.55

The asynchronous parallel implementation on the parallel computer has very good scalability in terms of number of processors and dimension of the power system model. Most of the communication time is concentrated on the initial broadcast of data. The communication cost is very high in the evaluation of small scale system models like IEEE-RTS.

The number of analyzed system states in the parallel methodology may be different from that of the sequential execution due to the asynchronous convergence detection criterion and the extra simulation. The reliability indices calculated may be not exactly the same as the sequential ones but are statistically equivalent and with the same or greater level of accuracy.

In the sequential MCS approach, a DC power flow model and a LP-based remedial actions scheme were used for the adequacy analysis of each state within a simulated year. The efficiencies obtained by the parallel methodologies I and II on 4, 6 and 10 processors for the test systems, together with the CPU time of sequential execution, are shown on Table 5.

Table 5. Sequential MCS - Parallel Results

System	CPU time	Efficiency (%)					
		Methodology I			Methodology II		
	p=1	p=4	p=6	p=10	p=4	p=6	p=10
RTS	1.95 min	97.23	86.81	78.83	95.42	89.81	75.96
BSO	13.02 min	67.61	71.50	63.51	93.57	93.27	82.23
BSE	15.30 hour	95.69	97.02	87.72	97.81	97.41	91.20

The efficiency of methodology I is dependent on the number of simulated years and on the number of processors of the parallel platform. If the number of years is large or a multiple of the number of processors, the efficiencies are very good. Alternatively, if the number of simulated years is small, what may occur for systems with low reliability like the BSO, or not a multiple of the number of processors, the efficiency degrades considerably. This means that this methodology is not scalable.

Methodology II is scalable with both the number of years and the number of processors. Although in some cases the efficiency is smaller than in methodology I due to the higher complexity of the parallelization strategy, there is no scalability limitation imposed by the methodology.

The sequential MCS result is very relevant since its parallelization is more sequentially constrained than the non-sequential one. The time saved by using parallel processing can be used to improve the solution, for example by using an AC power flow model in the sequential MCS approach and nonlinear programming techniques for the solution of OPF.

7 Dynamic Security Assessment

This section reports a preliminary work developed by Jardim et al. [2000] aiming to the implementation of a parallel version of Dynamic Security Assessment software in Beowulf class of computer cluster.

7.1 Problem Description

Reliable power system operation requires on-line security assessment, one of the main functions in energy management systems. In particular, Dynamic Security Assessment (DSA) is of primary concern on those systems constrained by stability limits. The main benefits of security assessment are improved reliability, better use of system resources, and maintenance schedule flexibility. For commercial purposes, the calculation of transmission transfer capability requires effective security assessment. On-line static security assessment is a well-established technology and is widely used on energy management environment. On-line dynamic security assessment, on the other hand, has found difficulties in being implemented in control centers because of the lack of sufficiently fast tools and the poor monitoring conditions.

Effective dynamic security assessment requires step-by-step time simulation, which is computationally intensive and time consuming. That is one of the reasons why it has been considered unsuitable for on-line applications. This motivated the development of fast tools. The best results have been obtained with the hybrid methods. However, those methods are based on many simplifying assumptions, which depending on the operating conditions, are not always true. On the other hand, faster computer systems and parallel processing have turned step-by-step time simulation into an attractive method for on-line dynamic security assessment.

The DSA system used in the tests reported in this section is based on the unified (voltage and angular stability) methodology. The system takes advantage of high performance cluster computing and advanced numerical methods to make step-by-step time domain simulation very efficient and compatible with online requirements. Each simulation is performed until the system reaches the steady state condition. This allows the assessment of all stability phenomena. Variable-step-variable-order numerical integration algorithm is employed to make the simulation process more efficient and allow for medium term simulation with small additional computational cost.

7.2 Parallel Implementation

The on-line dynamic security assessment system developed is based on procedures similar to those well established for the off-line environment (i.e., a power flow case is the base for the simulation of various contingencies). Those practices are standard and considered as the reference of accuracy. The basic difference in the on-line environment is the need of automation of the whole process. Also, the requirements on the overall time for the assessment are quite different. The overall DSA scheme is shown in Figure 17.

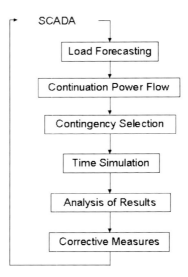

Fig. 17. Dynamic Security Assessment Scheme

The base case is obtained running a load flow using as input data the forecasted load conditions in the near future (from a few minutes to a few hours). The load forecasting module uses data produced by the SCADA system and processed by the state estimator. The continuation method is used to guarantee the convergence of the load flow algorithm. A list of previously defined contingencies is then processed by the contingency selection module which, in the version of the DSA package used in the tests reported in this paper, is based on the tangent vector technique. As a result of this selection, a list of contingencies prone to cause emergency conditions in the system is formed. A time domain simulation is, then, performed for each one of the selected contingencies using a variable step variable order dynamic simulator. The DSA method also contains a module for the analysis of results and a module for corrective actions which were not implemented in the version used in this work.

The most time consuming steps in the DSA scheme shown in Figure 17 are contingency selection and time domain simulation. In the first, a large number of continuation load flow cases need to be performed whereas in the former several relatively long time domain simulation cases need to be run. However, both the load flow and simulation cases are totally independent from each other and can be performed in parallel using the available processors on the cluster.

The parallelization strategy adopted in this case is based on the master-slave programming model, where the master process is responsible for controlling processes allocation, for scattering and gathering data and for ensuring good load balancing. The slave processes, on the other hand, are responsible for solving the different continuation load flow and then solving the different time domain simulation allocated to them by the master dynamic process scheduler.

7.3 Results

The dynamic security assessment parallel application described above was implemented on a Cluster composed of 8 Pentium personal computers interconnected by a Fast-Ethernet (100 Base-T) network via a 12 ports 100 Mbps switch. The test system used is a model of the Brazilian southeastern utility power system composed of approximately 700 buses, 1000 circuits and 80 generators. A list of 96 pre-selected contingencies has been used for evaluation purposes.

The execution time of the sequential and the parallel full optimized code, executed on 1 PC and on 4, 6 and 8 PCs of the cluster, together with their efficiencies, are shown in Table 6.

Table 6. Dynamic Security Assessment - Parallel Results

p=1	p=4		p=8		p=10	
Time (min)	Time (min)	Effic. (%)	Time (sec)	Effic. (%)	Time (sec)	Effic. (%)
3.82	1.22	78.50	49.30	77.91	37.21	76.99

The system is able to assess the security for the selected contingencies in less than 38 seconds on 8 PCs. Note that if faster results are required, one can use faster computers and/or add more computers to the cluster. This is a fair example of cluster scalability benefit.

8 Distribution Network Reconfiguration

This section describes the work developed by Borges et al. [2003] in which the results of a parallel implementation of a Distribution Network Reconfiguration application in a network of PCs is reported. The solution methodology uses Parallel Genetic Algorithms.

8.1 Problem Description

The problem of network reconfiguration for losses minimization deals with changing the network topology by opening and closing switching devices, in order to reduce the electrical losses of a given operation condition, taking into consideration quality, reliability and safety restrictions.

The distribution networks usually operate in a radial configuration. The switching facilities available on the network allow transferring part of the load of a feeder to another. Network reconfiguration can also be used to improve the load distribution between the feeders, to increase the system robustness in relation to fails, to enhance reliability, to facilitate the restoration of de-energized areas with the operation of a smaller number of devices, etc.

The reconfiguration problem can be formulated as a non-linear integer programming problem as shown below (Nara et al., 1992):

$$\text{Minimize} \quad \sum_{j \in \Omega_A} L_j$$
$$C_k$$

$$\text{subject to} \quad \sum S_{j,i} = D_i, \quad i \in \Omega_N$$
$$S_j \leq S_{j,max}, \quad j \in \Omega_A$$
$$\Delta V_i \leq \Delta V_{max}, \quad i \in \Omega_N$$
$$\prod \delta_i = 1, \quad i \in \Omega_N$$

where:

L_j: active losses in branch j;
C_k: k-th configuration;
Ω_A: set of feeders branches;
$S_{j,i}$: power flow in branch j connected to load point i;
D_i: power demand at load point i;
Ω_N: set of load ponts;
S_j: power flow in branch j;
ΔV_i: voltage drop at node i;
ΔV_{max}: maximum voltage drop at node i;
δ_i: binary variable indicating if the load point i is energized (1) or not (0).

The objective function represents the sum of the active losses in all feeders' branches, that is, the total active losses. The first restriction establishes the power balance at the load points. The next two restrictions establish limits in the capacity of each feeder branch and in the voltage drop of each load point. The last restriction imposes that the demand at all load points are attended. Besides the restrictions above, it is also required that the network maintains its radial topology. Other restrictions can be included in the problem formulation as necessary.

The solution of the problem defined above has been tried using a large number of different techniques. In many of them, the solution process is based on an algorithm of power flow which is responsible for the calculation of the active losses, the voltage drop and the power flow on feeders and transformers. The process consists, essentially, in the generation of potential solutions for the problem (by opening and closing devices) and then performing the adequacy verification of each potential solution in order to identify the best one. In this paper, intelligent system techniques based on Genetic Algorithms are used for that purpose. The adequacy evaluation is performed by a power flow algorithm with the representation of dispersed generation.

8.2 Parallel Implementation

A parallel implementation of the problem of network reconfiguration has been implemented based on the parallel multi-population genetic algorithm (PGA), in which a different population is allocated to each processor of the parallel platform and they interchange good candidate solutions to enhance the optimal solution search.

Figure 18 shows a layout of the reconfiguration problem implementation using Multi-Population PGA on a cluster composed of four PCs.

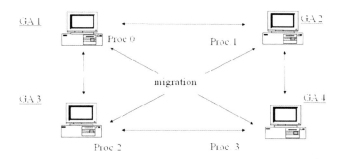

Fig. 18. Multi-Population PGA

8.3 Results

The reconfiguration problem based on the multi-population PGA was implemented on a cluster composed of four Intel Pentium III 1.0 GHz microcomputers, interconnected by a switched Fast-Ethernet. Two different test systems were used. The first test system is composed of 5 distribution feeders arranged in two substations, with a total of 26 switching devices that allow several different operative configurations. The second test system represents an actual substation of the Brazilian distribution system and is composed of 7 feeders with a total of 121 switching devices for reconfiguration.The speedup obtained by the parallel solution was almost linear with respect to the number of processor, requiring less than 8 sec for the first test system, and less than 3 min for the second test system.

9 Concluding Remarks

Electrical energy systems planning, operation, and control requires large amount of computational work due to large mathematical models and fast time response, in certain applications. The use of HPC in this field has been experiencing a steady progress in the last decades and a relatively steep rise in the last few years. This chapter presented a review of potential application areas and described several actual implementations of parallel processing to electrical energy systems problems. In particular, the following applications have been presented: hydrothermal systems optimization, tuning of power system stabilizers, security constrained optimal power flow, composite reliability evaluation, dynamic security assessment, and distribution network reconfiguration.

In the foreseen scenario of smart grid, it will be necessary the manipulation of large amount of data collected from the distributed meters and sensors. For this purpose it will be necessary to apply high performance processors like the modern Graphics Processing Units (GPU) and geographically distributed processing techniques like Cloud Computing.

Appendices

A- Benders Decomposition

Benders Decomposition allows handling separately the base problem and each of the N sub-problems. To represent the possible unfeasibility of each sub-problem, penalty variables are added to the base problem in order to represent the amount of violation associated with the each sub-problem. Therefore, the minimization of the constraints violations can be defined as a new objective function and the sub-problem can be formulated as:

$$w(z_o) = \min d^r \cdot r$$
$$s.t.$$
$$a(z_0) \leq b \tag{5.2}$$

Where $r \geq 0$ is the vector of penalty variables for the group of operative constraints and d^r is the cost vector. From this formulation, it can be seen that if $w(z_0) = 0$, the sub-problem is feasible and if $w(z_0) > 0$, the sub-problem is unfeasible.

The global problem can, then, be re-written in terms of z_o as follows, where the scalar functions $w_i(z_o)$ are the solutions of the sub-problems (5.2) for the given operation point z_o.

$$\min f(z_o)$$
$$s.t.$$
$$a_0(z_0) \leq b_0$$
$$w_i(z_0) \leq 0 \tag{5.3}$$
$$for \quad i = 1, 2, \ldots, N$$

The Benders Decomposition method involves in obtaining an approximation of $w_i(z_o)$ based on an iterative solution of the base problem and the N sub-problems. The Lagrange multipliers associated with the solution of each sub-problem are used to form a linear constraint, known as Benders Cut, which are added to the base problem solution.

B- Genetic Algorithms

Genetic Algorithms (GA) are heuristics optimization methods based on Darwin's specimens evolution theory, in which the best fitted individuals have higher probability of maintaining their genes on the next generations. GA works with the codification of the problem variables in a structure called chromosome. For an ample range of technologic and scientific problems, GA has demonstrated to be a good global optimization method. However, the computational effort required by solving large dimensions optimization problems based on GA may become prohibitive in actual engineering applications. One solution to avoid that inconvenience is the use of parallel processing by the implementation of Parallel Genetic Algorithms (PGA) [8.2]. In this approach, parts of the global problem are solved on different processors that communicate and collaborate for the overall problem solution.

The Multi-Population PGA is a GA parallelization technique in which different populations are assigned to different processors and usually presents the best performance (higher fitness value in smaller computation time). In the Multi-Population PGA, the initial populations are created by randomly generating their chromosomes using different seeds for the random number generator algorithm in each processor. After a fixed number of generations, the populations interchange a certain number of individuals in a process called Migration [Cantú-Paz 1999]. The migration establishes the communication and integration strategy between the populations. The migration operator selects the best chromosome of each processor and spread them to the others, as shown in Figure A1.

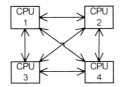

Fig. A1. Simple example of parallelization

The new parameters introduced by the migration operator are:

- Migration interval: establishes the number of generations between successive migrations;
- Migration rate: indicates the number of individuals to be communicated to the other populations at each migration;
- Individuals' Selection strategy: establishes the rules for choosing the individuals that are chosen to migrate from one population to the others;
- Reception strategy: establishes the rules for incorporating the individuals migrated from other populations into the population that receives them.

Acknowledgments. The authors acknowledge the decisive participation of R.J. Pinto (Cepel) and M.E.P. Maceira (Cepel) in the research work reported in section 3; S.L. Escalante (Coppe) and A.L.B. Bomfim (Eletrobras) in the research work reported in section 4; J.M.T. Alves (Cepel) and A.L.O. Filho (Cepel) in the research work reported in section 5; J.L. Jardim (Nexant) and C.A.S. Neto (ONS) in the research work reported in section 7.

References

Allan, R., Billinton, R.: Probabilistic assessment of power systems. Proceedings of the IEEE 88(2), 140–162 (2000)

Alves, J.M.T., Borges, C.L.T., Filho, A.L.O.: Distributed Security Constrained Optimal Power Flow Integrated to a DSM Based Energy Management System for Real Time Power Systems Security Control. In: Daydé, M., Palma, J.M.L.M., Coutinho, Á.L.G.A., Pacitti, E., Lopes, J.C. (eds.) VECPAR 2006. LNCS, vol. 4395, pp. 131–144. Springer, Heidelberg (2007)

Billinton, R., Li, W.: Reliability Assessment of Electric Power Systems Using Monte Carlo Methods. Plenum Press, New York (1984)

Bomfim, A.L.B., Taranto, G.N., Falcao, D.M.: Simultaneous Tuning of Power System Damping Controllers Using Genetic Algorithms. IEEE Trans. on Power Systems 15(1), 163–169 (2000)

Bomfim, A.L.B., Escalante, S.L., Taranto, G.N., Falcao, D.M.: Parallel genetic algorithm to tune multiple power system stabilizers in large interconnected power systems. In: Proceedings of the VIII IREP Symposium - Bulk Power System Dynamics and Control (2010)

Borges, C.L.T., Falcao, D.M., Mello, J.C.O., Melo, A.C.G.: Composite reliability evaluation by sequential monte carlo simulation on parallel and distributed processing environments. IEEE Transactions on Power Systems 16(2), 203–209 (2001-a)

Borges, C.L.T., Falcao, D.M., Mello, J.C.O., Melo, A.C.G.: Concurrent composite reliability evaluation using the state sampling approach. Electric Power Systems Research 57, 149–155 (2001-b)

Borges, C.L.T., Manzoni, A., Viveros, E.R.C., Falcao, D.M.: A parallel genetic algorithm based methodology for network reconfiguration in the presence of dispersed generation. In: Proceedings of 17th CIRED - International Conference on Electricity Distribution (2003)

Borges, C.L.T., Taranto, G.N.: A unified online security assessment system. In: Proceedings of the 38th CIGRÉ Session - Session (2000)

Cantú–Paz, E.: Topologies, migration rates, and multi-population parallel genetic algorithms. Illinois Genetic Algorithms Laboratory – IlliGAL Report No. 99007, University of Illinois, Urbana–Champaign (1999)

Chai, J., Bose, A.: Bottlenecks in parallel algorithms for power system stability analysis. IEEE Transactions on Power Systems 8(1), 9–15 (1993)

Decker, I.C., Falcao, D.M., Kaszkurewicz, E.: Conjugate gradient methods for power system dynamic simulation on parallel computers. IEEE Transactions on Power Systems 11(3), 1218–1227 (1996)

Fan, J., Borlase, S.: The Evolution of Distribution. IEEE Power & Energy Magazine 7(2), 63–68 (2009)

Falcao, D.M., Kaszkurewicz, E., Almeida, H.L.S.: Application of parallel processing techniques to the simulation of power system electromagnetic transients. IEEE Transactions on Power Systems 8(1), 90–96 (1993)

Falcao, D.M.: High Performance Computing in Power System Applications. In: Palma, J.M.L.M., Dongarra, J. (eds.) VECPAR 1996. LNCS, vol. 1215, Springer, Heidelberg (1997)

Faria Jr, H., Binato, S., Falcao, D.M.: Power transmission network design by greedy randomized adaptive path relinking. IEEE Transactions on Power Systems 20(1), 43–49 (2005)

Gen, M., Cheng, R.: Genetic Algorithms and Engineering Design. John Wiley & Sons (1997)

Granville, S., Lima, M.C.A.: Application of decomposition techniques to var plan-ning: methodological & computational aspects. IEEE Transactions on Power System 9(4), 1780–1787 (1994)

Ipakchi, A., Albuyeh, F.: Grid of the future. IEEE Power and Energy Magazine 7(2), 52–62 (2009)

Jardim Jardim, J., Neto, C.A.S., Alves da Silva, A.P., Zambroni de Souza, A.C., Falcao, D.M., Borges, C.L.T., Taranto, G.N.: A unified online security assessment system. In: Proceedings of the 38th CIGRÉ Session - Session (2000)

Kim, J.W., Kim, S.W., Park, P.G., Park, T.J.: On The Similarities between Binary-Coded GA and Real-Coded GA in Wide Search Space. IEEE Press (2002)

LAPACK - Linear Algebra PACKage, http://www.netlib.org/lapack/

Latorre, G., Cruz, R.D., Areiza, J.M., Villegas, A.: classification of publications and models on transmission expansion planning. IEEE Transactions on Power System 18(2), 938–946 (2003)

Rodriguez, J.I.R., Falcao, D.M., Taranto, G.N., Almeida, H.L.S.: Short-term transmission ex-pansion planning by a combined genetic algorithm and hill-climbing technique. In: Proceedings of the International Conference on Intelligent Systems Applications to Power Systems (2009)

Maceira, M.E.P., Terry, L.A., Costa, F.S., et al.: Chain of Optimization Models for Setting the Energy Dispatch and Spot Price in the Brazilian System. In: Proccedings of the Power System Computation Conference - PSCC 2002 (2002)

Monticelli, A., Pereira, M.V.F., Granville, S.: Security-constrained optimal power flow with post-contingency corrective rescheduling. IEEE Transactions on Power System 2(1), 175–182 (1987)

NACAD, Center for Service in High Performance Computing - COPPE/UFRJ, http://www.nacad.ufrj.br/index.php

Nara, K., Shiose, A., Kitagawa, M., Ishihara, T.: Implementation of genetic algorithm for the distribution systems loss minimum re-configuration. IEEE Transactions on Power Systems 7(3), 1044–1051 (1992)

ONS, Electric System National Operator, http://www.ons.org.br/

Ott, A.L.: Evolution of computing requirements in the PJM market: past and future. In: Proceedings of the IEEE PES General Meeting (2010)

PacDyn, ELETROBRAS CEPEL – User Manual, Rio de Janeiro, Brazil

Pereira, M.V.F., Pinto, L.M.V.G.: Multi-stage Stochastic Optimization Applied to Energy Planning. Mathematical Programming 52, 359–375 (1991)

Pinto, R.J., Duarte, V.S., Maceira, M.E.P., Borges, C.L.T., Falcao, D.M., Aveleda, A.A., Hinrichsen, A.C.P.A.: Methodology for the applications of distributed processing to the expan-sion and operational planning. In: Proceedings of the XXXIX Brazilian Symposium of Operational Research (2007) (in Portuguese)

Pinto, R.J., Borges, C.L.T., Maceira, M.E.P.: Distributed processing applied to hydrothermal system expansion and operating planning. In: Proceedings of the 30th Iberian Latin American Congress on Computational Methods in Engineering (2009)

Rodriguez, J.I.R., Falcao, D.M., Taranto, G.N., Almeida, H.L.S.: Short-term transmission ex-pansion planning by a combined genetic algorithm and hill-climbing technique. In: Proceedings of the International Conference on Intelligent Systems Applications to Power Systems (2009)

Stanley, H.H., Phadke, A.G., Renz, A.B.: The future of power transmission. IEEE Power & Energy Magazine 8(2), 34–40 (2010)

Trachian, P.: Machine learning and windowed subsecond event detection on PMU data via hadoop and the open PDC. In: Proceedings of the IEEE PES General Meeting (2010)

Tomim, M.A., Martí, J.R., Wang, L.: Parallel solution of large power system networks using the multi-area Thévenin equivalents (MATE) algorithm. International Journal of Electrical Power & Energy Systems 31(9), 497–503 (2009)

Tylavsky, D.J., Bose, A., Alvarado, F., Betancourt, R., Clements, K., Heydt, G.T., Huang, I.M., La Scala, M., Pai, M.A.: Parallel processing in power systems computation. IEEE Transactions on Power Systems 7(2), 629–638 (1992)

Zhang, L., Hoyt, A., Calson, B.: A review of high performance technical computing in Midwet ISO. In: Proceedings of the IEEE PES General Meeting (2010)

Zhu, J., Chow, M.: A review of emerging techniques on generation expansion planning. IEEE Transactions on Power System 12(4), 1722–1728 (1997)

High Performance Computing for Power System Dynamic Simulation

Siddhartha Kumar Khaitan and James D. McCalley

Electrical and Computer Engineering,
Iowa State University, Iowa, USA
{skhaitan,jdm}@iastate.edu

Abstract. High-speed extended term (HSET) time domain simulation (TDS) is intended to provide very fast computational capability to predict extended-term dynamic system response to disturbances and identify corrective actions. The extended-term dynamic simulation of a power system is valuable because it provides ability for the rigorous evaluation and analysis of outages which may include cascading. It is important for secure power grid expansion, enhances power system security and reliability, both under normal and abnormal conditions. In this chapter the design of the envisioned future dynamic security assessment processing system (DSAPS) is presented where HSET-TDS forms the core module. The power system is mathematically represented by a system of differential and algebraic equations (DAEs). These DAEs arise out of the modeling of the dynamic components such as generators, exciters, governors, automatic generation control, load tap changers, induction motors, network modeling and so on. To provide very fast computational capability within the HSET-TDS, this chapter motivates the need for high performance computing (HPC) for power system dynamic simulations through detailed modeling of power system components and efficient numerical algorithms to solve the resulting DAEs. The developed HSET-TDS is first validated for accuracy against commercial power simulators (PSSE, DSA Tools, Power-World) and then it is compared for computational efficiency. The chapter investigates some of the promising direct sparse linear solver for fast extended term time domain simulation and makes recommendation for the modern power grid computations. The results provide very important insights with regards to the impact of the different numerical linear solver algorithms for enhancing the power system TDS.

1 Introduction

HSET-TDS is intended to provide very fast computational capability to predict extended-term dynamic system response to disturbances and identify corrective actions. The design of HSET-TDS is motivated by the low-probability, high-consequence events to which the power systems are continuously exposed. Such events, usually comprised of multi-element (so-called "N-k") outages, often cause additional cascading events spanning minutes or even hours, and are typically perceived to be unlikely and therefore undeserving of preventive action. These preventive actions are also associated increased costs due to off-economic dispatch. Yet, such events do occur in today's energy control centers. The operational personnel have no decision-support function available to assist them in identifying effective corrective action, or even in becoming familiar with system performance under such potentially high-consequence events. There is indication in other industries (e.g., airline, nuclear, process control) that they employ a computational capability which provides operators with ways to predict system response and identify corrective actions. We believe that the power system operators should also have a similar capability. With this motivation, we present the design essential features of the HSET-TDS.

1.1 Dynamic Security Assessment Processing System (DSAPS)

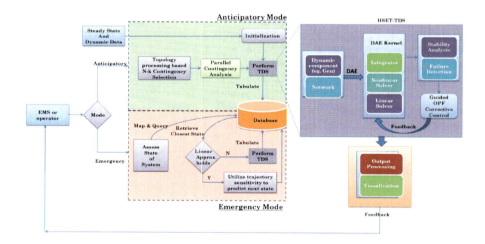

Fig. 1. Overall framework of the dynamic security assessment processing system

HSET is intended to be a part of the energy management system (EMS), and serve as a decision support tool. Fig. 1 shows the overall architecture of the HSET-TDS as the future DSAPS [15–18].

The simulator is envisioned to operate in two modes namely the anticipatory mode and the emergency mode. HSET-TDS will be usually in an anticipatory mode, continuously computing responses to many contingencies (contingency selection will be based on the topological processing of the power grid which is represented by the node breaker model rather than bus branch model) and storing preparatory corrective actions that would be accessed by operational personnel should one of the contingencies occur. The goal of anticipatory computing is to cover as much of the event-probability space as possible within a particular window of computing time. Results can be archived and reused when similar conditions are met. Since the number of N-k contingencies explodes with the system size, we employ load-balanced contingency parallelization (discussed in a separate chapter of the book).

HSET-TDS will be available in an emergency mode where it is run following initiation of a severe disturbance. In this mode, HSET-TDS will compute corrective control actions based on direct integration or previously computed linear trajectory sensitivities if the current state of the system is within the region of accuracy of the closest retrieved system state. In case it performs direct integration, the results would be archived.

The power system is mathematically represented by a system of differential and algebraic equations (DAEs). These DAEs arise out of the modeling of the dynamic components such as generators, exciters, governors, automatic generation control, load tap changers, induction motors and from the network modeling. The network is represented by algebraic equations whereas the other components comprise of both differential and algebraic equations. Together they form the DAE. The power system dynamic response to disturbances is decided not only by fast dynamics of its machines, but also by the action of slow processes such as tap changers and load dynamics. They often cause voltage problems and/or thermal loading problems after an extended period. To capture such effects, simulators must be capable of modeling both fast and slow dynamics and also capable of lengthening time steps when fast dynamics are inactive. This requres an adaptive time stepping integration algorithm. The differential equations are discretized using the integration algorithm. The resulting system of nonlinear equations are solved using a variant of the Newton method. The linear system subproblem of DAE forms a major part of the solution and is highly ill conditioned for TDS and thus quite challenging. With recent advances in sparse linear solver algorithms, it is possible to significantly speedup the solution of the DAE. We take advantage of these algorithms to enhance the computational efficiency of the HSET-TDS. Later in the chapter we give more detailed information about the choice of different linear solvers and their impact on the computational efficiency of simulations.

The simulator must also have the necessary intelligence to recognize, based on the evolution of the state variables, when failure conditions are encountered, including detection of the operation of as-designed protective relays. Once such conditions are encountered it would determine appropriate corrective actions and restart the simulation at some particular point in time (in advance of the failure condition) with the corrective action modeled.

Corrective controls will be identified by a guided optimal power flow (GOPF). Following identification of a failure condition, HSET-TDS will access embedded intelligence to select from a group of possible actions; the amount of corrective action will be determined by the optimal power flow (OPF). The philosophy is to prepare and revise, track and defend (see Fig. 1).

2 Modeling

Recent advances in HPC have fueled the growth of several computation-intensive fields(e.g. [2, 14, 32]). The need for HPC is strongly motivated by the large and complex interconnected infrastructure of the modern power system with increased renewable sources, electronic controls and associated requirement of detailed modeling to accurately capture the operation of the grid and to control it. Adequate simulation models are indispensable to perform various studies for maintaining system reliability at local and regional level. To capture the multi-scale dynamics, the components could be modeled to different degrees of detail according to the phenomenon studied. As the complexity of the models increase the dimension of the DAEs increase and computation burden increases nonlinearly. Therefore it is a common practice to do model reduction to gain computational efficiency. However, it comes with a cost of losing accuracy and very fast transients. Therefore, on the one hand detailed modeling is important and one the other hand it requires very efficient numerical algorithms on fast hardware architecture for enabling very fast near real-time simulations. This is very computationally challenging especially for time frame from minutes to hours and HPC would play a major role in such simulation studies. In this chapter we particularly utilize very fast linear solver algorithms to gain computational efficiency. This is discussed later in the chapter.

HSET-TDS is designed to seamlessly interface with the components, modeled at different levels of granularity and complexity. HSET-TDS has an extensive library of generator (2nd, 3rd, 4th, 6th order, GENROU etc , exciter (IEEET1,IEEEX1 etc) and governor (TGOV1,GAST, HYGOV etc) models. In this chapter we present the modeling details of the network and the 6th order generator, IEEET1 exciter and the TGOV1 governor. Other modeling details are left out for simplicity. Interested readers can refer to the [1], [19],[25],[5] for further details of the models.

2.1 Reference Frame

The voltage equations developed for an elementary synchronous machine have machine inductances which are functions of the rotor speed. The time varying inductances in the synchronous machine is due to varying magnetic reluctances of electric circuits in relative motion. This makes the coefficients of differential equations, which describe the dynamic behavior of the synchronous machine, as time varying quantities except when the rotor is stalled. In order to reduce the complexity of solving these time varying differential equations, the time varying parameters are transformed to another frame of reference where they behave as constants. The transformation of variables to a frame of reference rotating at an arbitrary angular velocity can be thought of as a change of coordinates from one reference to another one. By choosing different speed of the rotation of the reference frame, different transformations can be obtained. In the analysis of power system components and electric machines, the following reference frames are generally used

1. Arbitrary reference frame : ω=Arbitrary value
2. Stationary reference frame : ω=0
3. Rotor reference frame : $\omega=\omega_r$
4. Synchronous reference frame : $\omega=\omega_s$

For the analysis of the synchronous machine, the time varying inductances are eliminated only if rotor reference frame is chosen. This transformation is commonly known as Park's transformation where the stator variables are referred to a reference frame fixed in rotor [5]. However in the analysis of induction machine, synchronous reference frame would be more beneficial, where both the stator and rotor variables are transformed to a reference frame rotating in synchronism with the rotating magnetic field.

The details of winding inductances and voltage equations derivations and their transformation to a suitable reference frame can be found in [19],[25],[5]. Each one of these references chooses a particular alignment of q-d axis which leads to differences in the final equations' notations and signs, though all of them are perfectly correct and valid and one can choose any one of them for analysis.

2.2 Network Modeling

The power system network consists of transformers and interlinked transmission lines each of which are modeled using π equivalent circuit. The whole power system network is represented by the nodal network equation as shown in (1).

$$\begin{bmatrix} \widetilde{I}_1 \\ \vdots \\ \widetilde{I}_i \\ \vdots \\ \widetilde{I}_n \end{bmatrix} = \begin{bmatrix} \overline{Y}_{11} & \cdots & \overline{Y}_{1i} & \cdots & \overline{Y}_{1n} \\ \vdots & \ddots & \vdots & & \vdots \\ \overline{Y}_{i1} & \cdots & \overline{Y}_{ii} & \cdots & \overline{Y}_{in} \\ \vdots & & \vdots & \ddots & \vdots \\ \overline{Y}_{n1} & \cdots & \overline{Y}_{ni} & \cdots & \overline{Y}_{nn} \end{bmatrix} \begin{bmatrix} \widetilde{V}_1 \\ \vdots \\ \widetilde{V}_i \\ \vdots \\ \widetilde{V}_n \end{bmatrix} \quad (1)$$

In compact form, the network equations are represented by $\widetilde{\underline{I}} = \overline{\underline{Y}} * \widetilde{\underline{V}}$, where the matrix $\overline{\underline{Y}}$ represents the nodal admittance matrix, $\widetilde{\underline{I}}$ and $\widetilde{\underline{V}}$ are current and voltage phasors at 60 Hz. [Note: Phasors are represented with a tilde

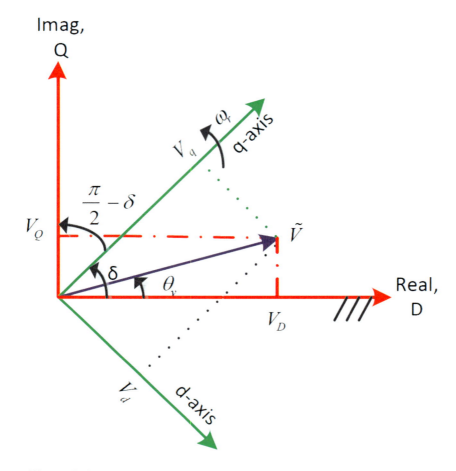

Fig. 2. Relationship between the generator and the network reference frame

sign at the top, whereas complex numbers are represented by bar sign at the top].

2.3 Need for Change of Reference Frame

The power system network equation (1) is in network's complex coordinates say (D,Q) while the synchronous generator equations are in rotor reference frame (q,d orthogonal coordinates). The relationship between the two system of coordinates is shown in Fig. 2. Here, the q-axis of the synchronous generator is lagging with respect to the network coordinate Q-axis by an angle ($\frac{\pi}{2} - \delta$). The coordinate D is along positive x axis direction and Q is along positive y axis direction. The relationship between the network coordinates and the generator coordinates is given by (2)

$$\begin{bmatrix} F_D \\ F_Q \end{bmatrix} = \begin{bmatrix} sin(\delta) & cos(\delta) \\ -cos(\delta) & sin(\delta) \end{bmatrix} \begin{bmatrix} F_d \\ F_q \end{bmatrix}$$
$$F_{DQ} = T * F_{dq} \tag{2}$$

where T is the transformation matrix and F denotes voltage and current quantities. There are as many (d,q) system of coordinates as the number of generators, as each generator may operate at a particular rotor angle δ.

2.4 Power System Component Modeling

Generator Modeling. The differential equations for a 6th order generator model, given in (3) is obtained by considering d-axis lagging the q-axis by $\frac{\pi}{2}$ and aligned at an angle δ with respect to the network coordinates [25], as shown in Fig. 2.

$$\frac{2H_i}{\omega_s}\frac{d\omega_i}{dt} = T_{mi} - T_{ei} \tag{3a}$$

$$\frac{d\delta_i}{dt} = \omega_i - \omega_s \tag{3b}$$

$$T'_{doi}\frac{dE'_{qi}}{dt} = -E'_{qi} - (X_{di} - X'_{di})I_{di} + E_{fdi} \tag{3c}$$

$$T'_{qoi}\frac{dE'_{di}}{dt} = -E'_{di} - (X_{qi} - X'_{qi})I_{qi} \tag{3d}$$

$$T''_{doi}\frac{dE''_{qi}}{dt} = \frac{(X''_q - X_{ls})}{(X'_d - X_{ls})}\dot{E}'_{qi} + E'_{qi} - E''_{qi} + (X'_d - X''_d)I_{qi} \tag{3e}$$

$$T''_{qoi}\frac{dE''_{di}}{dt} = \frac{(X''_q - X_{ls})}{(X'_d - X_{ls})}\dot{E}'_{di} + E'_{di} - E''_{di} + (X'_q - X''_q)I_{di} \tag{3f}$$

$$T_{ei} = E_{qi}I_{di} + E_{di}I_{qi} + (X'_{qi} - X'_{di})I_{di}I_{qi} + D_i(\omega_i - \omega_s) \tag{3g}$$

The algebaric equations of the generator are as follows:

$$I_{Di} = I_{di}sin(\delta_i) + I_{qi}cos(\delta_i) \tag{4a}$$
$$I_{Qi} = -I_{di}cos(\delta_i) + I_{qi}sin(\delta_i) \tag{4b}$$
$$V_{di} = V_{Di}sin(\delta_i) - V_{Qi}cos(\delta_i) \tag{4c}$$
$$V_{qi} = V_{Di}cos(\delta_i) + V_{Qi}sin(\delta_i) \tag{4d}$$
$$E''_{di} = V_{di} + R_{si}I_{qi} - X''_q I_{qi} \tag{4e}$$
$$E''_{qi} = V_{qi} + R_{si}I_{qi} + X''_d I_{di} \tag{4f}$$

Exciter. The IEEET1 [1] is a standard exciter model (Fig. 3).

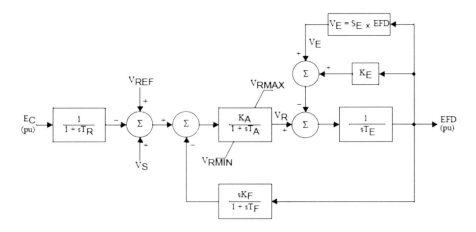

Fig. 3. IEEET1 Model[1]

The differential equations for IEEET1 exciter are as follows:

$$T_{\text{Ei}} \frac{dE_{\text{fdi}}}{dt} = -(K_{\text{Ei}} + S_{\text{Ei}}(E_{\text{fdi}}))E_{\text{fdi}} + V_{\text{Ri}} \tag{5a}$$

$$T_{\text{Fi}} \frac{dR_{\text{fi}}}{dt} = -R_{\text{fi}} + \frac{K_{\text{Fi}}}{T_{\text{Fi}}} * E_{\text{fdi}} \tag{5b}$$

$$T_{\text{Ai}} \frac{dV_{\text{Ri}}}{dt} = -V_{\text{Ri}} + K_{\text{Ai}}R_{\text{Fi}} - \frac{K_{\text{Ai}}K_{\text{Fi}}}{T_{\text{Fi}}} + K_{\text{Ai}}(V_{\text{refi}} - V_{\text{i}}) \tag{5c}$$

In steady state the equation becomes,

$$-(K_{\text{Ei}} + S_{\text{Ei}}(E_{\text{fdi}}))E_{\text{fdi}} + V_{\text{Ri}} = 0 \tag{6a}$$

$$\text{Where, } S_{\text{Ei}}(E_{\text{fdi}}) = A_{\text{x}} e^{B_{\text{x}} E_{\text{fdi}}} \tag{6b}$$

$$R_{\text{fi}} = \frac{K_{\text{Fi}}}{T_{\text{Fi}}} * E_{\text{fdi}} \tag{6c}$$

$$V_{\text{refi}} = V_{\text{i}} + \frac{V_{\text{Ri}}}{K_{\text{Ai}}} \tag{6d}$$

Governor. The block diagram of TGOV1 [1] governor model in s-domain is shown in Fig. 4.

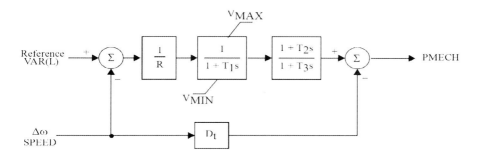

Fig. 4. TGOV1 Model[1]

The differential and algebraic equations for the TGOV1 model are given below.

The differential equations are:

$$\frac{dP_{m0}}{dt} = \frac{\frac{(\omega_{ref}-\omega+1)}{R} - P_{m0}}{T_1} \tag{7a}$$

$$\frac{dP_{m1}}{dt} = \frac{\frac{(1-\omega)T_2}{T_1 R} + \frac{\omega_{ref}T_2}{T_1 R} + \frac{(1-T_2)(P_{m0})}{T_1} - P_{m1}T_1}{T_3} \tag{7b}$$

The algebraic equations are:

$$P_{mech} = P_{m1} - \Delta\omega D_t \tag{7c}$$

2.5 Initialization Process for Dynamic Simulation

The overall procedure for initialization of generator and other components for the dynamic simulation in HSET-TDS is shown in Fig. 5. The detailed steps and equations are enumerated below.

1. Data conversion
   The parameters of all the dynamic models should be converted to a common base. Often a base value of 100 MVA is used.
2. Calculation of generator currents in network reference frame
   The first step in the initialization process is to find out the current injected from the generator node into the network. In order to find out the injected current from a particular generator, the real and reactive power of the corresponding generator and the complex voltage of the bus to which the generator is connected are required. These informations are

obtained from the power flow solution. Equation (8) is used to calculate the injected currents from a generator in network reference frame (D-Q).

$$I_i \angle \theta_{I_i}{}^* = \frac{P_{Gi} + jQ_{Gi}}{V_i \angle \theta_{V_i}}$$
$$P_{Gi} = P_i - P_{Li}, \; Q_{Gi} = Q_i - Q_{Li} \tag{8}$$
$$I_{Di} + jI_{Qi} = I_i \angle \theta_{I_i}$$

P_{Gi} is the real power injection from the generator at bus 'i'. P_{Li} is the real power consumed by load at bus 'i' and P_i is the total real power injected into the network from the bus node 'i'. Similarly, Q_{Gi} specifies the reactive power produced by the generator at bus 'i'. Q_{Li} is the reactive power consumed by load at bus 'i' and Q_i is the total reactive power injected into the network from the bus node 'i'. The angle $\angle \theta_{I_i}$ specifies how much the current phasor I_i is leading or lagging with respect to D-axis of the network coordinate reference.

As a check for the calculated generator injected currents, (9) can be used. If there are static loads connected at that particular generator node, then those loads should be converted into equivalent impedances or admittance using Eq. (10).

$$I_i \angle \theta_{I_i} = \sum_{j=1}^{N} \overline{Y}_{ij} \widetilde{V}_j \tag{9}$$

$$\overline{Y}_{L,i} = \frac{P_{L,i}}{V_{L,i}^2} - j\frac{Q_{L,i}}{V_{L,i}^2} \tag{10}$$

3. Compute generator torque angle or machine internal angle δ

   Figure (6) shows the representation of stator algebraic equations in steady state condition. The torque angle δ is the angle of the internal voltage behind the impedance $(R_s + jX_q)$ and is computed using (11)

$$\delta_i = \text{angle}(V_i \angle \theta_{V_i} + (R_{si} + jX_{qi})I_i \angle \theta_{I_i}) \tag{11}$$

4. Calculation of generator currents in machine reference frame

   The currents injected from the generator node calculated in network coordinates (D-Q) are transformed into machine reference frame (d-q) using (12)

$$I_{di} + jI_{qi} = I_i \angle (\theta_{I_i} - \delta_i - \frac{\pi}{2})$$
$$I_{di} = I_{Di} \sin(\delta_i) - I_{Qi} \cos(\delta_i) \tag{12}$$
$$I_{qi} = I_{Qi} \sin(\delta_i) + I_{Di} \cos(\delta_i)$$

5. Calculation of generator voltages in machine reference frame

   Similarly, the voltage at the generator node in network coordinates (D-Q) are transformed into machine reference frame (d-q) using (13)

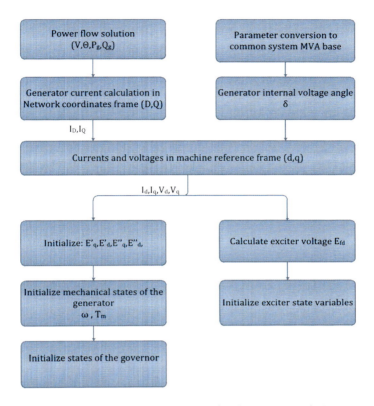

Fig. 5. Initialization procedure for dynamic simulation

$$V_{di} + jV_{qi} = V_i \angle (\theta_{V_i} - \delta_i - \frac{\pi}{2})$$
$$V_{di} = V_{Di} sin(\delta_i) - V_{Qi} cos(\delta_i) \quad (13)$$
$$V_{qi} = V_{Qi} sin(\delta_i) + V_{Di} cos(\delta_i)$$

6. Compute E'_{di} and E'_{qi} phasors of each generator

The initial values for E'_{di} and E'_{qi} are calculated using (14) and (15) respectively. As a check for the initial calculated value of E'_{di}, (16) is used. Equation (16) is obtained from (3d) by setting the differential equation to zero.

$$E'_{di} = V_{di} + R_{si} I_{di} - X'_{qi} I_{qi} \quad (14)$$

$$E'_{qi} = V_{di} + R_{si} I_{qi} + X'_{di} I_{di} \quad (15)$$

$$E'_{di} = (X_{qi} - X'_{qi}) I_{qi} \quad (16)$$

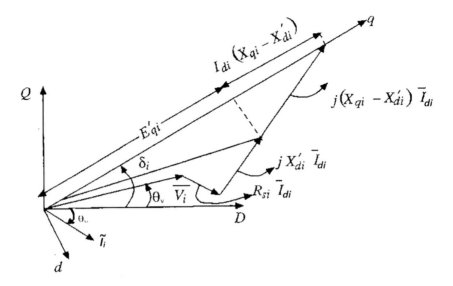

Fig. 6. Steady state representation of generator equations

7. Compute $E_{di}^{''}$ and $E_{qi}^{''}$ phasors of each generator
   Using (4e) and (4f), the initial values for $E_{di}^{''}$ and $E_{qi}^{''}$ are calculated respectively as shown in

$$\begin{aligned} E_{di}^{''} &= V_{di} + R_{si}I_{qi} - X_q^{''}I_{qi} \\ E_{qi}^{''} &= V_{qi} + R_{si}I_{qi} + X_d^{''}I_{di} \end{aligned} \tag{17}$$

8. The exciter initialization
   The excitation voltage E_{fd} is initialized using (18a) which is obtained by setting the derivative of (3c) equal to zero. Once E_{fd} is known, by setting the differential equations (5a) - (5c) equal to zero gives the exciter initialization equations (18b) - (18d)

$$E'_{\text{fd}i} = E'_{\text{q}i} + (X_{\text{d}i} - X'_{\text{d}i})I_{di} \tag{18a}$$

$$V_{\text{R}i} = (K_{\text{E}i} + S_{\text{E}i}(E'_{\text{fd}i}))E'_{\text{fd}i} \tag{18b}$$

$$R_{\text{f}i} = \frac{K_{\text{F}i}}{T_{\text{F}i}}E'_{\text{fd}i} \tag{18c}$$

$$V_{\text{ref}i} = V_i + \frac{V_{\text{R}i}}{K_{\text{A}i}} \tag{18d}$$

9. Compute the mechanical states ω and T_m of the generator
   The generator mechanical states ω and T_m are computed using (19a) and (19b) respectively, which are obtained by setting the derivatives of (3b) and (3a) as equal to zero.

$$\omega_i = 1 \tag{19a}$$

$$T_{mi} = E'_{di} I_{di} + E'_{qi} I_{qi} + (X'_{qi} - X'_{di}) I_{qi} I_{di} \tag{19b}$$

10. Compute the governor initial states
    From the algebraic equation of governor (7c), the variable $P_{m1,i}$ is initialized. By setting the differential equation (7a) equal to zero, the variable $P_{m0,i}$ is initialized as shown in (20b)

$$P_{m1,i} = T_{mi}^{p.u} \tag{20a}$$

$$P_{m0,i} = \frac{\omega_{ref}}{R} \tag{20b}$$

Proper initialization is a must for dynamic simulation. A quick check of proper initialization is to simulate a flat run and see that all the variables remain constant.

3 Solution Methodology

A set of differential equations represented by (21) describes the dynamic behavior of the power system components which includes the generator, exciter, governor, and dynamic loads like induction motor.

$$\dot{x} = f(x, y) \tag{21}$$

$$0 = g(x, y) \tag{22}$$

The differential equations of generator, exciter and governor given in (3), (5) and (7) respectively together form (21). Similarly all the algebraic equations are collected together in (22) which mainly includes the network equations specified by (1), and the generator and the governor algebraic equations as in (4) and (7c) respectively. Equations (21) and (22) defines the electromechanical state of power system at any instant of time and these two sets of equations forms the differential and algebraic model (DAE) of the power system. A system of differential equations are referred to be a stiff system if they have both slow varying and fast dynamics present in them. If the generator dynamic models include the sub transient equations (fast varying components) and slow rotor dynamics, then the DAE of the power system becomes stiff. The integration method which is used to solve the DAE of the power system should be capable of handling stiff systems.

3.1 Strategy

The effectiveness of solving DAE of the power system depends on the numerical algorithms and the type of generator models used. The DAE of the power system can be solved either by the partitioned solution strategy (also called alternating solution strategy) or the simultaneous solution approach [21]. In *partitioned solution* or *alternating solution* strategy, for a given time step, both the differential equations and the algebraic equations are solved separately. The solution of the differential equations x, is used as a fixed input while solving the the algebraic equations and similarly the solution of the algebraic equations y, is used as a fixed input while solving the differential equations. This process is repeated iteratively until the solution converges. In the case of simultaneous or combined solution approach, the differential equations are discretized using an integration methods to form a set of non linear equations which are then combined with the algebraic equations. Equation (23) represents the full set of non linear equations that describes the DAE of the power system.

$$0 = F(x, y) \tag{23}$$

The partitioned solution approach is suited for simulations that covers a shorter time interval. This method allows a number of simplifications like partial matrix inversion, triangular factorization to speed up the solution. However, if proper care is not taken in handling these simplifications it will lead to large interfacing errors. In the case of simultaneous appraoch, Newton's method is usually employed to solve the set of algebraic equations. It ensures no convergence problem even if large integration step length is used with stiff systems like power system. Since both differential and algebraic equations are solved together; there are no interfacing problems. This method also allows for large and variable time steps which is not feasible for partitioned strategy. In this research simultaneous approach is adopted which is suitable for both short-term and long-term dynamic simulations.

3.2 Numerical Methods for Simultaneous Solution Strategy

Computational efficiency is a major bottleneck in the deployment of online TDS in today's control centers even for a short amount of simulated time, e.g., 10 seconds, for a limited number of contingencies. Online simulation of minutes to hours (extended-term) for a very large number of contingencies requires computational efficiency several orders of magnitude greater than what is today's state-of-the-art. We are not aware of any company/control center that has implemented extended-term simulation capability for online purposes. This motivates the current research to achieve high computational efficiency for online deployment of extended-term TDS. In this chapter this is addressed through algorithmic improvements and software

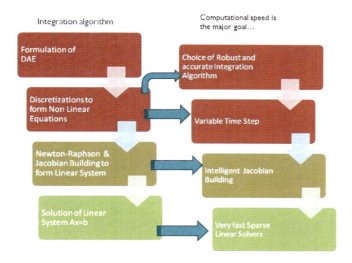

Fig. 7. Integration process

implementation for solution of DAE through variable time step integrators and sparse linear solvers for solving the linear system of equations that arise in the TDS. However we focus only on the linear solvers in this chapter.

Fig. 7 shows, on the left side, the steps involved in the solution of the DAE. The discretizations of the DAE with any implicit integration scheme results in a set of nonlinear equations. The core of the resulting nonlinear equations is the solution of a sparse linear system. On the right side of Fig. 7 the potential areas within the integration process where computational efficiency can be enhanced are shown. We have addressed all of them in this research. Specifically, the solution of the linear system is the most computationally intensive part of a DAE solver. This is exploited in the work described here, via implementation of the state-of-the-art sparse linear solver algorithms. When combined with a robust integration scheme, this achieves very fast TDS.

3.3 Integration Method

There have been several efforts to develop robust and fast numerical integration methods to deal with stiffness and ensure A-stability [3, 6–8, 17, 24]. Different numerical integration schemes differ in their convergence, order, stability, and other properties. In HSET-TDS we have implemented several different explicit and implicit integrators such as backward differentiation formula (BDF), Trapezoidal, Euler etc.

3.4 Nonlinear Solver

The equation (23) is the set of the nonlinear equations which are solved iteratively by the Newton method at each time step. Newton methods have quadratic convergence. However it is a common practice to keep the jacobian constant over a large number of time steps to alleviate the need to refactorize it at every time step. This is commonly called as very dishonest Newton method (VHDN). VHDN in general does not have quadratic convergence. However, VHDN results in significant computational gain. The Newton iterations terminate when the residual vectors are smaller than pre-specified tolerances based on norm and rate of convergence. Computation of the correction vectors requires the solution of the set of linear equations which is discussed next.

3.5 Linear Solvers

There have been significant advances in the field of sparse linear solvers. Linear solvers can be categorized into direct linear solvers and the iterative linear solvers. There are a number of algorithms available for both the direct and the iterative solvers. However not all algorithms are suitable for all applications like in power system computations. The choice of the algorithm depends on the problem at hand and the numerical characteristics of the jacobian involved. Some of the characteristics that impact the choice of the methods are a) diagonal dominance (b) numerical stability (c) symmetry (d) conditioning (e) structure (banded, tridiagonal, etc.) and so on. In this research we focus on the direct linear solvers and present the impact of the three different sparse solvers (UMFPACK,KLU and PARDISO) [9, 13, 29] on the computational efficiency of the short-term transient simulations. However there are other sparse solvers (like WSMP, HSL MA78 etc) available which would evaluated in our future research.

In general, the algorithms for sparse matrices are more complicated than for dense matrices. The complexity is mainly attributed to the need to efficiently handling fill-in in the factor matrices. A typical sparse solver consists of four distinct phases as opposed to two in the dense solver algorithms:

1. The ordering step minimizes the fill-in and exploits special structures such as block triangular form.
2. An analysis step or symbolic factorization determines the nonzero structures of the factors and creates suitable data structures for the factors.
3. Numerical factorization computes the factor matrices.
4. The solve step performs forward and/or backward substitutions.

The section below presents a brief overview of the sparse direct solvers that are being used in this study.

UMFPACK. In the present study, UMFPACK v5.6.0 [12] is used. UMFPACK consists of a set of ANSI/ISO C routines for solving unsymmetric sparse linear systems using the unsymmetric multifrontal method. It requires the unsymmetric, sparse matrix to be input in a sparse triplet format. Multifrontal methods are a generalization of the frontal methods developed primarily for finite element problems [4] for symmetric positive definite systems which were later extended to unsymmetric systems [11]. UMFPACK first performs a column pre-ordering to reduce the fill-in. It automatically selects different strategies for pre-ordering the rows and columns depending on the symmetric nature of the matrix. The solver has different built in fill reducing schemes such as COLAMD [12] and approximate minimum degree (AMD) [10]. During the factorization stage, a sequence of dense rectangular frontal matrices is generated for factorization. A supernodal column elimination tree is generated in which each node in the tree represents a frontal matrix. The chain of frontal matrices is factorized in a single working array.

KLU. It is a software package, written in C, for solving sparse unsymmetric linear systems of equations that typically arise in circuit simulation applications [13]. It performs a permutation to block triangular form (BTF), and employs several methods for fnding a fill-reducing ordering and implements Gilbert/Peierls sparse left-looking LU factorization algorithm to factorize each block. KLU is found to be particularly suitable for power system dynamic simulation applications and outperforms other linear solvers used in this study.

PARDISO. The PARDISO package is high-performance, robust, memory efficient and easy to use software for solving large sparse symmetric and unsymmetric linear systems of equations. The solver uses a combination of left and right-looking supernode techniques [26–30]. To improve sequential and parallel sparse numerical factorization performance, the algorithms are based on a Level-3 BLAS update. Preprocessing via nonsymmetric row permutations are used to place large matrix entries on the diagonal to obtain a better scaled and diagonally dominant jacobian (A in Fig. 7) matrix. This preprocessing reduces the need for partial pivoting, thereby speeding up the factorization process. Complete block diagonal supernode pivoting allows dynamical interchanges of columns and rows during the factorization process. In this research PARDISO is used as the in-core sequential solver.

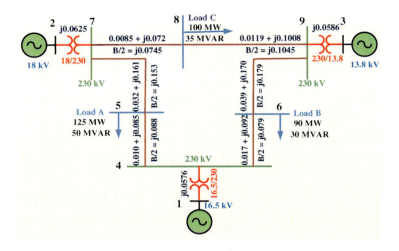

Fig. 8. WSCC nine bus system [5]

4 Validation Results against Commercial Software

4.1 Test Systems

It is important to validate the developed software against well acknowledged and widely used commercial power system simulators to establish its accuracy and to estimate the error in modeling and solution strategy.

Towards this end, two systems are selected namely WSCC nine bus system as shown in Fig. 8 (with 3 generators, 3 loads and 9 buses) and a large system which has 13029 buses, 431 generators, 12488 branches and 5950 loads. Each generator is provided with an exciter and a governor model.

The test systems are validated and the performance comparison is performed against commercial power system dynamic simulation packages like PSSE [31] , PowerWorld [23], DSA Tools (TSAT) [20] and GE PSLF [22].

4.2 Nine Bus System

WSCC system as shown in Fig. 8 has three generators, three loads and six transmission lines. Each generator is represented by a GENROU model along with IEEET1 exciter model and TGOV1 governor model. However, the results shown in this section are simulated with governor modeled as constant. Loads are represented as constant admittance during dynamic simulation. A three phase fault is created at bus 7 at time t=5s and the fault is removed after 6 cycles (0.1 seconds).

Figures (9)-(11) shows the comparison of results among different simulation packages (PSSE, DSA and PowerWorld). From Fig. (9)-(10), it is clear

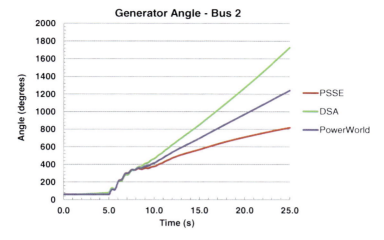

Fig. 9. Comparison of generator angle at bus 2 for a fault at bus 7 with PSSE, Power World and DSA Tools

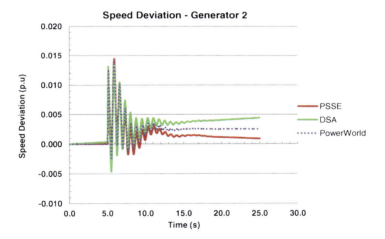

Fig. 10. Comparison of generator speed deviation at bus 2 for a fault at bus 7 with PSSE, Power World and DSA Tools

that there are significant differences in the results of these softwares. Although one may expect that all these commercial power simulators produce the exact same results for this small system and with the same models available in these packages, this is not found to be true. These are mainly due the internal differences in the implementaions of the power system components and could also be due the numerical algorithms implemented to handle the DAE.

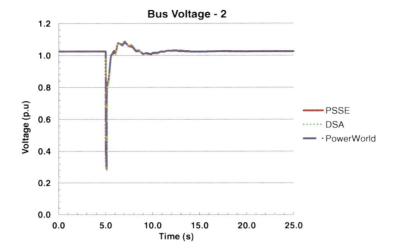

Fig. 11. Comparison of voltage at bus 2 for a fault at bus 7 with PSSE, Power World and DSA Tools

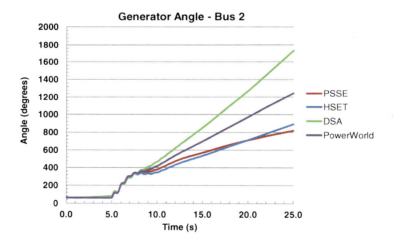

Fig. 12. Comparison of generator angle at bus 2 for a fault at bus 7 with HSET, PSSE, Power World and DSA Tools

Numercial methods play an important role in the accuracy of the results. In our experience we have also found that initialization could also be important factor in the evolution of the system trajectory and also in computational efficiency.

High Performance Computing for Power System Dynamic Simulation 63

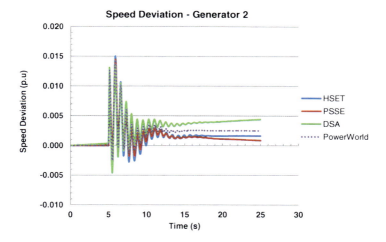

Fig. 13. Comparison of generator speed deviation at bus 2 for a fault at bus 7 with HSET, PSSE, Power World and DSA Tools

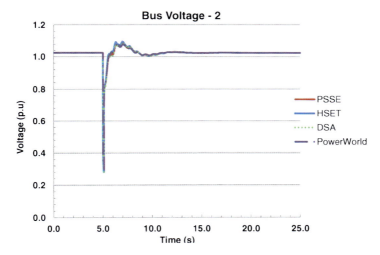

Fig. 14. Comparison of voltage at bus 2 for a fault at bus 7 with HSET, PSSE, Power World and DSA Tools

Figures (12)-(14) compares HSET-TDS results with these software packages. From figures (12)-(14), we find a close match between PSSE and the results obtained via HSET-TDS.

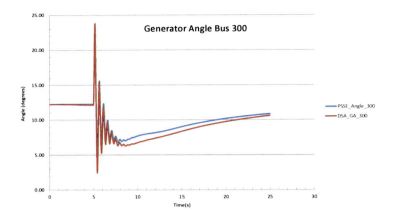

Fig. 15. Comparison of generator angle at bus 300 for a fault at bus 300 with PSSE and DSA Tools

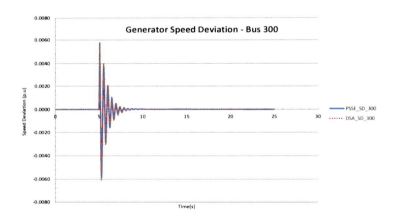

Fig. 16. Comparison of generator speed deviation at bus 300 for a fault at bus 300 with PSSE and DSA Tools

4.3 Large Test System

HSET-TDS simulation package is also tested on a large system as described in section 4.1. Each generator (GENROU) is provided with an exciter (IEEET1) and a governor (TGOV1) model. A three phase fault is created at the generator bus 300 at time t=5s and the fault is cleared after 0.1 seconds (6 cycles).

High Performance Computing for Power System Dynamic Simulation   65

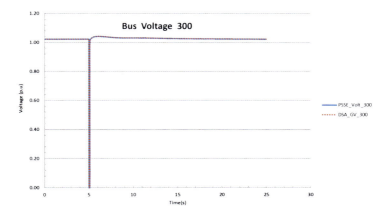

Fig. 17. Comparison of bus voltage 300 for a fault at bus 300 with PSSE and DSA Tools

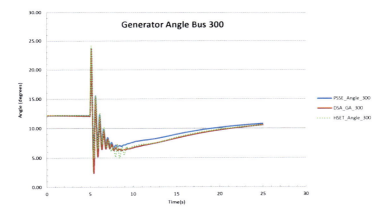

Fig. 18. Comparison of generator angle at bus 300 for a fault at bus 300 with PSSE, DSA Tools and HSET

The simulation results are compared with that of PSSE and DSA tools[1]. Figures (15)-(17) shows the comparison of results among different simulation packages, where as figures (18)-(20) compares HSET results with other software packages. The results show a excellent match with PSSE and DSA packages.

---
[1] Since we do not have the license for the PowerWorld and the evaluation version does not allow simulation of such large systems; therefore we do not present results with PowerWorld package for the large system.

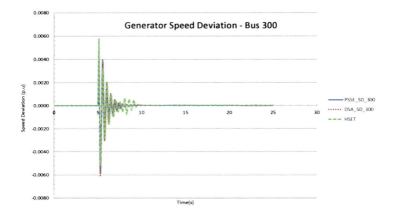

Fig. 19. Comparison of generator speed deviation at bus 300 for a fault at bus 300 with PSSE, DSA Tools and HSET

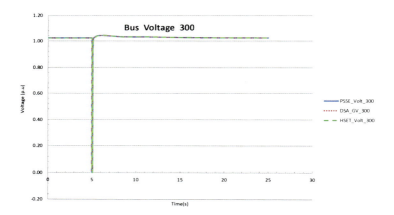

Fig. 20. Comparison of bus voltage 300 for a fault at bus 300 with PSSE, DSA Tools and HSET

5 Performance Comparison

As discussed in section 3.5, linear system solution is most computationally intensive. The power system dynamic jacobian are highly sparse and direct sparse algorithms are used for computational efficiency. Different sparse solvers significantly impact the solution times of the dynamic simulations. In this section we present the comparison results of three different sparse solvers used in this study (see section 3.5). The results for two different cases are presented.

Table 1. Simulation times with different linear solvers and speed-ups

Contingency	UMFPACK	KLU	PARDISO	UMFPACK/KLU	PARDISO/KLU
C1	34.66	9.15	23.77	3.79	2.60
C2	27.04	8.00	24.13	3.38	3.02

1. C1: There is bus fault on bus number 13182 at time t=5 seconds and it is cleared at time t=5.3 seconds. Then there is a branch fault on branch 13181-13182 at time t=8.5 seconds and it is cleared at time t=8.8 seconds. The simulation is run for 25 seconds.
2. C2: There is bus fault on bus number 300 at time t=5 seconds and it is cleared at time t=5.1 seconds. The simulation is run for 25 seconds.

Table 1 shows simulation times with three linear solvers, namely UMFPACK, KLU, PARDISO for C1 and C2. It also shows the speedup of KLU over UMFPACK and PARDISO. Clearly, KLU is nearly 3 times faster than UMFPACK and PARDISO solvers.

Table 2 shows simulation times with PSSE and HSET. Clearly HSET is nearly 4 times faster than commercial software PSSE.

Table 2. Simulation times with PSSE with HSET and speed-ups

Contingency	PSSE	HSET	Speedup
C1	37.00	9.15	4.04
C2	40.00	8.00	5.00

6 Conclusion

Some control centers have the real-time model which represents more than 32,000 buses and 42,000 branches. It is likely that this model contains over 3000 generators. Thus magnitude of transient simulation would be in terms of tens of thousands of differential algebraic equations assuming 12th order differential equations corresponding to each generator, governor and exciter combination. Thus computational speed is a main concern, together with data measurement, network processing, and detailed modeling. The magnitude of transient simulation is in terms of tens of thousands of differential algebraic equations. Decreasing computational time for any heavily used tool is advantageous in increasing the operators' ability to take quick, apt and accurate action and analyze more scenarios. Our particular interest is for on-line deployment of long-term time-domain simulation. Initial results presented in this chapter show significant performance improvements over commercial software like PSSE. In this context, HPC will play a very crucial role in decreasing the wall clock time of the dynamic simulations through the development and implementation of optimized numerical algorithms on modern high-speed hardware architecture (both serial and parallel platforms) for power system specific applications.

References

1. Siemens PTI Power Technologies Inc., PSS/E 33, Program Application Guide, vol. II (May 2011)
2. Agrawal, A., et al.: Parallel pairwise statistical significance estimation of local sequence alignment using message passing interface library. Concurrency and Computation: Practice and Experience (2011)
3. Alvarado, F., Lasseter, R., Sanchez, J.: Testing of trapezoidal integration with damping for the solution of power transient problems. IEEE Transactions on Power Apparatus and Systems (12), 3783–3790 (1983)
4. Amestoy, P., et al.: Vectorization of a multiprocessor multifrontal code. International Journal of High Performance Computing Applications 3(3), 41–59 (1989)
5. Anderson, P., Fouad, A.: of Electrical, I., Engineers, E.: Power system control and stability. IEEE Press power engineering series, IEEE Press (2003), http://books.google.com/books?id=8-xSAAAAMAAJ
6. Astic, J., Bihain, A., Jerosolimski, M.: The mixed adams-bdf variable step size algorithm to simulate transient and long term phenomena in power systems. IEEE Transactions on Power Systems 9(2), 929–935 (1994)
7. Berzins, M., Furzeland, R.: An adaptive theta method for the solution of stiff and nonstiff differential equations. Applied Numerical Mathematics 9(1), 1–19 (1992)
8. Brenan, K., Campbell, S., Campbell, S., Petzold, L.: Numerical solution of initial-value problems in differential-algebraic equations, vol. 14. Society for Industrial Mathematics (1996)
9. Davis, T.: Algorithm 832: Umfpack v4. 3—an Unsymmetric-Pattern Multifrontal Method. ACM Transactions on Mathematical Software (TOMS) 30(2), 196–199 (2004)
10. Davis, T.: A column pre-ordering strategy for the unsymmetric-pattern multifrontal method. ACM Transactions on Mathematical Software (TOMS) 30(2), 165–195 (2004)
11. Davis, T., Duff, I.: A combined unifrontal/multifrontal method for unsymmetric sparse matrices. ACM Transactions on Mathematical Software (TOMS) 25(1), 1–20 (1999)
12. Davis, T., Gilbert, J., Larimore, S., Ng, E.: A column approximate minimum degree ordering algorithm. ACM Transactions on Mathematical Software (TOMS) 30(3), 353–376 (2004)
13. Davis, T., Stanley, K.: Klu: a" clark kent" sparse lu factorization algorithm for circuit matrices. In: 2004 SIAM Conference on Parallel Processing for Scientific Computing (PP04) (2004)
14. Honbo, D., Agrawal, A., Choudhary, A.: Efficient pairwise statistical significance estimation using fpgas. In: Proceedings of BIOCOMP 2010, pp. 571–577 (2010)

15. Khaitan, S., Fu, C., McCalley, J.: Fast parallelized algorithms for on-line extended-term dynamic cascading analysis. In: Power Systems Conference and Exposition, PSCE 2009, IEEE/PES, pp. 1–7. IEEE (2009)
16. Khaitan, S., McCalley, J.: A class of new preconditioners for linear solvers used in power system time-domain simulation. IEEE Transactions on Power Systems 25(4), 1835–1844 (2010)
17. Khaitan, S., McCalley, J., Chen, Q.: Multifrontal solver for online power system time-domain simulation. IEEE Transactions on Power Systems 23(4), 1727–1737 (2008)
18. Khaitan, S., McCalley, J., Raju, M.: Numerical methods for on-line power system load flow analysis. Energy Systems 1(3), 273–289 (2010)
19. Krause, P., Wasynczuk, O., Sudhoff, S.: Analysis of electric machinery and drive systems. IEEE Press series on power engineering. IEEE Press (2002), http://books.google.com/books?id=8m4vAQAAIAAJ
20. Kundur, P.: http://www.dsatools.com/
21. Machowski, J., Bialek, J., Bumby, J.: Power system dynamics: stability and control. Wiley (2011)
22. Murdoch, C.: Pslf/plot manuals. GE PSLF. Modified 20 (2005)
23. Overbye, T.: http://www.powerworld.com/
24. Sanchez-Gasca, J., D'Aquila, R., Price, W., Paserba, J.: Variable time step, implicit integration for extended-term power system dynamic simulation. In: Proceedings of Power Industry Computer Application Conference, pp. 183–189. IEEE Computer Society Press, Los Alamitos (1995)
25. Sauer, P., Pai, A.: Power system dynamics and stability. Prentice-Hall (1998), http://books.google.com/books?id=dO0eAQAAIAAJ
26. Schenk, O.: Scalable parallel sparse LU factorization methods on shared memory multiprocessors. Hartung-Gorre (2000)
27. Schenk, O., Gärtner, K.: Sparse factorization with two-level scheduling in paradiso. In: Proceedings of the 10th SIAM Conference on Parallel Processing for Scientific Computing, p. 10 (2001)
28. Schenk, O., Gärtner, K.: Two-level dynamic scheduling in pardiso: Improved scalability on shared memory multiprocessing systems. Parallel Computing 28(2), 187–197 (2002)
29. Schenk, O., Gärtner, K.: Solving unsymmetric sparse systems of linear equations with pardiso. Future Generation Computer Systems 20(3), 475–487 (2004)
30. Schenk, O., Gärtner, K., Fichtner, W.: Efficient sparse lu factorization with left-right looking strategy on shared memory multiprocessors. BIT Numerical Mathematics 40(1), 158–176 (2000)
31. Siemens, http://www.energy.siemens.com/
32. Zhang, Y., et al.: Efficient pairwise statistical significance estimation for local sequence alignment using GPU. In: 2011 IEEE 1st International Conference on Computational Advances in Bio and Medical Sciences (ICCABS), pp. 226–231. IEEE (2011)

Distributed Parallel Power System Simulation

Mike Zhou

Chief Architect, InterPSS Systems
Canton, Michigan, USA
mike@interpss.org

Abstract. The information technology (IT) world has changed fundamentally and drastically from running software applications on a single computer with a single CPU to now running software as services in distributed and parallel computing environment. Power system operation has been also shifted from being solely based on off-line planning study to more and more real-time market-driven. Facing these challenges, we will discuss in this chapter how to design power system analysis and simulation software to take advantage of the new IT technologies and to meet the real-time application requirements, using the InterPSS project as a concrete example.

1 Introduction

Due to technology advance, business competiveness and regulation changes, it is becoming more and more demanded to carry out large amount of computation in real-time in power system operation. The author started power system simulation programming 30 years ago, when studying as a PhD student at Tsinghua University (Beijing, China). In 1985, China was planning to build the Three Gorges Power Plant, which would be the largest hydro power plant in the world. At that time, there was no commercial packaged power system simulation software available in China. The author was assigned the task to develop a power system simulation program for the feasibility study, including loadflow, short circuit and transient stability with HVDC transmission lines. The simulation program was completed in about 3 months and used for the study. It was written in FORTRAN language in the procedure programming style, and running then on a single computer with a single Central Processing Unit (CPU).

The information technology world has changed fundamentally and drastically since then. Today, a typical enterprise computing environment consists of a set, sometime a farm, of computers connected by high-speed and reliable computer networks. The computers are usually equipped with multiple multi-core processors, capable of running many, sometimes massive numbers of, computation tasks in parallel to achieve highly-available and fault-tolerant computations. Distributed and parallel computing is the foundation for today's highly-available SOA (Service-Oriented Architecture) enterprise computing architecture.

Also, the way how power system is operated has changed from planned to real-time and market-driven. Power system operations used to be planned according to the off-line planning study many months ahead. Today, power system operations are moving to market-driven, allowing transmission of power from one node to another to be arranged real-time on-demand by the market participants. High performance computing (HCP) technology-based real-time power system analysis tools will play a more and more import role to ensure reliability and stability of future power system operations.

1.1 Challenges

Instead of increasing the CPU clock speed and straight-line instruction throughput, the computer industry is currently turning to the multi-core processor architecture and the distributed computing approach for high performance computing (HPC). The amount of performance gain by the use of a multi-core processor depends very much on the software algorithms and implementation. In particular, the possible gains are limited by the fraction of the software that can be parallelized to run on multiple cores/CPU/computers simultaneously. Therefore distributed and parallel computing is one of the most important issues, which needs to be addressed, for today and tomorrow's HPC in power and energy systems.

For those who wrote FORTRAN code for power system analysis and simulation before, the following code snippet might not be foreign:

```
INTEGER N, BusNo(10000)
REAL P(10000), Q(10000)                                    (1)
COMMON /BUSDATA/ N, BusNo, P, Q
```

where, N is the number of buses; BusNo bus number array; and P, Q for storing bus p and q. BUSDATA is a common data block for passing data between subroutines, which is accessible globally in the application. There are at least two major problems when trying to run the above code in a distributed and parallel computing environment.

1.1.1 Multi-thread Safe

The objective of parallel computing is to reduce the overall computation time of a computation task by dividing the task into a set of subtasks (simulation jobs) that can be processed independently and concurrently. Multiple subtasks can concurrently read the same memory location (the common data black), which is shared among them. However, concurrent writes to the same memory location generated by multiple subtasks would create conflicts. The order in which the subtasks would perform their write is non-deterministic, making the final state unpredictable. In computer science, the common data block is called static global variable. It has been a known fact that using static global variable makes the code multi-thread unsafe. Therefore the above code set (1) could not be safely executed in a computation process in parallel without causing data integrity issues, because of the common data block usage.

1.1.2 Serialization/De-serialization

In distributed computing, there are often situations where application information in the memory of a computer, needs to be serialized and sent to another computer via the network. On the receiving end at the target computer, the serialized information needs to be de-serialized to reconstruct the original data object graph in the memory to continue the application logic execution. For example, for a mission-critical software application process, if an active computer running the application crashes for some reason, the application process is required to be migrated to a backup computer and continue the execution there. This is the so-called fault-tolerance design. One of the key requirements for a successful fault-tolerant implementation is that the information in the memory used by an application process should be able to be easily and seamlessly serialized and de-serialized. Using the above coding style (1), in large-scale power system simulation software, as one can imagine, there might be hundreds or even thousands of individual variables and arrays, scattered over many places. Serializing these variables and arrays in the memory, and then de-serialized to re-construct the data object graph at a different computer would be extremely difficult, if not impossible.

1.2 Near Real-Time Batch Processing

In computer science, a thread of execution is the smallest unit of processing that can be scheduled by a computer's operating system. A thread is contained inside a computer application process. Multiple threads can co-exist within the same application process and share resources such as the memory, while different processes cannot share these resources. In general, threads are "cheap" and application processes are "expensive", in terms of system resources usage. A parallel power system simulation approach was presented in [Lin, 2006], where a job scheduler is used to start multiple single-thread application processes to perform power system simulation in parallel. It is a known fact among computer science professionals that this kind of approach, which is sometimes called near real-time batch processing, is not efficient and has many side-effects.

"We argue that objects that interact in a distributed system need to be dealt with in ways that are intrinsically different from objects that interact in a single address space. ... Further, work in distributed object-oriented systems that is based on a model that ignores or denies these differences is doomed to failure, and could easily lead to an industry-wide rejection of the notion of distributed object-based systems."[Waldo 2004] In our opinion, FORTRAN, along with other procedural programming language-based legacy power system software packages, which were designed to function properly on a single computer with a single CUP (single address space), in general cannot be evolved and adapted to take full advantage of the modern multi-processor, multi-core hardware architecture and the distributed computing environment. The legacy power system simulation software packages cannot be easily deployed in, and able of taking advantage of, today's highly-available enterprise computing environment.

1.3 InterPSS Project

The discussion material presented in this chapter is based on an actual power system simulation software implementation – the InterPSS project [Zhou 2005]. The question, the InterPSS development team want to address, is not so much how to develop or adapt a distributed parallel computing system to run some existing legacy power system simulation software, but rather on how to develop a completely new power system simulation engine in certain way and with certain features, that will enable it to run in most modern enterprise computing environments available today, and being able to be adapted to possible future HPC innovations.

The InterPSS project is based on modern software development methodology, using the object-oriented power system simulation programming approach [Zhou, 1996] and Java programming language, with no legacy FORTRAN code. Its open and loosely coupled software architecture allows components developed by others to be easily plugged into the InterPSS platform to augment its functionalities or alter its behaviors, and equally important, allows its components to be integrated into other software systems. The project is currently developed and maintained by a team of developers living in the United States and China.

In summary, the focus of this chapter is not on a particular distributed and parallel computing technology, but rather on how to architect and design power system analysis and simulation software to run in modern distributed and parallel computing environment to achieve high performance.

2 Power System Modeling

Power system simulation software, except for those targeted for university teaching or academic research, is very complex and usually takes years to develop. When starting a large-scale software development project, the first important step is to study how to model the problem which the software is attempted to solve. This section discusses power system modeling from power system analysis and simulation perspective. A model, in the most general sense, is anything used in any way to represent anything else. There are different types of models for different purposes.

- Physical model - Some models are physical, for instance, a transformer model in a physical power system laboratory which may be assembled and made work like the actual object (transformer) it represents.
- Conceptual model - A conceptual model may only be drawn on paper, described in words, or imagined in the mind, which is commonly used to help us get to know and understand the subject matter it represents. For example, the IEC

Common Information Model (CIM) is a conceptual model for representing power system at the physical device level.
- Concrete implementation model - A conceptual model could be implemented in multiple ways by different concrete implementation models. For example, a computer implementation (concrete) model is a computer software application which implements a conceptual model.

2.1 Power System Object Model

Matrix has been used to formulate power system analysis since the computer-based power system simulation was introduced to power system analysis. Almost all modern power system analysis algorithms are formulated based the [Y]-matrix - the power network Node/Branch admittance matrix. [Y]-matrix essentially is a conceptual model representing power network for the simulation purpose. The [Y]-matrix model has been commonly implemented in FORTRAN using one or two dimensional arrays as the fundamental data structure. One would often find that power system analysis software carries a label of 50K-Bus version, which really means that the arrays used in the software to store bus related data have a fixed dimension of 50,000. This FORTRAN-style computer implementation of the [Y]-matrix power network model has influenced several generations of power engineers and been used by almost all commercial power system analysis software packages today.

It is very important to point out that a conceptual model could be implemented in multiple ways. This has been demonstrated by an object-oriented implementation of the very same power system [Y]-matrix model in [Zhou, 1996], where C++ Template-based linked list, instead of array, was used as the underlying data structure. In this implementation, a set of power system modeling vocabulary (Network, Bus, and Branch) and ontology (relationship between the Network, Bus and Branch concepts) are defined for the purpose of modeling power system for analysis and simulation. The power system object model could be illustrated using the UML – Unified Modeling Language [OMG, 2000] representation as shown in Figure-1.

In the object model, a Bus object is a "node" to which a set of Branch objects can be connected. A Branch object is an "edge" with two terminals which can be connected between two Bus objects. A Network object is a container where Bus and Branch objects can be defined, and Branch objects can be connected between these Bus objects to form a network topology for power system simulation. In UML terminology, [0..n] reads zero to many and [1..n] one to many. A Network object contains [1..n] Bus objects and [1..n] Branch objects, while a Bus object can be associated with [0..n] branches. A Branch object connects (associates) to [1..n] Bus objects. Typically, a Branch object connects to 2 Bus objects. However, the number might be 1 for grounding branch, or 3 for three-winding transformer.

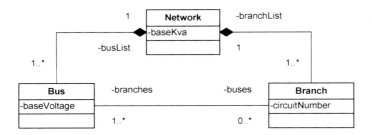

Fig. 1. Power System Object Model

2.2 Model-Based Approach

Model-driven architecture (MDA) [OMG, 2001], is a design approach for development of software systems. In this approach, software specifications are expressed as conceptual models, which are then "translated" into one or more concrete implementation models that the computers can run. The translation is generally performed by automated tools, such as the Eclipse Modeling Framework (EMF) [Eclipse 2010]. Eclipse EMF is a modeling framework and code generation facility for building tools and other applications based on structured data models. From a model specification, Eclipse EMF provides tools and runtime support to produce a set of Java classes (concrete implementation) for the model.

MDA-based software development approach is sometimes called model-based approach. In this approach, the model is placed at the center of software development and the rest evolve around the model. InterPSS uses this approach and is based on Eclipse EMF [Zhou, 2005]. The immediate concern, one might have, is the performance, since power system simulation is computation-intensive. For an 11865-bus test system, it took the Eclipse EMF generated Java code 1.4 sec to load the test case data to build the model, and 15.4 sec to run a complete loadflow analysis using the PQ method. The testing was performed on a Lenovo ThinkPad T400 laptop. It has been concluded that the code generated by computer from the Eclipse EMF-based power system simulation model out-performs the code manually written by the InterPSS development team. The following is a summary of the model-based approach used by InterPSS:

- The power system simulation object model [Zhou, 2005] is first expressed in Eclipse EMF as a conceptual model, which is easy to be understood and maintained by human beings.
- Using Eclipse EMF code generation, the conceptual model is turned to a set of Java classes, a concrete implementation model, which can be used to represent power network object model in computer memory for simulation purposes.
- The two challenges presented in Section-1.1 are addressed. Since the conceptual model is represented by a set of UML classes, it is quite easy to change and adjust, to make the generated concrete implementation model multi-thread safe.

- Eclipse EMF includes a powerful framework for model persistence. Out-of-the-box, it supports XML serialization/de-serialization. Therefore, the Eclipse EMF-based InterPSS power system simulation object model could be serialized to and de-serialized from Xml document automatically.

2.3 Simulation Algorithm Implementation

When the author implemented his first loadflow program about 30 years ago, the first step was implementing loadflow algorithms, i.e., the Newton-Raphson (NR) method and the Fast-Decoupled (PQ) method. Then, input data routines were created to feed data into the algorithm routines. While this approach is very intuitive and straightforward, it was found late that the program was difficult to maintain and extend, since the program data structure was heavily influenced by the algorithm implementation. In the model-based approach, the model is defined and created first. The algorithm is developed after, often following the visitor design pattern, which is described below.

In software engineering, the visitor design pattern [Gamma, et al, 1994] is a way of separating an algorithm from the model structure it operates on. A practical result of this separation is the ability to add new algorithms to existing model structures without modifying those structures. Thus, using the visitor pattern decouples the algorithm implementation and the model that the algorithm is intended to apply to, and makes the software easy to maintain and evolve. The following are some sample code, illustrating an implementation of loadflow algorithm using the visitor pattern:

```
// create an aclf network object
AclfNetwork aclfNet = CoreObjectFactory.createAclfNetwork();
// load some loadflow data into the network object
SampleCases.load_LF_5BusSystem(aclfNet);
// create an visitor - a Loadflow algorithm
IAclfNetVisitor visitor = new IAclfNetVisitor() {
   public void visit(AclfNetwork net) {
      // define your algorithm here
      . . . }
};
// run loadflow calculation by accepting the visitor visiting
// the aclf network object
aclfNet.accept(visitor);
```

The first step is to create a power system simulation object model – an `AclfNetwork` object. Next, some data is loaded into the model. At this time, the exact loadflow algorithm to be used has not been defined yet. All it says is that necessary information for loadflow calculation have been loaded into the simulation model object. Then a loadflow algorithm is defined as a visitor object and loadflow calculation is performed by the visitor visiting the simulation model object. The loadflow calculation results are stored in the model and ready to be output in any format, for example, by yet another visitor implementation. In this way, the power system simulation model and simulation algorithm are decoupled, which makes it possible for the simulation algorithm to be replaced with another implementation through software configuration using dependency injection [Fowler 2004], see more details in the upcoming Section-3.2.

2.4 Open Data Model (ODM)

ODM [Zhou 2005] is an Open Model for Exchanging Power System Simulation Data. The model is specified in XML using XML schema [W3C 2010]. The ODM project, an open-source project, is currently sponsored by the IEEE Open-Source Software Task Force. InterPSS is the ODM schema author and a main contributor to the open-source project.

One might be familiar with the IEEE common data format [IEEE Task Force 1974]. The following is a line describing a bus record:

```
2 Bus 2       HV  1  1  2 1.045   -4.98       21.7      12.7     ->
  40.0    42.4    132.0  1.045    50.0   -40.0    0.0     0.0              0
```

Using the ODM model, the same information could be represented in an XML record, as follows:

```
<aclfBus id="Bus2" offLine="false" number="2"
         zoneNumber="1" areaNumber="1" name="Bus 2      HV">
   <baseVoltage unit="KV" value="132.0"/>
   <voltage unit="PU" value="1.045"/>
   <angle unit="DEG" value="-4.98"/>
   <genData>
      <equivGen code="PV">
         <power unit="MVA" im="42.4" re="40.0"/>
         <desiredVoltage unit="PU" value="1.045"/>
         <qLimit unit="MVAR" min="-40.0" max="50.0"/>
      </equivGen>
   </genData>
   <loadData>
      <equivLoad code="CONST_P">
         <constPLoad unit="MVA" im="12.7" re="21.7"/>
      </equivLoad>
   </loadData>
</aclfBus>
```

There are many advantages to represent data in XML format. In fact, XML currently is the de facto standard for representing data for computer processing. One obvious benefit is that it is very easy to read by human. Anyone with power engineering training can precisely interpret the data without any confusion. There is no need to guess if the value 40.0 is generation P or load P, in PU or Mw.

The key ODM schema concepts are illustrated in Figure-2. The Network complex type (NetworkXmlType) has a busList element and a branchList element. The branch element in the complex type (branchTypeList) is of the Branch complex type (BaseBranchXmlType). The fromBus and toBus elements are of complex type BusRefRecordXmlType, which has a reference to existing Bus record. The tertiaryBus is optional, used for 3-winding transformer.

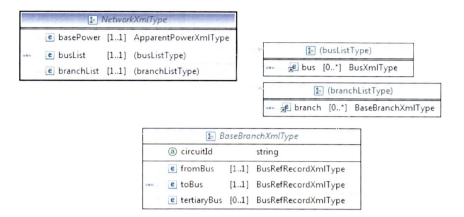

Fig. 2. Key ODM Schema Concepts

IEEE common data format specification has two main sets of records – bus record and branch record. Bus number is used to link these two sets of records together. In certain sense, it can be considered as a concrete implement model of the [Y]-matrix model. Similarly, the key ODM schema concepts, shown in Figure-2, are intended to describe the [Y]-matrix in an efficient way. Therefore, we can consider the ODM model is another concrete implementation model of the [Y]-matrix model.

2.5 Summary

A conceptual model could be implemented in multiple ways by different concrete implementation models. Power network sparse [Y]-matrix is a conceptual model, which has been implemented in two different ways in the InterPSS project: 1) the InterPSS object model using Eclipse EMF and 2) the ODM model using XML schema. InterPSS object model is for computer in memory representation of power network object relationship, while the ODM model is for power network simulation data persistence and data exchange. The relationship between these two models and the simulation algorithms is illustrated in Figure-3. At the runtime, power network simulation data stored in the ODM format (XML document) is loaded into computer and mapped to the InterPSS object model. Simulation algorithms, such as loadflow algorithm, are applied to the object model to perform, for example, loadflow analysis.

In the model-based approach, power system simulation object model is at the center of the software development and the rest evolve around the model in a decoupled way, such that different simulation algorithm implementations can be associated with the model through software configuration using the dependency injection approach. Since the underlying data model object and the simulation algorithms are decoupled, it is not difficult to make the data model multi-thread safe and to implement data model serialization/de-serialization for distributed parallel computation.

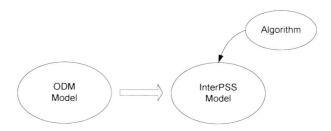

Fig. 3. Model and algorithm relationship

3 Software Architecture

The question to be addressed in this chapter is not so much about how to develop a new or adapt to an existing distributed and parallel computing system to run the legacy power system simulation software, but rather on how to architect power system simulation software that will enable it to run in today's enterprise computing environments, and to easily adapt to possible future HPC innovations. The problem is somewhat similar to that on how to build a high-voltage transmission system for scalability. When building a local power distribution system, the goal is usually to supply power to some well-defined load points from utility source bus. Flexibility and extensibility is not a major concern, since the number of variation scenarios in power distribution systems is limited. When we were studying the Three Gorges plant, there were so many unknowns and possible variation scenarios. In fact the plant was built more than 20 years after the initial study. At that time, an "architecture" of parallel high-voltage HVAC and HVDC transmission lines were suggested for reliability and stability considerations, and for flexibility and extensibility, without knowing details of the generation equipment to be used to build the plant and the power consumption details in the eastern China region (the receiving end). The concept of designing scalable system architecture with flexibility and extensibility is not foreign to the power engineering community.

In this section, the InterPSS project will be used as a concrete example to explain the basic software architecture concepts. The discussion will stay at the conceptual level, and no computer science background is needed to understand the discussion in this section.

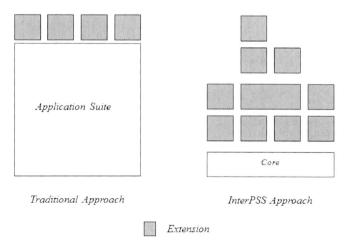

Fig. 4. Comparison of Different Software Architecture

3.1 InterPSS Software Architecture

Traditional power system simulation software is monolithic with limited extension capability, as illustrated in Figure-4. InterPSS on the other hand has an open architecture with a small core simulation engine. The other parts, including input and output modules, user interface modules, and controller models are extension plug-ins, which could be modified or replaced. For example, InterPSS loadflow algorithms, sparse equation solution routines, transient stability simulation result output modules and dynamic event handlers for transient stability simulation are plugins, which could be taken away and replaced by any custom implementation.

InterPSS core simulation engine is very simple. It consists of the following three parts:

- Power system object model – The power system object model discussed in Section-2. It, in a nutshell, is a concrete implementation of the power network sparse [Y]-matrix model (conceptual model). The implementation is in such a generic way that it could be used or extended to represent power network for any type of [Y]-matrix model based power system simulation.
- Plug-in framework – A dependency injection based plug-in framework allowing other components, such as loadflow simulation algorithm, to be plug-in or switched through software configuration.
- Default (replaceable) algorithm – The core simulation engine has a set of default implementation of commonly used power system simulation algorithms, such as loadflow Newton-Raphson method, fast-decoupled method and Gauss-Seidel method. However, these default algorithm implementations are plug-ins, meaning that they could be taken away and replaced.

InterPSS has a simple yet powerful open architecture. It is simple, since in essence, it only has a generic power network sparse [Y]-matrix solver. It is powerful, since using the plug-in extension framework, it could be extended and used to solve any [Y]-matrix based power system simulation problem. InterPSS core simulation engine is designed to be multi-thread safe, without static global variable usage in the power system object model. The object model, which is based on the Eclipse EMF, could be seamlessly serialized to and de-serialized from XML document. The core simulation engine is implemented using 100% Java with no legacy FORTRAN or C code. It can run on any operating system where a Java virtual machine (JVM) exists. It is easy to maintain since sparse [Y]-matrix for power system simulation is well understood and is unlikely to change for the foreseeable future, which enables the core simulation engine to be quite stable and simple to maintain.

3.2 Dependency Injection

Dependency injection [Fowler 2004] in object-oriented programming is a design pattern with a core principle of separating behavior from dependency (i.e., the relationship of this object to any other object that it interacts with) resolution. This is a technique for decoupling highly dependent software components. Software developers strive to reduce dependencies between components in software for maintainability, flexibility and extensibility. This leads to a new challenge - how can a component know all the other components it needs to fulfill its functionality or purpose? The traditional approach was to hard-code the dependency. As soon as a simulation algorithm, for example, InterPSS' loadflow implementation, is necessary, the component would call a library and execute a piece of code. If a second algorithm, for example, a custom loadflow algorithm implementation, must be supported, this piece of code would have to be modified or, even worse, copied and then modified.

Dependency injection offers a solution. Instead of hard-coding the dependencies, a component just lists the necessary services and a dependency injection framework supplies access to these service. At runtime, an independent component will load and configure, say, a loadflow algorithm, and offer a standard interface to interact with the algorithm. In this way, the details have been moved from the original component to a set of new, small, algorithm implementation specific components, reducing the complexity of each and all of them. In the dependency injection terms, these new components are called "service components" because they render a service (loadflow algorithm) for one or more other components. Dependency injection is a specific form of inversion of control where the concern being inverted is the process of obtaining the needed dependency.

There are several dependency injection implementations using Java programming language. The Spring Framework [Spring 2010] is the most popular one, which is used in InterPSS.

3.3 Software Extension

At the conceptual level, almost all InterPSS simulation related components could be considered as an implementation of the generic bus I-V or branch I-V function, as shown in Figure-5, where,

  I - Bus injection or branch current;
  V - Bus or branch terminal voltage;
  t - Time for time-domain simulation;
  dt - Time-domain simulation step.

t and dt only apply to time-domain dynamic simulation.

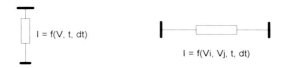

Fig. 5. Extension Concept

The following is a simple example to implement a constant power load (p = 1.6 pu, q = 0.8 pu) through bus object extension.

```
AclfBus bus = (AclfBus)net.getBus("1");
bus.setExtensionObject(new AbstractAclfBus(bus) {
    public boolean isLoad() { return true; }
    public double getLoadP() { return 1.6; }
    public double getLoadQ() { return 0.8; }
});
```

The behavior of AclfBus object (id = "1") in this case is overridden by the extension object. This example, of course, is over-simplified and only for illustrating the extension concept purpose. In the real-world situation, dependency injection is used to glue custom extension code to any InterPSS simulation object to alter its behavior.

3.4 Summary

"A good architecture is not there is nothing to add, rather nothing could be taken away". InterPSS has a simple yet powerful open architecture. "Change is the only constant" is what Bill Gates said when he retired from Microsoft. The main reason, such a simple yet flexible architecture is selected by the InterPSS development team, is for the future changes and extension. InterPSS core simulation engine is multi-thread safe, and its simulation object could be seamlessly serialized/de-serialized. In the next two sections, we will discuss InterPSS extension to grid computing and cloud computing.

4 Grid Computing

InterPSS grid computing implementation will be used in this section as an example to discuss key concepts in applying grid computing to power system simulation. Until recently, mainstream understanding of computer simulation of power system is to run a simulation program, such as a loadflow, Transient Stability or Voltage Stability program, or a combination of such programs on a single computer. Recent advance in computer network technology, and software development methodology has made it possible to perform grid computing, where a set of network-connected computers, forming a computational grid, located in local area network (LAN), are used to solve certain category of computationally intensive problems.

The Sparse matrix solution approach [Tinney, 1967] made simulation of power system of (practically) any size and any complexity possible using a single computer. The approach forms the foundation for today's computer-based power system study. The Fast Decoupled approach [Stott, 1974] made high-voltage power network analysis much faster. The [B]-matrix approximation concept used in the Fast Decoupled approach forms the foundation for today's power system on-line real-time security analysis and assessment. However, the power engineering community seems to agree that, by using only one single computer, it is impossible, without some sort of approximation (simplification/reduction/screening), to meet the increasing power system real-time online simulation requirements, where thousands of contingency analysis cases need to be evaluated in a very short period of time using a more accurate AC loadflow power network model.

4.1 Grid Computing Concept

Grid computing was started at the beginning to share unused CPU cycles over the Internet. However, the concept has been extended and applied recently to HPC, using a set of connected computers in a LAN (Local Area Network). Google has used grid computing successfully to conquer the Internet search challenges. We believe grid Computing can also be used to solve power system real-time on-line simulation problems, using the accurate AC loadflow model.

4.1.1 Split/Aggregate

The trademark of computational grids is the ability to split a simulation task into a set of sub-tasks, execute them in parallel, and aggregate sub-results into one final result. Split and aggregate design allows parallelizing the processes of subtask execution to gain performance and scalability. The split/aggregate approach sometimes is called map/reduce [Dean 2004].

4.1.2 Gridgain

GridGain (www.gridgain.org) is an open-source project, providing a platform for implementing computational grid. Its goal is to improve overall performance of computation-intensive applications by splitting and parallelizing the workload. InterPSS grid computing solution currently is built on top of the Gridgain platform. However, the dependency of InterPSS on Gridgain is quite loose, which allows InterPSS grid computing solution to be moved relative easily to other grid computing platform in the future, if necessary.

4.2 Grid Computing Implementation

With grid computing solution, power system simulation in many cases might not be constrained by the computation speed of a single computer with one CPU any more. One can quickly build a grid computing environment with reasonable cost to achieve HPC power system simulation. Using InterPSS core simulation engine and Gridgain, a grid computing solution has been developed for power system simulation. A typical grid computing environment setup is shown in Figure-6. InterPSS is installed on a master node. Power system simulation jobs, such as loadflow analysis run or transient stability simulation run, are distributed to the slave nodes. InterPSS core simulation engine software itself is also distributed through the network from the master node to the slave nodes one time when the simulation starts at the slave node. InterPSS simulation model object is serialized at the master node into an XML document and distributed over the network to the remote slave node, where the XML document is de-serialized into the original object model and becomes ready for power system simulation.

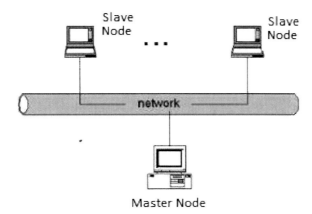

Fig. 6. Typical Grid Computing Setup

4.2.1 Terminology

The following are terminologies used in this section's discussion:

- Slave Gird Node - A Gridgain agent instance, running on a physical computer, where compiled simulation code can be deployed from a master grid node remotely and automatically through the network to perform certain simulation job. One or more slave grid node(s) can be hosted on a physical computer.
- Master Grid Node - A computer with InterPSS grid computing edition installation. One can have multiple master grid nodes in a computation grid. Also, one can run one or more slave grid node(s) on a master grid node computer.
- Computation Grid - One or more master grid nodes and one or many slave grid node(s) in a LAN form an InterPSS computational grid.
- Grid Job - A unit of simulation work, which can be distributed from a master grid node to any slave grid node to run independently to perform certain power system simulation, such as loadflow analysis or transient stability simulation.
- Grid Task – Represent a simulation problem, which can be broken into a set of grid jobs. Grid task is responsible to split itself into grid jobs, distribute them to remote grid nodes, and aggregate simulation sub-results from remote grid nodes for decision making or display purpose.

4.2.2 Remote Job Creation

Computer network has a finite bandwidth. Excessive communication between a master grid node and slave grid nodes cross the network during the simulation process may slow down the simulation. There are two ways to create simulation jobs, as shown in Figure-7.

- Master node simulation job creation - This is the default behavior. From a base case, simulation jobs are created at amaster node and then distributed to remote slave nodes to perform simulation. If there are a large number of simulation jobs, sending them in real-time through the network may take significant amount of time and might congest the network.
- Remote node simulation job creation - In many situations, it may be more efficient to distribute the base case once to remote grid nodes. For example, in the case of N-1 contingency analysis, the base case and a list of contingency description could be sent to remote slave grid node. Then simulation job for a contingency case, for example, opening branch Bus1->Bus2, can be created at a remote slave grid node from the base case.

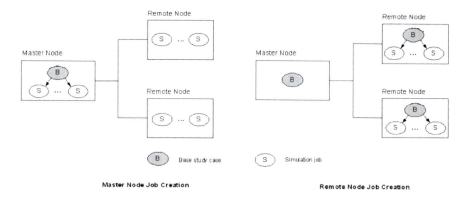

Fig. 7. Simulation Job Creation

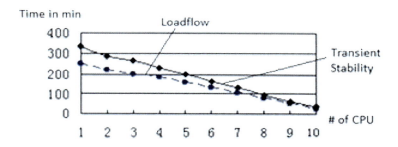

Fig. 8. InterPSS Grid Computing Performance Study

4.2.3 Performance Study

The result of an experiment by a research group at the South China University of Technology [Hou 2009] using InterPSS grid computing is shown in Figure-8.

- loadflow study: Based on a base case of 1,245 buses and 1,994 branches, perform 2,500 N-1 contingency analysis by running 2,500 full AC loadflow runs in parallel on remote grid nodes;
- Transient Stability study: Perform 1,000 transient stability simulation runs in parallel on remote grid nodes.

The results in Figure-8 indicate that InterPSS grid computing approach can achieve almost linear scalability, meaning doubling the computing resource (CPU) can approximately speed-up the computation two times.

With grid computing, for certain categories of computation-intensive power system simulation problems, where the split/aggregate technique can be used to solve the problem in parallel, computation speed might not be the deciding or limiting factor in selecting simulation algorithm any more. The computation speed constraint can be much relaxed. In the following section, an AC loadflow model and grid computing based security assessment implementation will be discussed.

4.3 Security Assessment

Power system security assessment is the analysis performed to determine whether, and to what extent, a power system is reasonably safe and reliable from serious interference to its normal operation. From the power system simulation perspective, security assessment is based on contingency analysis, which runs in energy management system in power system control center to give the operators insight on what might happen to the power system in the event of an unplanned or un-scheduled equipment outage. Typical contingency on a power system consists of outages such as loss of generators or transmission lines. The security assessment concept and terminology, defined in [Bula 1992], are used in the following discussion.

Real-time on-line security assessment algorithms used in today's power system control centers are mostly designed to be run on a single computer. Because of the computation speed limitation, most of them are based on the [B]-matrix approximation [Stott, 1974]. The following approaches are commonly used in the existing security assessment algorithms to speed-up computation:

- Simplified model - Because of the computation speed limitation, on-line analysis normally treats a power network as a DC network using $[\Delta P] = [B] \cdot [\Delta \theta]$. Reactive power and voltage are normally not considered in the economic dispatch calculation in power control center.
- Linear prediction - Linear prediction approach is commonly used to predict, for example, the power flow increase on branch-A, if there is a branch-B outage. Again the [B]-matrix is used to calculate the LODF - Line Outage Distribution Factor. LODF assumes that the power network under contingency conditions can be reasonably accurately represented by a network of impedance and constant voltage sources, which is no longer accurate in today's flexible power network with many electronic control and adjustment equipments, such as HVDC transmission line, FACTS and SVC (Static Var Compensator).
- Screening - Contingency analysis requires running thousands study cases in a short period of time, which is not feasible using a single computer and the accurate AC loadflow power network model. Therefore, the common practice today is to use some kind of screening process to pre-select a sub-set of the study cases to run. The selection is normally based on sensitivity analysis using the [B]-matrix.

The limitation of the above approaches and the need for AC Loadflow power network model based security assessment were discussed in [Wood 1984], Section-11.3.3. An AC loadflow model and grid computing based security assessment has been implemented in InterPSS, including three main features: 1) N-1 contingency analysis, 2) security margin computation, and 3) preventive/corrective action evaluation.

4.3.1 N-1 Contingency Analysis

Using the grid computing approach, full AC loadflow-based contingency analysis cases are run at remote grid nodes in parallel. The analysis results are sent back to master grid node in real-time through the network. The master grid node collects all analysis results and selects the largest branch MVA flow from all contingency cases for each branch to perform the security assessment.

Table 1. Branch MVA Rating Violation Report

```
Branch Id        Mva      Mva      Violation   Description
                 Flow     Rating
=================================================================
0001->0002(1)    241.1    200.0    21%         Open branch 0001->0005
0005->0004(1)    103.2    100.0     3%         Open branch 0002->0003
0006->0013(1)     51.3     50.0     3%         Open branch 0007->0008
0003->0004(1)    101.2    100.0     1%         Open branch 0002->0003
```

Table 2. Bus Voltage Security Margin

```
                      Bus Voltage Limit: [1.10, 0.90]
Bus Id   High V-Margin      Low V-Margin      Description
=================================================================
0002     1.0500    5.0%     1.0247   12.5%    Generator outage at bus 0002
0001     1.0600    4.0%     1.0600   16.0%    Open branch 0001->0005
0014     1.0426    5.7%     0.9970    9.7%    Open branch 0009->0014
0013     1.0552    4.5%     0.9980    9.8%    Open branch 0006>0013
0012     1.0590    4.1%     1.0000   10.0%    Open branch 0007->0008
0011     1.0635    3.7%     1.0000   10.0%    Open branch 0007->0008
0009     1.0689    3.1%     1.0000   10.0%    Open branch 0007->0008
0010     1.0620    3.8%     1.0000   10.0%    Open branch 0007->0008
0008     1.0900    1.0%     1.0366   13.7%    Generator outage at bus 0008
0007     1.0680    3.2%     1.0000   10.0%    Open branch 0007->0008
0006     1.0700    3.0%     1.0451   14.5%    Generator outage at bus 0006
0005     1.0288    7.1%     1.0000   10.0%    Open branch 0007->0008
0004     1.0210    7.9%     0.9983    9.8%    Open branch 0002->0003
0003     1.0100    9.0%     0.9537    5.4%    Open branch 0002->0003
```

Table 3. Branch Rating Security Margins

```
Branch Id        Mva      Mva       P + jQ           Margin   Description
                 Flow     Rating
=================================================================================
0002->0003(1)     98.4    100.0    ( 98.4+j  2.4)      2%     Open branch 0003->0004
0006->0012(1)     26.4     50.0    ( 11.4+j23.8)      47%     Open branch 0007->0008
0009->0014(1)     27.4     50.0    ( 27.3-j  1.8)     45%     Open branch 0005->0006
0002->0004(1)     93.8    100.0    ( 93.7+j  3.5)      6%     Open branch 0002->0003
0006->0013(1)     51.3     50.0    ( 23.2+j45.7)      -3%     Open branch 0007->0008
0010->0009(1)     33.7     50.0    ( 33.5-j  3.8)     33%     Open branch 0005->0006
0002->0005(1)     78.0    100.0    ( 77.9-j  1.9)     22%     Open branch 0001->0005
0013->0014(1)     16.4     50.0    ( 15.3+j  5.9)     67%     Open branch 0009->0014
0005->0004(1)    103.2    100.0    (102.3-j13.4)      -3%     Open branch 0002->0003
0004->0007(1)     59.8    100.0    (-57.4+j16.5)      40%     Open branch 0005->0006
0001->0002(1)    241.1    200.0    (240.1-j21.8)     -21%     Open branch 0001->0005
0003->0004(1)    101.2    100.0    (101.1-j  4.7)     -1%     Open branch 0002->0003
0005->0006(1)     62.9    100.0    ( 47.3+j13.5)      37%     Open branch 0010->0009
0004->0009(1)     33.5    100.0    (-32.9+j  6.4)     66%     Open branch 0005->0006
0007->0008(1)     23.3    100.0    (  0.0+j23.3)      77%     Open branch 0002->0003
0012->0013(1)     13.6     50.0    ( 13.1+j  3.8)     73%     Open branch 0006->0013
0001->0005(1)     95.5    100.0    ( 95.0+j  9.1)      5%     Open branch 0002->0003
0007->0009(1)     57.6    100.0    (-57.4+j  4.2)     42%     Open branch 0005->0006
0006->0011(1)     34.0     50.0    ( 14.6+j30.7)      32%     Open branch 0007->0008
0011->0010(1)     26.4     50.0    (-23.7+j11.6)      47%     Open branch 0005->0006
```

Using the IEEE-14 bus system as an example, N-1 contingency analysis is performed to perform security assessment and determine security margins. There are 23 contingency cases in total - 19 branch outages and 4 generator outages. The analysis results are listed in Table-1. Out of the 23 contingencies, 4 resulted in branch MVA rating violations. The most severe case is the "Open branch 0001-0005" case, where the branch 0001->0002 is overloaded 21%.

4.3.2 Security Margin Computation

In addition to violation analysis, security margin is computed based on analysis of all contingencies. The result reveals how much adjustment room are available with regard to bus voltage upper and low limits, and branch MVA rating limits for all concerned contingencies.

- Bus Voltage Security Margin - Using the IEEE-14 bus system, bus voltage security margin computation results are listed in Table-2. For all the 23 contingencies, the highest voltage happened at Bus 0008 (1.0900) for the contingency "Generator outage at bus 0008", and the lowest voltage happened at Bus 0003 (0.9537) for the contingency "Open branch 0002->0003".
- Branch Rating Security Margin - Using the IEEE-14 bus system, branch MVA rating security margin computation results are listed in Table-3. As shown in the table, Branch 0002->0003, although with no MVA rating violation, only has 2% room for increasing its MVA flow before hitting its branch rating limit, while Branch 0006-0012 has 47% room (margin) for scheduling more power transfer increase.

The bus voltage and branch MVA rating security margin results in Table-2 and Table-3 are obtained by running 29 full AC Loadloaw runs in parallel using the grid computing approach. There is no simplification or approximation in calculating the margin.

4.3.3 Preventive/Corrective Action Evaluation

When there are violations in contingency situations, sometimes, one might want to evaluate preventive or corrective actions, such as adjusting transformer turn ratio or load shedding. For example, the following is a sample corrective action rule:

```
Under contingency conditions, if branch 0001->0002 has
branch rating violation, cut Bus 0014 load first (Rule
Priority=1). If the condition still exits, cut Bus 0013
load (Rule Priority=2).
```

The priority field controls the order in which the rule actions are applied.

After applying the corrective actions, cutting load at bus 0014 and then 0013 in the event of Branch 0001->0002 overloading, the branch MVA rating violation has been reduced from 21% to 4%, as compared with Table-3.

Distributed Parallel Power System Simulation

Table 4. Branch MVA Rating Violation Report With Corrective Action

```
Branch Id       Mva     Mva     Violation   Description
                Flow    Rating
===================================================================
0001->0002(1)   207.8   200.0   4%          Open branch 0001->0005
0005->0004(1)   103.2   100.0   3%          Open branch 0002->0003
0006->0013(1)    51.3    50.0   3%          Open branch 0007->0008
0003->0004(1)   101.2   100.0   1%          Open branch 0002->0003
```

4.4 Summary

Before grid computing is available, one of the major concerns in real-time on-line power system simulation in power control center is the computation speed due to the usage of a single computer for the simulation. This has influenced the research of simplified models, liner prediction and/or screening approaches for real-time on-line power system simulation over the last 40 years. With the introduction of grid computing, the computation speed constraint can be much relaxed when selecting the next generation of on-line power system simulation algorithms for future power control center. It has been demonstrated that the speed of the grid computing based power system security assessment can be approximately increased linearly with the increase of computing resource (CPU or computer machines).

InterPSS grid computing solution is currently implemented on top of the Gridgain grid computing platform. Because of the modern software techniques and approaches used in InterPSS, it was quite easy and straightforward to run the InterPSS core simulation engine on top of the Gridgain grid computing platform. However, this really does not mean that we endorse Gridgain as the best grid computing platform for power system simulation. In fact, the dependency of InterPSS on Gridgain is quite decoupled, which will allow the underlying grid computing platform to be switched relative easily in the future, if necessary. During the InterPSS grid computing development process, the main objective was to architect and structure InterPSS core simulation engine so that it can be integrated into any Java technology-based grid computing platform easily.

5 Cloud Computing

Cloud computing is an evolution of a variety of technologies that have come together to alter the approach to build information technology (IT) infrastructure. With cloud computing, users can access resources on demand via the network from anywhere, for as long as they need, without worrying about any maintenance or management of the actual infrastructure resources. If the network is the Internet, it is called public cloud; or if the network is a company's intranet, it is called private cloud.

5.1 Cloud Computing Concept

Cloud computing is commonly defined as a model for enterprise IT organizations to deliver and consume software services and IT infrastructure resources. cloud service enables the user to quickly and easily build new cloud-enabled applications and business processes. Cloud computing has many advantages over the traditional IT infrastructure approach. Cloud services might not require any on premise infrastructure. It supports a usage cost model – either subscription or on-demand, and can leverage the cloud infrastructure for elasticity and scalability. There are commonly three core capabilities that cloud software management provides to deliver scalable and elastic infrastructure, platforms and applications:

- Infrastructure Management: The ability to define a shared pool (or pools) of IT infrastructure from physical resources or virtual resources;
- Application Management: The ability to encapsulate and migrate existing software applications and platforms from one location to another so that the applications become more elastic;
- Operation Management: A simple and straight way to deliver the service to the user. The service might be virtual, meaning the user does not know where is service is physically located.

The cloud computing concept – putting computing resources together to achieve efficiency and scalability should not be foreign to power engineers. Power system in certain sense can be considered as a cloud of generation resources. The power supply is 1) virtual, meaning power consumer really does not know where the power is actually generated, and 2) elastic, meaning the power supply can be ramped from 1 MW to 100MW almost instantly.

5.2 Cloud Computing Implementation

A cloud computing solution has been developed using the InterPSS core simulation engine. The solution is currently hosted, for demonstration purpose, within Google cloud environment using Google App Engine. As shown in Figure-9, Google App Engine provides a Java runtime environment. InterPSS core simulation engine has been deployed into the cloud and running there 24x7, accessible from anywhere around the world, since October 2009.

InterPSS cloud computing solution has a Web interface, as shown in Figure-10. After uploading a file in one of the supported formats, such as PSS/E V30, UCTE or IEEE CDF, one can perform a number of analyses.

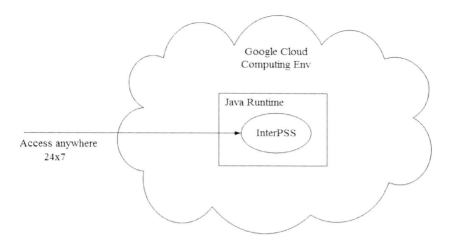

Fig. 9. InterPSS Cloud Computing Runtime

5.2.1 Loadflow Analysis

There are currently 3 loadflow methods [NR, NR+, PQ] available in InterPSS cloud Edition. NR stands for the Newton-Raphson method, PQ for the Fast Decouple method. The +sign indicates additional features, including: 1) non-divergence; 2) add swing bus if necessary; and 3) Turn-off all buses in an island if there is no generation source.

Using the loss allocation algorithm, transmission or distribution system losses can be allocated along the transmission path to the load points or generation source points [Zhou 2005].

5.2.2 Contingency Analysis

Three types of contingency analysis are available: N-1 Analysis; N-1-1 Analysis and N-2 Analysis.

- N-1 Analysis - Open each branch in the power network and run full AC loadflow analysis for each contingency.
- N-1-1 Analysis - First open each branch in the power network and run full AC loadflow analysis for each contingency. Then for each N-1 contingency, open each branch with branch MVA rating violation and run another full AC loadflow analysis for each N-1-1 contingency.
- N-2 Analysis - Open all combination of two branches in the power network and run full AC loadflow analysis for each N-2 contingency.

Fig. 10. InterPSS Cloud Computing User Interface

5.3 Summary

The InterPSS cloud computing solution is currently hosted in Google cloud computing environment. Because of the modern software techniques and approaches used in InterPSS architecture and development, it was quite easy and straightforward to run the InterPSS core simulation engine in the Google cloud. Since it was deployed into the cloud and went live October 2009, we found that the computer hardware and software maintenance effort has been almost zero. In fact, the InterPSS development team does not know, and really does not care, where the InterPSS core simulation engine is physically running and on what hardware or operating system. This is similar to our power consumption scenario, when someone starts a motor and uses electric energy from the power grid, he/she really does not know where the power is actually generated.

The relationship between InterPSS and the underlying cloud computing environment is quite decoupled, which allows the underlying cloud computing environment to be switched relative easily in the future. Hosting InterPSS core simulation engine in the Google cloud really does not mean that we recommend Google as the best cloud computing environment for power system simulations. During the InterPSS cloud computing solution design and development process, the main objective was to architect and structure InterPSS core simulation engine in such a way so that it can be hosted in any cloud computing environment where Java language runtime is available.

6 Trajectory-Based Power System Stability Analysis

With grid computing, certain types of power system simulation can be performed in parallel. In Section 4, application of grid computing to power system security assessment was discussed, where grid computing was used to implement a well-known subject in a different way. In this section, an on-going InterPSS research project will be discussed, where grid computing is used to formulate a new stability analysis approach, which to our knowledge, has not been tried before.

6.1 Power System Stability

In large-scale, multi-machine power networks, it is often observed that under certain conditions, the system might become unstable following a large disturbance, such as a fault, as illustrated in Figure-11. The exact reason causing the instability in large-scale multi-machine power networks has not been fully understood in general, since an adequate approach to fully analyze the dynamic behavior near the stability limit region is not available to date - at least to our knowledge. The following three approaches are currently often used for power system stability study:

- Time-domain simulation: Time-domain based transient stability simulation can very accurately simulate power system dynamic behavior following a large disturbance. The simulation results are displayed as a set of time-domain curves for power engineers to examine. Some trend analysis, using for example, data-mining techniques, might be performed in addition, for pattern identification. Yet, by examining the time domain points, it is not possible to pin-point the exact reason that caused the system instability.
- Small-signal stability analysis: Linear small-signal stability analysis is sometimes performed at certain operation point(s). Today, to our knowledge, analysis in this category are performed at steady-state, pre-disturbance operation point. Power system is non-linear, especially when it is approaching the stability limit region. It is known that this type of linear approach, applied at pre-disturbance operation point, in general, is not capable of predicting, with any certainty, power system behavior in the stability limit region following a large disturbance.
- Transient energy function (TEF) analysis: TEF is also performed around pre-disturbance steady-state condition. Its accuracy to predict power system dynamic behavior in the stability limit region following a large disturbance has not been proven and convinced the power engineering community.

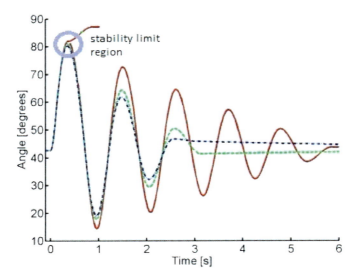

Fig. 11. Stability Limit Region

In summary, the time-domain simulation approach is quite accurate, yet, the simulation result - currently recorded as a set of time-domain points, could not be used for in-depth power system dynamic behavior analysis. The linear small-signal analysis approach is very powerful in terms of analysis functionality, yet, it is currently only applied to pre-disturbance steady-state condition, which is known to be unable to accurately predict power system dynamic behavior following a large disturbance. A research project is currently underway at InterPSS to combine these two approaches together to perform continuous small signal analysis along the trajectory following a large disturbance, calculated using the time-domain simulation approach. Similar approach has been used in voltage stability analysis, where a load change trajectory is manually picked, and multiple loadflow calculations and Jacobian matrix analysis are performed along the load-changing trajectory to determine voltage stability.

6.2 *Trajectory-Based Stability Analysis*

It would be interesting to systematically analyze large-scale multi-machine power system dynamic behavior around the stability limit region, using the combination of time-domain simulation and small-signal stability analysis. The difference between the trajectory-based approach and the commonly used voltage stability approach is that in the trajectory-based approach, the trajectory is accurate, obtained by time-domain transient stability simulation, following a large disturbance, while in voltage stability analysis, the trajectory, obtained by increasing load to find voltage stability limit, is somewhat artificial. In the trajectory-based stability analysis, voltage stability analysis along the trajectory could be also performed, since deteriorating voltage and reactive power conditions in the stability limit region might be, in certain situations, the cause or a major contributing factor to the system instability. It is our belief that the dynamic behavior analysis around the stability limit region, obtained by time-domain simulation, following a large disturbance can provide insight into large-scale power system stability problems, which might be unknown to the power engineering community to date.

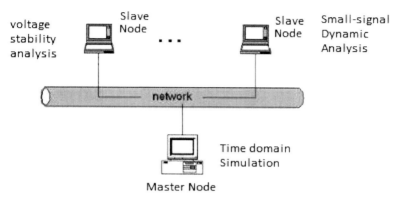

Fig. 12. Trajectory-based Stability Analysis

Grid computing is used to perform the combined analysis, as shown in Figure-12. Power system simulation objects, capable of small-signal stability and voltage stability analysis, are serialized along the trajectory during the time-domain transient stability simulation into XML documents and distributed over the network to the remote slave grid nodes. Then the XML document is de-serialized at the remote grid nodes into the original object model and becomes ready for small-signal stability and voltage-stability analysis. This is still an on-going InterPSS research project. Future research result will be posted at InterPSS Community Site [Zhou 2005].

6.3 PMU Data Processing

One situation, where the trajectory-based stability analysis approach might be applicable in the future, is in the area of PMU (Phasor Measurement Unit) data processing. PUM measurement data describes in real-time power system operation states, which could be considered equivalent to the points on the trajectory describing the system dynamic behavior. The points could be sent in real-time to a grid computing-based distributed parallel power system simulation environment to perform small-signal stability and voltage stability analysis to evaluate system stability to identify potential risky conditions and give early warning if certain patterns, which might have high rick to lead to system instability based on historic data analysis, are identified. This approach is similar to how large investment banks in US, such as Goldman Sachs, process the market data in real-time using their high-performance distributed parallel computing environment to make their investment decisions.

7 Summary

There are many challenges when attempting to adapt the existing legacy power system simulation software packages, which were created many years ago, mostly using FORTRAN programming language and the procedural programming approach, to the new IT environment to support the real-time software application requirements. Among other things, the legacy power system simulation software packages are multi-thread unsafe. The simulation state in computer memory is difficult to be distributed from one computer to another. How to architect and design power system simulation software systems to take advantage of the latest distributed IT infrastructure and make it suitable for real-time application has been discussed in this chapter, using the InterPSS software project as a concrete example.

A model-based power system software development approach has been presented. In this approach, a power system simulation object model is first created and placed at the center of the development process, and the rest of the software components evolve around the object model. Power system simulation algorithms are implemented in a de-coupled way following the visitor software design pattern with regard to the model so that different algorithm implementations can be accommodated.

Power system software systems are very complex. They normally take many years to evolve and become mature. Open, flexible and extensible software architecture, so that others can modify the existing functionalities and add new features, is very important for the software evolution. The InterPSS core simulation engine is such an example. It was first developed for a desktop application. Because of its simplicity and open architecture, it has been late ported to a grid computing platform and a cloud computing environment.

Application of InterPSS grid computing to power system security assessment has been discussed. Accurate full AC loadflow model is used in the security assessment computation. It has been shown that InterPSS grid computing approach can achieve linear scalability. Application of InterPSS to power system stability analysis using the trajectory-based stability analysis approach, which is currently an on-going InterPSS research project, has been outlined.

8 Looking Beyond

All the above has been focusing on how to get the simulation or analysis results quickly and in real-time. Naturally the next question would be, after you get all the results, how to act on them. This is where Complex Event Processing (CEP) [Luckham, 2002] might be playing an important role in the future. Natural events, such as thunder storm, or human activities (events), such as switching on/off washing machine, happen all the time. Most events are harmless, since the impacted natural or man-made systems, such as power system, are robust enough to handle and tolerate the event, and transition safely to a new steady-state. However, there are cases that certain event or combination of events might lead to disaster. Power system blackout is such an example, where large disturbance (event) causes cascading system failures. In power system operation, monitoring the events and correlating them to identify potential issues to prevent cascading system failure or limit the impact of such failure is currently conducted by human system operator.

CEP refers using computer system to process events happening across different layers of a complex system, identify the most meaningful event(s) within the event cloud, analyze their impact, correlate the events, and take subsequent action(s), if necessary, in real time. Event correlation is most interesting, since it might identify certain event pattern, which, based on historic data or human experience, might lead to disaster. If such pattern is identified in real-time, early warning could be sent out and immediate action taken to prevent disaster happening or limit its impact. We believe that CEP has the potential to be used in power system control center to assist system operator to:

- Monitor system operation and recommend preventive or corrective action based on system operating rules when certain system situation (event) occurs;
- Correlate system events and give early warning of potential cascading failure risk;
- Synthesis real-time on-line power system analysis information and give recommendations for preventive correction to power system operation.

CEP application to power system operation will require significant effort to modernize current power system control center IT infrastructure to be ready for real-time IT concept and technology. In 2005, the author spent about a year at Goldman Sachs, working with their middleware messaging IT group to support the firm's real-time stock trading infrastructure. The goal of stocking trading is to maximize potential profit while controlling the risk (exposure) of the entire firm. Power System operation has a similar goal – to optimize power generation, transmission and distribution while maintaining the system reliability, stability and security margin. It is our belief that the real-time information technology, developed in the Wall Street over the last 20 years, could be applied to power system operation to make power system a "Smart" grid in the future. To apply the real-time information technology to future power control center, we see a strong need for research and development of distributed parallel power system simulation technology using modern software development concept and methodology. The InterPSS project is such an attempt, using modern software technology to solve a traditional classic power engineering problem in anticipating the application of real-time information technology and CEP to power system operation in the near future.

Acknowledgement. The author would like to acknowledge the contribution to the InterPSS project from the power system group at South China University of Technology, under the leadership of Professor Yao Zhang, with special thanks to Professor Zhigang Wu of South China University of Technology; Stephen Hao, currently a Ph.D. student at Arizona State University; Tony Huang, currently a graduate student at South China University of Technology.

References

Balu, N., et al.: On-line Power System Security Analysis. Proc. of the IEEE 80(2) (February 1992)
Dean, F.: MapReduce: Simplified Data Processing on Large Clusters (2004) (available on the Web)
Eclipse, "Eclipse Modeling Framework" (2004), http://www.eclipse.org/modeling/emf/ (2010)
Flower, M.: Inversion of Control Containers and the Dependency Injection pattern (2004) (available on Web)
Gamma, E., et al.: Design Patterns: Elements of Reusable Object-Oriented Software. Addison-Wesley (1994)
Hou, G., et al.: A New Power System Grid Computing Platform Based on Internet Open Sources. Power System Automation (Chinese) 33(1) (2009)
IEEE Task Force, Common Data Format for the Exchange of Solved Load Flow Data. IEEE Trans. on PAS PAS-92(6), 1916–1925 (1973)
Lin, Y., et al.: Simple Application of Parallel Processing Techniques to Power Sys-tem Contingency Calculations. N. American T&D Conf. & Expo., Montreal, Canada, June 13-15 (2006)
Luckham, D.: The Power of Events: An Introduction to Complex Event Processing in Distributed Enterprise Systems. Addison-Wesley (2002)
OMG, Model-Driven Architecture (2001), http://www.omg.org/mda/
OMG, Unified Modeling Language (2000), http://www.omg.org/spec/UML/ (2002)
SpringSource, The Spring Framework (2010), http://www.springsource.org
Stott, B., et al.: Fast Decoupled Load Flow. IEEE Trans. on PAS PAS-93(3), 859–869 (1974)
Tinney, W.F., et al.: Direct Solution of Sparse Network Equation by Optimal Ordered Triangular Factorization. Proceeding of IEEE 55(11), 1801–1809 (1967)
W3C, XML Schema (2010), http://www.w3.org/XML/Schema
Waldo, J., et al.: A Note on Distributed Computing (November 1994) (available on the Web)
Wood, A.J., et al.: Power Generation Operation and Control. John Wiley & Sons, Inc. (1984)
Zhou, E.: Object-oriented Programming, C++ and Power System Simulation. IEEE Trans. on Power Systems 11(1), 206–216 (1996)
Zhou, M., et al.: InterPSS Community Site (2005), http://community.interpss.org

A MAS-Based Cluster Computing Platform for Modern EMS

Boming Zhang[1], Wenchuan Wu[1], and Chuanlin Zhao[2]

[1] Department of Electrical Engineering, Tsinghua University, Beijing 100084, China
zhangbm@tsinghua.edu.cn,
wuwench@tsinghua.edu.cn
[2] School of Electrical Engineering & Computer Science, Washington State University,
PO Box 642752, Pullman, Washington
cheka008@gmail.com

Abstract. To meet the requirement of dispatch and control in Smart Grid environment, a MAS-Based cluster computing platform is developed for modern Energy Management System(EMS). Multi-Agent technology is used to develop the high-performance computing platform. "Double-Dimensional Federal Architecture" is proposed and used to implement the advanced application software of modern EMS. A middleware named "WLGrid" is implemented to simplify software development and to integrate distributed computers in Electric Power Control Center(EPCC) into a "Super Computer". Methods of "Task Partition", "Concurrent Dispatching Strategy", "Dynamic Load Balancing" and "Priority-based Resource Allocation" are introduced to accelerate the computation and to improve the real-time property of the whole system. Onsite tests are done and the results are reported to show the effectiveness of these methods.

1 Introduction

Real time power grid operations heavily rely on computer simulation which should be fast enough to meet the requirement for real-time decision making. In order to speed-up computer simulation, High Performance Computing (HPC) technology is needed which has been introduced into power grid operation[Huang Z, Nieplocha J. 2008]. HPC technology used to adopt a parallel computing mechanism to make full use of advantages of multi-core processors and cluster computers. With the technical development of multi-core processors and cluster computers, HPC is improving today's power grid operation functions and will transform power grid operation in the future[Huang Z, Nieplocha J. 2008]. In recent years, HPC has become a hot topic in power system area. Green II RC et al(2011) presents a review on applications and trends of HPC for electric power systems. Zhang L et al(2010) reports HPC applications to Midwest ISO.

In Electric Power Control Center(EPCC), execution of Energy Management System(EMS) application functions can be speeded up by using parallel computing technology. To implement the parallel computing technology, Multi Agent System(MAS) is indispensable no mater to use multi-core microprocessors with tightly-coupled shared memory architecture or to use cluster computers with distributed memory architecture. Task partitioning and coordination to achieve static load balancing or dynamic load balancing are the key in this application which is easy to implement by using MAS technology.

1.1 Agent and Multi-Agent System

Agent and Multi-Agent System(MAS) are two important concepts in computer science and distributed artificial intelligence areas, which represent a new method to conceptualize, to describe and to implement complex systems. In recent years, the application of Multi-Agent technology in power system field has been paid more and more attention. A wide range of applications in power system has been developed with this technology, including fault diagnostics [Davidson et al 2006], condition monitoring system[McArthur S.D.J et al 2004], power market simulation [Widergren et al 2004], micro grid control [Dimeas et al 2005], substation automation [Buse et al 2003], voltage control [Sheng et al 2002], agent modeling[Liu CC 2008] etc. The IEEE Power Engineering Society's (PES) Intelligent System Subcommittee (within the PSACE Committee) has formed a Working Group to investigate the benefits gained by the application of Multi-Agent technology in power system field, and provide recommendation and guidance on the proper use of this technology[McArthur S.D.J et al 2007-Part I][McArthur S.D.J et al 2007-Part II].

Although computer science community has proposed myriad definition about Agent, many of them are either too "strong" to be fulfilled by current technology, or too "week" to differentiate from existing system. To solve this problem, IEEE MAS Group takes the requirement of modern power system into consideration, and provides a definition about intelligent agent which is suitable for power system usage [McArthur S.D.J et al 2007-Part I].

The definition emphasizes that an intelligent agent must have the feature called "Flexible Autonomy", which covers three characteristics:

- **Reactivity:** An intelligent agent is able to react to changes in its environment in a timely fashion, to take actions based on those changes and to implement the function it is designed to achieve.
- **Pro-activeness:** An intelligent agent exhibits goal-directed behavior. Goal-directed behavior connotes that an agent will dynamically change its behavior in order to achieve its goals.
- **Social ability:** An intelligent agent should have the ability to communicate with other agents. The communication is not just a simple and conventional data transmission, but includes the ability to execute complex protocols, such as cooperation, negotiation and competition.

And Multi-Agent System is not only the simple collection of multiple agents, but also involves the communication, cooperation, coordination among these agents. In MAS, data is distributed, and calculation is asynchronous and concurrent.

By complex interaction among agents, MAS can solve a complicated problem in a far more flexible and reliable way, while several individual agents can't do so successfully if they do separately and independently.

1.2 Why Multi-Agent Technology Is Needed in a Smart Grid?

Smart Grid has raised great technical challenges to modern power system. By taking the advantages of "Flexible Autonomy" feature[McArthur S.D.J et al 2007-Part I], Multi-Agent technology can be helpful to design and implement a highly efficient, flexible and reliable software and hardware system, facilitating to overcome those challenges. Distributed Autonomous Real-Time(DART) is a basic requirement for modern Energy Management System(EMS) in Smart Grid[Moslehi K 2004].

Advantages of Multi-Agent technology applied to modern EMS in Electric Power Control Center(EPCC) are as follows.

1.2.1 Autonomy Advantages

Agent's autonomy feature indicates that it has fully control over its state and behavior. In conventional methodology, different modules inside a complex system interact with each other by "Method Invocation". The module to be called has no choice but to execute.

Things get changed in Multi-Agent world, where "Message Passing" replaces "Method Invocation". Instead of sending a command, one agent only sends message requiring for another agent to take actions. The service provider agent makes its own decision on whether to fulfill the request, or reject it, or forward to other agents, or discriminate them by different priority. This is a very important ability for an online security analysis system in modern EMS in EPCC because a calculation has to be finished in a limited time so that computing resource can be released for responding to other more emergent calls.

Standard message interface can also help to enhance the flexibility and extensibility of the system. In old system, different modules are generally tightly coupled. Upgrading such system should first stop current system entirely, and then update the current system by the new one, and re-compile and re-start the new system.

In MAS, since each agent is autonomous and not directly linked to other agents, so the whole system is loose-coupled. Then it is easy to replace the old agent with the new one, without affecting other running agents. Agent platform will direct the request shifting from the old agent to the new one, and the request is fulfilled without any awareness about the change of service provider. This feature makes that components can be plugged in and out dynamically without stopping the whole system, which guarantees the "7 x 24" availability of online security analysis system.

1.2.2 Distribution Advantages

Smart Grid requires high performance computation for EMS advanced applications in EPCC.

- It must deal with a large network model which includes the entire interconnected grid.
- To evaluate the system's comprehensive security status, power grid should be examined from multiple aspects, including static security, transient stability, voltage stability, etc. Then intensive calculation is caused by solving lots of algebra equations for power flow calculation and differential algebra equations for dynamic stability analysis.
- Online security analysis is to deal with a set of predefined contingencies. In order to get more trustable analysis result, a larger and more detailed contingency set is needed, resulting in a more heavy calculation burden.
- Online security analysis requires a high real-time property. To trace the variation of power system operation status in a timely fashion, all these heavy calculations must be completed in a short period so as to leave enough time for operator to make decision.

The distribution feature of MAS can help to solve the problems above. In MAS, multiple agents can be deployed on different machines across network and be run concurrently. An agent platform will connect all these machines seamlessly and integrate them into a "Super Computer" to complete the heavy calculation in short period, making the online security analysis feasible.

1.2.3 Pro-activeness Advantages

This feature will help to build a highly reliable software system in EPCC. For example, Agent-A sends request to Agent-B, while Agent-B is failing to provide such service for some reason. Since agent's behavior is Goal-directed, Agent-A will never drop, but search for other agents who can provide the same service, or execute a sub-optimal plan. Anyway, MAS reduces the risk of "Single Point Failure", and has a better fault tolerance ability, making it suitable for online security analysis application.

1.3 Problems to Be Solved in Applying Multi-Agent to Power System

Although the potential benefits of MAS have been far described, but several key technical challenges are yet to be overcome to develop MAS-based system efficiently, which includes[McArthur S.D.J et al 2007-Part I]:

- **Platform.** Multi-Agent platform will work as a container which hosts multiple agents, and provide services such as network communication, computing resource allocation and fault tolerance to them. The platform should have high reliability to continuously support online analysis, and also should be highly open and flexible to ease system upgrading and maintenance.
- **Toolkit.** Agent's advantages come from its features like Autonomy, Distribution, Pro-activeness and Social ability, but to implement those features is not an easy task. A toolkit, which reuses previous design and hides implementation details of various agents, is needed to popularize MAS in power system field.

- **Design Strategy.** This problem contains two sub-problems: "How to design and implement a single agent" and "how to form multiple agents into MAS". In this aspect, computer science community has provided a lot of solutions. Among these solutions, how to choose one that best suits the actual feature and requirement of power system is a problem worthy to study.

Following Sections will give our solutions to the problems listed above.

2 MAS-Based Software Architecture

Modern Energy Management System (EMS) in EPCC is a very complex system which is implemented by distributed computers and multiple functions. It must have a suitable architecture to make its implementation easier and to make the EMS extendable in the future. This Section designs a software architecture for modern EMS, which answers the following two questions:

- How to construct a single agent?
- How to integrate individual agents to a full-functional system?

This Section is divided into three Subsections: the first Subsection briefly discusses the theory on Multi-Agent architecture; the second Subsection classifies agents in modern EMS by three categories and introduces the functions of the agents of three categories; the third Subsection introduces a "Double-Dimensional Federal Architecture" which integrates agents from the three categories into a full-functional system.

2.1 Brief Introduction to Theory on Multi-Agent Architecture

2.1.1 Structure of Single Agent

Structure of single agent concerns the modules inside individual agent and the relationship between mules. It answers following questions: what modules are needed to compose an agent and how these modules interact with each other? How external changes affect agent's internal state and behavior? ..., etc.

General speaking, structure for single agent can be divided into following two categories:

1) Cognitive Agent

This kind of agent must have precise symbol presentation of itself and external world. Its decision making process is based on a range of classical artificial intelligence methods such as logical inference, pattern matching and symbol operation. Some structures, such as BDI (Belief-Desire-Intention) [Wooldridge M et al, 1999], even want to simulate human psychology, such as Belief, Desire and Intention.

This kind of structure faces many technical challenges, such as "how to translate complex real world into precise symbol model in limited time" and "how to perform logic inference in real-time fashion". These challenges make "Cognitive

Agent" hard to be implemented and operated, thus it is not suitable for use in complex environment, like EMS.

2) Reactive Agent

Different from cognitive agents, "Reactive Agent" neither needs to symbolize external world before it runs, nor needs to perform complex logic reasoning during runtime[McArthur S.D.J et al 2007 - Part II]. The theory supporting this structure believes that the complicated and intelligent behaviors of a Multi-Agent System are built upon the cooperation and coordination among the simple acts from individual agents, rather than relying on complex logic reasoning inside each one[Brooks RA, 1999].

Each Reactive Agent has no symbol model about external world, but is predefined some knowledge before it runs. During runtime, as long as external event satisfies some conditions, agent will act based on its predefined knowledge and affect its outside environment. Since no logic inference and reasoning are performed during runtime, Reactive Agents always have far-better real-time performance than cognitive ones.

2.1.2 Multi-Agent Architecture

Multi-Agent Architecture defines responsibility for each agent contained in the system, and makes rules for interaction between them. Roughly speaking, there are mainly two kinds of architectures, "Network" and "Federation".

1) Network Architecture

In this architecture, every agent should be aware of other agents' location, ability and state, and should also have communication knowledge and capacity. When a specific agent needs a service, it will find the best service provider itself, and send a request to that service provider. If the request is accepted, requester will communicate with service provider directly to fulfill its job as shown in Fig 1.

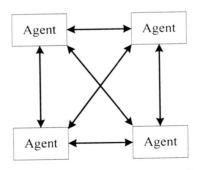

Fig. 1. Network architecture for MAS

This architecture is simple and straightforward. But when more agents enter the system, more communication channels are needed, which will cost a lot of network resources and deteriorate the performance. Another disadvantage is that it requires each agent to have communication knowledge and ability, which increase complexity in implementation. Therefore, "Agent Network" architecture only fits for some small system applications.

2) Federation Architecture

In this architecture, several related agents form a "Federation", and each federation should have a "Manager" as shown in Fig. 2.

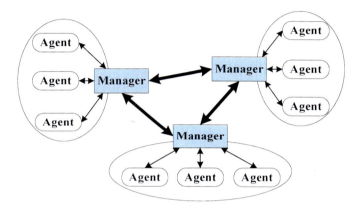

Fig. 2. Federation architecture for MAS

The manager has two responsibilities:

- **Internal responsibility:** It coordinates the agents inside the federation.
- **External responsibility:** Representing the agents inside the federation, it communicates with other federations.

Compared with "Agent Network" architecture, Federation architecture has two advantages:

- High-cost remote communication will only appear among Managers representing for different federations, while inside an individual federation, low-cost local communication will be enough, so communication overhead is highly reduced.
- Only the manager of each federation requires knowledge and ability for network communication, thus no network communication facilities are needed for agents inside the federation, which makes its implementation significantly easier.

Above advantages make "Federation Architecture" suitable for large-scale Multi-Agent System. Therefor "Federation Architecture" is adopted in our platform development.

2.2 Three Categories of Agents in the System

As described above, compared with "Cognitive Agent", "Reactive Agent" is easy to be implemented and operated, and also has far-better real-time performance, so "Reactive Agent" model is used in the proposed platform system.

According to their function and deployment position, agents in this system can be divided into three types: Client Agent, Master Agent and Slave Agent as show in Fig. 3.

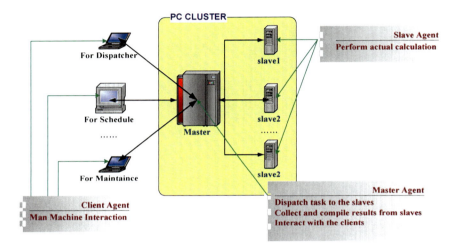

Fig. 3. Client Agent, Master Agent and Slave Agent

1) Client Agent

Client Agent is deployed on the workstations for man-machine interaction. Client Agent transfers users' command to the server, receive analysis results from the server and display them on the screen.

2) Master Agent

The responsibility of the Master Agent includes:
- Act upon the request from client agents or periodically;
- Split a large job into small pieces and distribute them among slave agents;
- Receive partial results from individual slave agent and compile them into a whole one after having received all;
- Send results to client agents for them to refresh display.

3) Slave Agent

Slave agents runs on computers in the PC cluster. It is the slave agent which performs the actual calculation. They receive tasks from the master agent, calculate them and send results back to master agent.

By dividing all these agents into three categories, two benefits can be achieved:

- In some conventional design plan, individual agent takes too much responsibility and always tries to finish a job all by itself, from foreground man-machine interaction to background calculation, making the implementation difficult. However, in this design, each agent only has single responsibility, and cooperates with others to complete a heavy and complex task, which makes system development easier.
- Client Agents, Master Agent, Slave Agents run on different computers, which can fully utilize the advantage of distributed computation to accelerate calculation, accommodating this design for online usage.

2.3 Double-Dimensional Federal Multi-Agent Architecture

2.3.1 Multi-Agent Interactions in This System

"Reactive Agent" structure is chosen to model each agent in this system. Some rules for multi-agent interaction are made in designing the Multi-Agent architecture. Because advanced applications of modern EMS in EPCC is a very complex system which involves a range of computers across network and multiple functions to implement the applications, in order to satisfy such an application requirement, following two interactions should be considered:

1) Interaction among Different Functions

In modern EMS, online security assessment should be performed considering multiple aspects, such as static security, transient stability and voltage stability, etc. All the analysis results on different aspects should be compiled to help operator to get a comprehensive view about the security status of the power grid under current operation condition.

Furthermore, assessment on a single security aspect also contains several sub-steps. Taking transient stability assessment for example, it contains four sub-steps like "Contingency Screening", "Detailed Simulation on Severe Contingencies", "Preventive Control" and "Total Transmission Capacity (TTC) calculation". All these sub-steps need interacting with each other. For instance, "Detailed Simulation" waits the result of "Contingency Screening" to get the most severe contingencies and its result determines whether to run "Preventive Control" or to run "TTC" in next step.

2) Interaction Inside a Single Function

In modern EMS, even a single function always involves a huge amount of calculation, which needs distributed computing and cooperation among client agents, master agents and slave agents in the distributed computing:

- Client agent has no calculation ability but knows how to display the result from other Slave agent on the screen.

- Master agent can figure out a task-dispatching strategy that can best use the computing resources in the cluster, but it doesn't know the algorithm to execute the task.
- Slave agent is the one who undertake the actual calculation.

Since each agent can only undertake part of the job, following cooperation is necessary:

- Master agent splits the huge job into small pieces and distributes them among multiple slave agents.
- Each slave agent calculates its own piece and sends result back to master agent.
- Master agent receives the results from all slave agents, and compiles the results once all results are received.
- Client agents receive compiled results from master agent, and refresh their display.

2.3.2 Federal Architecture Inside a Single Function

According to the description above, client agents, master agent and slave agents, which are deployed on multiple computers across the network, should cooperate with each other to accomplish a single function.

If "Agent Network" architecture as shown in Fig. 4 is chosen to model this interaction, it can be seen from Fig. 4 that too many remote communication channels are needed, and on the other side, each individual agent has to have network communication ability. It is too complicated to implement for a large system.

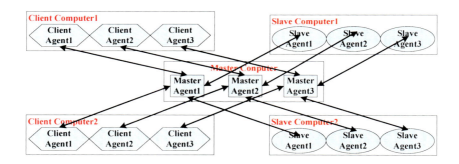

Fig. 4. Agent Network architecture inside single function

So "Federation Architecture" in Fig. 5 is adopted which is much more feasible for our applications.

A MAS-Based Cluster Computing Platform for Modern EMS 111

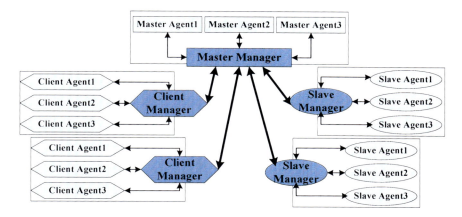

Fig. 5. Federation Architecture inside single function

In this architecture, agents running on the same computer are grouped as a federation, and each federation has a manager. These managers, according to their deployment location, are defined as "Client Manager", "Master Manager" and "Slave Manager". All these managers have the following responsibilities:

- Internal management, such as creating agents, destroying them, or invoking them by event or periodically.
- Manager, working as a representative for the federation, receives message from local agents, and forwards them into network.
- Router. After receiving messages from other federations, it routes them to specific local agents.

By using "Federation Architecture" to model the interaction within a single function, following benefits are achieved:

- First, only the managers require network communication ability and all the technical details are taken by individual agents and hided inside the agents, so this Federation Architecture is easy to implement.
- Second, high-cost network communication only appears among managers, while communication between manager and its members are low-cost local communication, which saves network resources and reduces overhead.

2.3.3 Federal Architecture among Multiple Functions

Just as described in previous Subsection, online security assessment and control in modern EMS in EPCC should compile analysis results based on multiple security aspects, while the analysis on each aspect includes several sub-steps, and each sub-step is accomplished by cooperation between agents distributed across the network. To model these interactions, a hierarchical architecture is designed, where the whole system is divided into three layers and each layer contains several agent federations as shown in Fig.6.

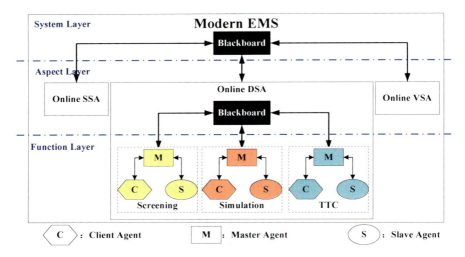

Fig. 6. Hierarchical architecture of Modern EMS in EPCC

1) Function Layer

This layer is on the bottom in Fig.6 which contains several Function Federations. Each federation undertakes a single sub-step of the analysis on a specific security aspect, such as Screening, Simulation and TTC calculation. Each federation in this layer is composed by client agents, master agent and slave agents who accomplish the single sub-step function.

It should be noticed that in such a federation:

- There may be several client agents to show results for different users
- Multiple slave agents share the huge workload in parallel way.
- Only one master agent works as the "Manager", who coordinates these client agents and slave agents.

Client agents and slave agents can join and exit the federation dynamically, without disrupting the whole system. For some functions without display requirement, there is no client agent in the federation. While for some functions without computing requirement, there is no slave agent in the federation.

2) Aspect Layer

The managers of several function federations form the "Aspect Federation", which analyzes the security of the power grid from a specific aspect, like Online SSA (Static Security Analysis), Online DSA (Dynamic Security Analysis) or Online VSA (Voltage Stability Analysis) as show in Fig. 6.

Members of an "Aspect Federation", which are the Master Agents of the "Function Federation", cooperate with each other by using "Blackboard" as the intermedia. When one "Function Federation" completes its calculation, the master agent of it writes its result on the "Blackboard", then other master agents in the federation read that result and decide whether to drive its "Function Federation" to act.

3) System Layer

"Aspect Federations" cooperate with each other with upper-level "Blackboard", which forms the "System Layer", or say, the core of modern EMS. When an "Aspect Federation" completes its analysis on a specific security aspect, it writes the security status or control strategy on the system-level blackboard. After all "Aspect Federation" finishes, modern EMS compiles all the security statuses and control strategies on the blackboard, and gives the operator a comprehensive view of current security status for him to make a decision.

2.3.4 Advantages of Double-Dimensional Federation Architecture

The previously proposed Federal Architecture can be named as "Double-Dimensional Federation Architecture", both inside a single function and among multiple functions. The proposed architecture has the advantages to deal with the complexities both inside a single function and among multiple functions.

Some earlier papers also discussed MAS-based architecture for modern EMS [Liu et al 2002], but they focus only on interaction among multiple functions ignoring details inside a single function. Multi-Agent architecture for a single function is always coarse-grained in the previous works, where the overall analysis toward one specific security aspect is always undertaken by a single agent. It is difficult for the Coarse-grained to implement because too much responsibility is taken by the agent, and it is hard to distribute the coarse-grained agents across network.

Different from previous works, Double-Dimensional Federation Architecture here will never let a single agent work alone, it always form multiple agents into the federation, and let them cooperate with each other to share a complex and huge task. Each member in the federation will never be worried about too much responsibility, but concentrate on a simple job, making them easy to implement. And members of the federations are distributed across network, so the benefits of distributed computing can be got for a better computing performance.

More importantly, the federation architecture is divided into three layers of "System layer", "Aspect layer" and "Function layer". In this layered architecture, cooperation not only happens inside a single federation, but also happens among multiple federations. There will never be too much work assigned to a single federation, benefiting to equalize computer loading.

And for interaction inside a single function, paper [Azevedo et al 2000] also divided agents into three categories, which are "interface agent", "mediator agent" and "application agent", but it doesn't give the details about how these agents communicate with each other. Just as described in previous Subsection, direct communication is complex and low-efficient. So Double-Dimensional Federation Architecture in our method uses "Federation Architecture" to import an indirect communication mechanism for client agents, master agents and slave agents inside a single function, which simplifies the implementation of each agent and reduces communication overhead.

3 MAS-Based Middleware for Cluster Computing – WLGrid

The advantages of the MAS are due to its advanced features, such as Autonomy, Pro-activeness and Distribution, but to realize these features is not an easy task. That's why IEEE PES's working group suggests that a toolkit that reuses previous design and implementation is necessary for popularizing MAS technology in power engineering field[McArthur S.D.J et al 2007 - Part I].

In recent years, Smart Grid has raised various technical challenges to modern EMS [T-FSM][Schainker R et al 2006], which includes:

- Heavy online calculation burden forces a modern EMS to be implemented by a highly- distributed way, while the location of the hardware, software and data should be transparent to the users, making the users feel that they are operating on a single Super Computer.
- Modern EMS should be highly flexible and extensible. Modules in modern EMS can be upgraded individually without halting the whole system. And flexibility also empowers the modern EMS to be applied to grids with different size and requirements, just by simply configuring different modules into the system, other than re-development.
- High availability. Online security assessment demands modern EMS to provide a non-stop service, which requires the system has a self-healing ability.
- ......

To satisfy all the requirements above, a middleware called "WLGrid" (abbreviation for "Wide Logic Grid") is developed in our method. This middleware serves two purposes:

- During development, the WLGrid works as a toolkit, which is used to implement those advanced features of agent and to hide implementation details to user. With the help of WLGrid, programmer can make autonomous, pro-active, distributed agents with less effort, which highly reduces the software development complexity.
- During runtime, the WLGrid works as a container which allows agents to be dynamically plugged in and out, and provides services like network communication, multi-processing and fault tolerance to them. With the help of WLGrid, computing resources in EPCC will be integrated into a "Super Computer", to conquer the challenges raised by Smart Grid.

This Section is arranged as follows. In first Subsection, some software design pattern is brief introduced. And then how to implement client agent, master agent and slave agent within a single function by using WLGrid is introduced in the second Subsection. After that, how to implement interaction among multiple functions based on WLGrid is given in the third Subsection. And the Subsection 4 reports some practical applications developed by using the WLGrid.

3.1 Software Design Pattern

Design pattern is a general and reusable solution to common problems in software design. By providing tested and proven paradigms, design pattern can help making the system more efficient, more flexible and more reliable. Since modern EMS is becoming more and more complex, design pattern is also being paid much more attention in power engineering field. During development of the WLGrid, "Template Method Pattern" and "Publish/Subscribe Pattern" are used and a good effect is achieved.

3.1.1 Template Method Pattern

Template Method Pattern uses base class to define a skeleton of an algorithm, and defers some steps to subclasses so as for them to provide their own concrete implementation. It allows subclasses redefine certain steps without changing algorithm's structure.

This pattern is just like a document template used in daily office work. Document template regulates procedures of an activity, which is full of highly-professional terms, but leaves those specific terms, such as name and date, blank. Using document template has the following advantages:

- One does not concern on those highly-technical terms, making a professional documenting work quick and easy;
- Since documents generated by templates have the same format, a unified method can be used to process them automatically.

By using "Template Method Pattern", same advantages will be achieved:
- During development, it reduces development complexity by hiding technical details
- During runtime, a framework can be generated in a unified and automatic manner, getting ready for handling different components. And a new component generated upon this template, can be easily integrated into the system without changing that framework.

3.1.2 Publish/Subscribe Pattern

In this pattern, message senders (publishers) do not proactively send their messages to specific receivers (subscribers). Instead, subscribers express their interest for some messages first, and then wait for those messages until the messages are delivered to the subscribers after they are generated by publishers.

In this pattern, publishers are loose-coupled with subscribers. They needn't ever know the existence of the counterpart, which makes any change at one side does not disturb the other. This pattern helps to build a highly-dynamical system, in which subscribers can join into or exit out of the system during runtime.

3.2 Agent Implementation for a Single Function

3.2.1 Template-Based Implementation Method

According to Double-Dimensional Federation Architecture described in Section 2, any individual function in WLGrid will be accomplished through cooperation among client agents, master agent and slave agents. So WLGrid is also composed by codes from the three sides, client side, master side and slave side, and each side mainly includes two parts, "Manager" and "Agent Template".

According to the federation architecture on "Single Function Dimension", each computer has one Manager, such as the WLClientManager, WLMasterManager and WLSlaveManager as shown in Fig. 7. These managers administer the agents on a local machine (create, destroy, or invoke them periodically), and perform network communication among federations.

After analyzing the characteristics of different functions in online security assessment and control, it has been found that many functions share the same distributed computation procedures, but only differ from each other in some specific steps. So it is suitable to apply "Template Method pattern" to develop these agents in a unified manner.

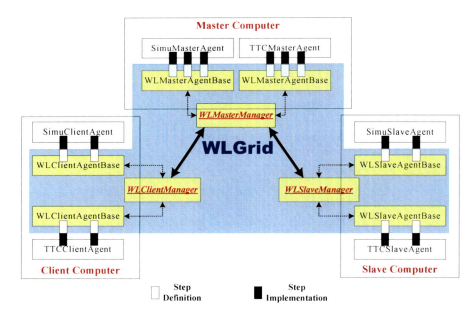

Fig. 7. Templation-based structure for WLGrid

A MAS-Based Cluster Computing Platform for Modern EMS 117

Base classes on different computers, such as WLClientAgentBase, WLMasterAgentBase and WLSlaveAgentBase, play the role as "Agent Template". These templates implement the general process for distributed computing, such as network communication, computing resource location and fault tolerance, but leave certain steps in the process "blank". Agents for specific functions, such as "Fault Screening", "Detailed Simulation on Severe Faults" and "TTC Calculation", will derive from these templates. By filling those "blanks", the concrete implementations can be made according to their algorithm details.

By applying "Template Method Pattern", two benefits can be achieved:

- Technical details about network communication, computing resource location, fault tolerance, etc. are all covered with these agent templates, thus agent implementation is highly simplified.
- The manager of certain federation can handle agents derived from the same template in a unified manner. When new agents enter the federation, the manager keeps unchanged and can still handle the new members as usual. This feature allows function modules being dynamically plugged in and out during runtime, without restarting the whole system.

3.2.2 Implementation for Client Agents

Client agents are generally deployed among computers in different offices in EPCC. Users in these offices have different interests in the information that modern EMS provides. For example, the operators mainly concern about the online security status of the current power grid, while operation planners care more about the results of off-line study. In order to serve for all staff in different offices in the EPCC flexibly, these client agents should be able to enter into and exit out of the system during runtime dynamically.

To meet such requirements, WLGrid uses a "Double Level Publish/Subscribe Pattern" to implement client agents shown in Fig. 8.

- **Subscription.** First, "Local Subscription", shown in (a) of Fig.8, is performed, that is, a client agent registers itself to its manager for the message of interest. The next step is "Remote Subscription", client manager collects all messages its members are interested in and forwards the messages to master agent manager as shown in (b) of Fig.8.
- **Publication.** When a message is generated at master side, such as a specific analysis has been finished and its result is ready to display, master manager first finds out the client managers which are interested in this message and forward the message to them as shown in (c) of Fig.8. After a client manager receives the message, it finds out the client agents who registered this message in advance and forwards the message to them. Finally, a specific client agent receives the message of interest and refreshes its display, which is shown in (d) of Fig.8.

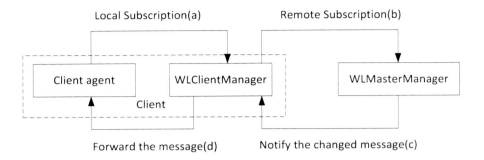

Fig. 8. Double Level Publish/Subscribe Pattern

By using this pattern, foreground man-machine interface is loose-coupled with background calculation service. When a new MMI workstation is deployed for displaying analysis result, the integration of the new MMI workstation to the system can be done during runtime instead of restarting the background server for this new client, thus "Non-Stop" services can be provided.

3.2.3 Implementation for Master Agents

Master agent's responsibility is to distribute a heavy task among slave agents and deal with the calculated results sent back. Although modern EMS includes various functions, it has been found that most of them follow the same procedure on the master side, which are:

- Master agent starts some initializing work;
- Master agent splits a heavy task into some small jobs and sends each job to an individual slave agent;
- Master agent waits and receives results sent back from each slave agent;
- After all slaves have sent their results back, master agent compiles these partial results into a whole one.

Many functions differ just on some concrete steps. For example, in the second step above, "small job" may differ for different functions. And in the final step, different functions may have different ways to deal with the compiled results, some may store them into database, some may publish them for client agent to display, while some others may transfer them to other modules for further analysis.

As a result, WLGrid here applies "Template Method Pattern" to get a better solution. WLGrid defines "WLMasterAgentBase" to work as the template, and its method "Process" represents the whole distributing computation procedure. "Process" has four sub-steps, including "Initialize", "Dispatch", "Collect" and "Compile", and works like the flow-chart as shown in Fig 9.

(1)"***Initialize***" method is invoked. Master agents derived from the template can do some preparation work in this method, such as loading predefined tasks from database, or parsing a binary buffer to get the task needed to do. "Initialize" can also check whether the current task satisfies its precondition, if not, current task will be abandoned.

If current task satisfies its precondition, master agent will call in slaves to do it. Master agent first broadcasts a message in the cluster to call for "bidding". After receiving the bidding invitation, each slave manager will check whether its federation has slave agent which is capable to do the job. If it finds such a member, the slave manager will send a message back to master agent to bid. Within a specific period, master agent will collect those slaves who have responded as "Available Slaves", and make "contract" with them.

By using this broadcast method to locate available computing resources, newly-added computers can be automatically discovered and the broken-down computers can be automatically got rid of, which improves the extensibility and the fault-tolerance of the system.

(2)Then master agent invokes "***Dispatch***" method to distribute a heavy job among "available slaves". In this step, different master agent will use different ways to split that heavy job. And after a small piece of job is sent to a specific slave, a copy of that job will be stored at master agent, in case that some slave agents may not be able to send back result in time.

(3)As long as any result is sent back, method "***Collect***" will be invoked to receive it. The results are always contained in a binary array, and different master agents use different method to parse them out. After getting the result from a slave, some agents will just cache the partial result for later compilation, while others may send another job to that slave so as to achieve a better load-balancing effect.

If some agent fails to send back its result within a specific period, then master agent will find out the copy of that failed job, and send it to another slave to do it again. This mechanism implements the Pro-activeness feature of an agent, which avoids "Single Point Failure" and makes the system self-healing.

(4)After all slaves have sent back results, "***Compile***" method will be invoked. Different master agents derived from template can override this method to deal with the results in different manner, such as saving them, or publishing them to client agents, or forwarding them to other agents for further analysis.

For some agents who are not suitable for running in a distributed manner, the whole template method "Process" can be overridden to implement a local-run method.

By applying "Template Method Pattern", technical details such as network communication and fault tolerance have been embedded in the template, making developing individual agent much easier.

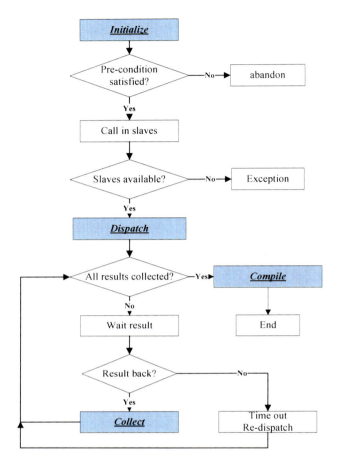

Fig. 9. Processing flow chart for master agent

3.2.4 Implementation for Slave Agents

Slave agents perform the actual calculation. Slave agents run in isolated processes, with a fault tolerant mechanism to ensure that failure on a single slave agent will not disturb others.

Similar with master agents, different slave agents share the same working process, with differences in a certain step. Fig. 10 shows the working flow chart between the Slave Manager and the Slave Process. In Fig. 10, the difference for different slave agents is just in the "Execute" step.

(1) Slave manager receives a task from master and writes the task into a file and then spawns a process to run slave agent;

(2) Slave agent reads task from the file first, and executes it, and then writes the result into a file and notifies the slave manager that the calculation has been finished;

(3) After slave manager receives the notification, it reads result from file, and sends result back to master.

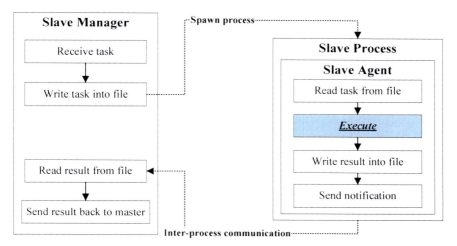

Fig. 10. Processing flow chart for slave agent

"Template Method Pattern" is used again here. WLGrid provides a base class "WLSlaveAgentBase" acting as the template. Building a slave agent is just to derive from the template, and to override its "Execute" method to implement certain algorithm, such as Time-Domain Simulation, Extend Equal Area Criterion (EEAC) calculation, and TTC calculation in Dynamic Security Analysis(DSA).

With the help of this template, programmers can focus on algorithm implementation, other than technical details. Because the file I/O and inter-process communication are all dealt with by template, thus software development is simplified and sped up.

3.3 Event Driven Blackboard Based Coordination among Multiple Functions

3.3.1 Difficulty in Multi-Agent Coordination

Above Subsection introduces how the WLGrid uses template-based method to implement client agent, master agent and slave agent, but it only covers the issue inside one single function. To get a full-functional modern EMS application, interaction among multiple functions is also needed.

Take online Dynamic Security Analysis (DSA) as an example. DSA consists of multiple function modules, such as "Contingencies Screening", "Detailed Simulation on Severe Contingencies", "TTC calculation" and "Preventive Control", as shown in Fig. 11.

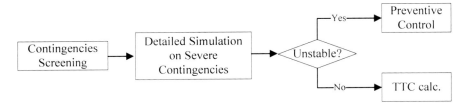

Fig. 11. Flow chart for Online DSA

Communication between agents is a critical problem in multi-agent coordination. Recent methods for multi-agent communication can be divided into two categories: direct method and indirect method. The direct method faces two problems if it is used in online DSA system:

- **Routing problem**, that is, which module should be invoked next is condition-based. In online DSA in Fig. 11, "Detailed Simulation" module should be aware of all the invoking preconditions of its subsequent modules such as TTC calculation and Preventive Control. TTC calculation can be only invoked when no unstable contingencies are discovered, otherwise "Preventive Control" should be invoked. If a new module is added following the "Detailed Simulation" module, the code of the "Detailed Simulation" module has to be modified.
- **Data compatibility problem**, that is, different followers requires different input. In the online DSA in Fig. 11, if "Detailed Simulation" module wants to invoke "Preventive Control", it should send out the critical contingencies to be remedied, but for TTC, only a "starting signal" is needed to send. So the "Detailed Simulation" module should know all the data requirements of its subsequent modules so as to send a specific message to the specific invoked module.

Both problems listed above, make multiple function modules linked by direct communication method, tightly coupled, deteriorating scalability of the system.

3.3.2 Event-Driven Blackboard

Because of shortages of direct communication methods, WLGrid uses a classical indirect method – "Blackboard model" to coordinate function modules in modern EMS.

The manner that "Blackboard model" works is like a group of experts sitting in front of a large blackboard. They work as a team to brainstorm a solution to a complex problem, using the blackboard as the workplace for cooperatively developing the solution. The session begins when the problem specifications are written onto the blackboard. The experts all watch the blackboard, waiting for an opportunity to apply their knowledge. Someone writes a contribution on the blackboard first and waits for further contribution from other experts, then the second expert can think and record his contribution on the blackboard subsequently, hopefully inspire other experts. This process of adding contributions to the blackboard continues until the problem is solved.

Blackboard model consists of three major components:

- **Knowledge Source (KS).** Like the human expert, Each KS is responsible for solving certain aspect of an overall problem. It reads information from the blackboard, processes it and writes a partial solution back onto the blackboard. KS has two major sub-components: Precondition and Action. Precondition specifies when it should be invoked, while Action specifies its modifications on blackboard. When the state on the blackboard has met the Precondition of a KS, its Action will be invoked to update the information on the blackboard, which may trigger other KSs. KSs cannot communicate with each other directly, and the blackboard is the only medium for them to interact with each other.
- **Blackboard.** It is a shared repository of the original problem, partial solution and the final solution. It can be thought of as a dynamic library of the contributions to the current problem that have been recently published by some Knowledge Sources.
- **Controller.** It controls the flow of problem-solving activity, by inspecting state changes on the blackboard and activates proper Knowledge Sources.

Blackboard model has a lot of variants, and WLGrid uses one of them, called "Event-driven Blackboard" here. "Event-driven Blackboard" uses "Publish/ Subscribe Pattern", whose working manner can be described as below:

- Before the system starts, each Knowledge Source should subscribe to the Controller to register its interested event which describes the changes on the Blackboard.
- During runtime, modification on the Blackboard will trigger events. All KSs who have registered the event will be notified. After receiving the notification, each KS will check current state of the Blackboard to decide whether it should act or not. If its precondition is satisfied, that KS will act and write its result on the Blackboard, which will trigger events to inspire other KSs.

By using "Event-Driven Blackboard" to coordinate multiple agents, both "routing problem" and "data compatibility problem" can be well solved:

- Each agent works as a Knowledge Source, when it receives notification, it will make its own decision to act, other than being commanded. This manner not only makes each agent more "autonomous", but also makes the whole system more extensible, because there is no need to modify the Controller when integrating a new agent.
- Different from direct communication method where one function module "pushes" data to its followers, in the Blackboard model, each agent "pulls" necessary data from Blackboard, which makes each agent more autonomous and loose-coupled with its peers.

3.3.3 Implementation of Event-Driven Blackboard

In our WLGrid, the "Event-Driven Blackboard" is developed based on an open-source embedded database called "SQLite"[SQLite]:

- An in-memory database is built with SQLite to act as the Blackboard. The database contains two kinds of tables. One is the static table, which stores fixed data such as the original input data, while the other is a dynamic table, which stores data to be updated by multiple KSs, such as partial solution and final solution.
- According to the layered federation architecture of WLGrid in Fig. 6, each master agent works as the KS in "Aspect Layer", while the Controller of "Aspect Blackboard" works as the KS in "System Layer". Before running, each agent should subscribe to the Controller, registering the event that the data in a specific table has changed to a specific state.
- Controller is developed based on a function "sqlite3_update_hook" provided by SQLite, which monitors the modification made in the database. According to the changing information returned by this function and the registration information, Controller will activate a proper agent.

3.4 Sample: Build Online DSA System with WLGrid

In order to demonstrate the effectiveness and convenience of WLGrid in developing online security assessment and preventive control system, a brief description about how to develop an Online DSA system with WLGrid is presented in this Subsection.

3.4.1 Making Tables on the Blackboard

In order to build an in-memory database to work as a Blackboard, following tables in the database should be constructed:

- **ScreenFault.** It is a static table, which stores the predefined contingencies to be screened each time.
- **SecurityThreshold.** It is also a static table, which stores some judging threshold, such as threshold values for CCT, etc.
- **ScreenRecord.** It is a dynamic table, which stores severe contingencies. This table have three columns:

    ◇ **State.** State of the severe contingency, indicating whether this contingency was preliminarily analyzed by Direct Method or accurately analyzed by Time Domain Simulation.
    ◇ **FaultId.** Identifier of the severe contingency. Detailed content of a fault can be fetched according to this id from table "ScreenFault".
    ◇ **CCT.** Critical clearing time of the severe contingency. A smaller CCT value indicates a more severe contingency.

3.4.2 Contingency Filtering Agent Federation

This Contingency Filtering Agent Federation performs a fast transient stability assessment for all the predefined contingencies, filters out stable ones and keeps the severe ones for further analysis.

FilterMasterAgent runs on master side. It derives from WLMasterAgentBase and provides concrete implementation for the four functions listed in Table 1.

Table 1. Functions provided by FilterMasterAgent

Name	Implementation Description
Initialize	Upload predefined contingencies from "ScreenFault" table on the blackboard.
Dispatch	Distribute all the contingencies among slaves in the cluster
Collect	Receive result. The result contains CCT calculated by a direct method such as EEAC, which should be refined further.
Compile	Sort these contingencies according to their CCTs, and insert severe faults with smaller CCT into "ScreenRecord" table on the Blackboard. The "State" columns of these faults are filled as "Preliminary", and their "CCT" columns are filled with the CCT calculated by EEAC.

FilterSlaveAgent runs on slave side. It is derived from WLSlaveAgentBase, and provides concrete implementation for the method listed in Table 2.

Table 2. Functions provided by FilterSlaveAgent

Name	Implementation Description
Execute	Parse contingencies from the input binary array, calculate their CCTs with EEAC algorithm, return a binary array with all those CCTs.

Contingency Filtering function module has no result to be shown on the screen, so there is no client agent in the federation.

3.4.2 Detailed Simulation Agent Federation

This federation uses "Time Domain Simulation" algorithm to calculate the accurate CCTs for those severe contingencies filtered out by Contingency Filtering.

SimuMasterAgent runs on master side. When it is started up, it should first subscribe to the Controller of the Blackboard for the events that new records with "Preliminary" state are inserted into table "ScreenRecord".

SimuMasterAgent is derived from WLMasterAgentBase, and provides concrete implementations for four functions listed in Table 3 below.

Table 3. Functions provided by SimuMasterAgent

Name	Implementation Description
Initialize	For each row in table "ScreenRecord", it loads contingency details according to "FaultId" column in table "ScreenFault", and takes the value from "CCT" column as initial value.
Dispatch	Send sub-tasks to each slave. Each sub-task contains several contingencies and their imprecise CCT values.
Collect	Receive the revised results. The results contain precise CCTs refined by "Time Domain Simulation" algorithm.
Compile	Update table "ScreenRecord", modify preliminary CCT in the "CCT" column by precise CCT, and modify "Preliminary" in the "State" column by "Accurate". Then submit the compiled results with severe contingencies and their precise CCTs to Master Manager.

SimuSlaveAgent runs on each slave computer. It is derived from WLSlaveAgentBase, and provides concrete implementation for the function listed in Table 4.

Table 4. Functions provided by SimuSlaveAgent

Name	Implementation Description
Execute	Parse out severe contingencies and their imprecise CCTs from input binary array. Taking imprecise CCTs as initial values, calculate precise CCTs by running time-domain simulation repeatedly, return a binary array including those precise CCTs.

SimuClientAgent runs on each client computer. It is derived from WLClientAgentBase, and shows the name and the CCT for each severe contingency on the screen.

3.4.3 TTC Agent Federation

This federation calculates the TTC for transmission interfaces.

TTCMasterAgent runs on master computer. Before it starts, it should first subscribe to the Blackboard Controller, registering the event that new records with "Accurate" state appear in table "ScreenRecord".

TTCMasterAgent is derived from WLMasterAgentBase, and provides concrete implementation for four functions listed in Table 5 below.

Table 5. Functions provided by TTCMasterAgent

Name	Implementation Description
Initialize	After being notified by Blackboard Controller, it checks all the records in table "ScreenRecord". If all their CCTs are larger than "Danger CCT", then it loads TTC tasks from the database, preparing for further calculation; otherwise, it neglects the notification.
Dispatch	Distribute sub-tasks among slaves in the cluster. Each sub-task contains a transmission interface and contingencies to be checked during power flow adjustment.
Collect	Receive the result. The result contains current power flow through that interface and its corresponding TTC under the unstable contingency.
Compile	Save all the TTC results into database, and submit the results to Master Manager.

TTCSlaveAgent runs on each slave computer. It is derived from WLSlaveAgentBase, and provides concrete implementation for the function listed in Table 6 below.

Table 6. Functions provided by TTCSlaveAgent

Name	Implementation Description
Execute	Parse out the transmission interface and the contingencies to be checked. Continually increase the power flow through the interface, and examine the stability under specified contingencies until an unstable case is found under a contingency. The power flow at this point is taken as TTC for that interface, and that unstable contingency is viewed as the restricting factor. Write these results into a binary array and return that array.

TTCClientAgent runs on client computer. It is derived from WLClientAgentBase, and displays the name, current power flow, TTC and restricting factor of each transmission interface on the screen.

3.4.4 Preventive Control Federation

PrectrlMasterAgent runs on master computer. When it runs, it should subscribe to Blackboard Controller first, registering the event that new records with the state "Accurate" appear in table "ScreenRecord".

Since Preventive Control requires fast response while its workload is not so heavy, running it in distributed manner is not a good choice, so Preventive Control run only on master computer. As a result, PrectrlMasterAgent should override the whole template method "Process" to implement the local-run fashion, working as below:

- After being notified by the Blackboard Controller, PrectrlMasterAgent first checks if there is any record in table "ScreenRecord" whose CCT is smaller than "Danger CCT". If there are no such an unstable contingency, notification is ignored, and no act is made.
- Otherwise, PrectrlMasterAgent loads the details of the unstable contingency from the table "ScreenFault" according to the "FaultId" column in table "ScreenRecord". Then it adjusts the power generation to improve current CCT for the unstable contingency until the CCT is larger than "Danger CCT".
- After adjustment, PrectrlMasterAgent submits the scheme to Master Manager, for the Master Manager to forward this control strategy to client computers.

Since Preventive Control runs locally on master computer, there is no slave agent in the federation.

PrectrlClientAgent runs on client computer. For each generator, it shows both the pre-adjusted and the post-adjusted generations on the screen.

3.4.5 Some Onsite Result

After all agents above are developed, they can be deployed on master computer, slave computers, and client computers separately. During runtime, managers on these computers will read configuration file and upload these agents automatically once the system is started up.

Online DSA system developed based on WLGrid has been successfully put into practice in two provincial EPCCs in China. Some screenshots of the system are shown below.

Fig. 12 shows MMI for the online contingency screening. Part (1) in the figure indicates the severe contingencies under current condition. Part (2) shows the calculation process of each slave agent. Parts (3), (4) and (5) plot the stability index (such as CCT, stability margin and max acceleration power of the critical generator) curves, which help dispatchers to trace stability level of the current power grid.

Fig. 13 gives MMI for online TTC monitoring. Part (1) in the figure lists out the actual power flow, maximum transmission capacity and the restricting factor of each transmission interface under current condition. Parts (2), (3), (4) and (5) plot TTC curves for four monitored transmission interfaces. Each curve pane corresponding to a transmission interface contains two curves, in which the lower blue one shows actual power flow flowing through the interface, while the upper red one shows the calculated TTC flowing through that interface. It can be seen from this figure that the TTC here is a variable curve instead of constant one as in traditional method.

A MAS-Based Cluster Computing Platform for Modern EMS 129

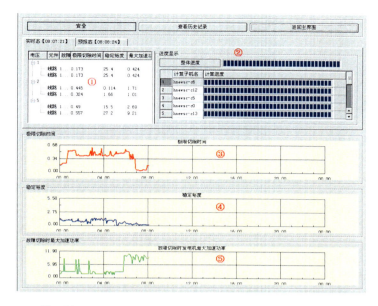

Fig. 12. MMI for Online Contingency Screening and Analysis

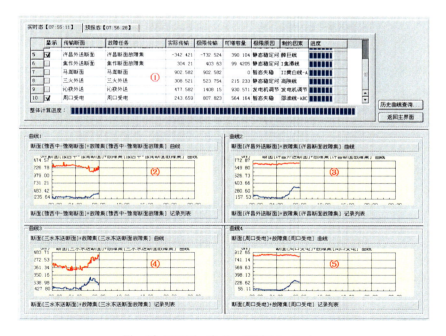

Fig. 13. MMI for Online TTC Monitoring

Fig. 14 is the MMI for preventive control. In part (1) of the figure, the bar charts represent the generation adjustment scheme of before(blue) and after(red) adjustments. In part (2) of the figure, the table lists CCT values before and after adjustments for a selected unstable contingency. It can be seen from this figure that the CCT has been increased from 0.254 sec. to 0.544 sec. , indicating that the power grid becomes more stable through this adjustment.

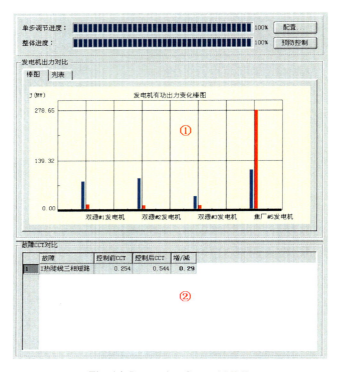

Fig. 14. Preventive Control MMI

4   Task Coordination and Dispatching Strategy in WLGrid

Real-time requirement to online security assessment and control system in EPCC is crucial issue. To speed up the calculation, the traditional "Main/Backup Machines" hardware framework has to be replaced by a "Computer Cluster" in modern EMS. But not the Cluster Computation will ensure the expected speed up, unless the communication overhead and the risk of failure can be well controlled. Therefore, software method is needed to coordinate and dispatch tasks among computers in the cluster to satisfy real-time requirement for online security assessment and preventive control.

The term "coordination" here refers to arranging the order of executing multiple tasks, while "dispatching" aims to distributing the workload properly among the computers in the cluster. This Section introduces the task coordination and dispatching strategy in WLGrid by following four Subsections:

(1) **Task Partition.** Task is partitioned to smaller pieces, getting ready for balancing both concurrency and communication overheads.
(2) **Concurrent Task Coordination.** By running multiple function modules concurrently and multiple tasks concurrently within a single module, the whole system is transformed into an assembly line, and online calculation is sped up.
(3) **Dynamic Load Balancing.** This strategy tries to distribute workload evenly among all computers in the cluster to reduce "Short-Board Effect".
(4) **Priority-Based Resource Allocation.** Multiple functions in the modern EMS are prioritized, and the computing resources are allocated according priority of a function, which guarantees that the most important and emergent task will always be allocated sufficient resource.

4.1 *Task Partition*

In order for computers in a cluster to share multiple tasks efficiently, a good task partition strategy is important.

Task Partition is composed with two aspects "Data Partition" and "Procedure Partition". "Data Partition" refers that a large dataset is distributed, and the same program is used to deal with different pieces of data on different computing resources. And "Procedure Partition" refers that the overall process is split into multiple sub-procedures, different sub-procedures run on different computing resources and cooperate with each other to complete the whole process.

The size of each subtask heavily affects the final calculation performance. Although small-grained tasks are more likely to run on different computing resources simultaneously and achieve a higher concurrency, but they also need to interact with others frequently, which may cause a high communication overhead. Coarse-grained tasks are more independent from each other, which need less communication during calculation, but they cannot take full advantages of abundant hardware resources.

According to Section 2, the WLGrid uses a Double-Dimensional architecture, including "Multi-Functions Dimension" and "Single-Function Dimension", so different task partition strategies for the two dimensions are needed and given in this Subsection.

4.1.1 Partition among Multiple Function Modules

Partition in this dimension belongs to "Procedure Partition". Online security assessment in power grid considers different types of security problems such as steady-state security, transient stability and voltage stability, etc. And the assessment even on a single aspect can also be divided into several sub-steps.

WLGrid divides the whole security assessment problem into several single security aspects, and for each security aspect, the function module is divided further into small-grained sub-function modules. Each sub-function module is implemented by a single sub-step, on different computing resources, to achieve a better concurrency

4.1.2 Partition within Single Function Module

Partition in this dimension belongs to "Data Partition", that is, each function module faces a large dataset which may contain a lot of predefined contingencies to be assessed for their stability or a lot of transmission interfaces to be checked for their TTC, and each slave agent within the function module will deal with only one subset of the dataset.

There are some fine-grained methods to do the partition. Taking DSA as an example, assessing the transient stability of a single contingency in a large power grid has to solve a large matrix. Network partitioned algorithm is to split the whole matrix into some smaller blocks with each slave calculating a single block. In this algorithm, high overhead on communication between slave and master is a critical problem. This is not suitable for our application.

Task partition instead of network partition is used in our method. By WLGrid, a relatively coarse-grained method is used to partition the task inside a single function module, that is, a single item in the dataset, e.g. a predefined contingency or a transmission interface, is the smallest unit and cannot be split any more.

For example, a slave will assess stability for a contingency, or calculate TTC for a transmission interface all by itself. During the calculation, the steps of the algorithm, or the matrix to be solved, are always processed by a single computer, other than distributed to other computer, to reduce the communication overhead.

4.2 Concurrent Task Coordination

Coordination strategy determines the execution order of each function module. Conventional EMS always applies a sequential coordination strategy. In this strategy, the whole system consists of several function modules, and these modules are executed one by one. Unless current modules have finished, the next module cannot start. And inside a module, the next task can only be dealt with unless current task is totally completed.

This sequential coordination strategy is suitable for the hardware with single CPU. If many programs run simultaneously on a single computer, they will cause serious contention over CPU and other computing resources, which will harm the execution performance of each program. As a result, a sequential coordination strategy is needed to avoid resource contention.

Multi-core computers with multiple CPUs in one computer are prevalent in current days. Multi-core computers can easily run more programs at the same time. In using the multi-core computer, avoiding "Short-Board Effect" becomes more concerned than saving in computing resources. So the conventional sequential coordination strategy cannot be efficient in the case of multiple CPU hardware.

"Short-Board Effect" refers that, if multiple agents run concurrently to share a big task, the time cost to finish this big task depends on the slowest agent. The fast agent finishes early, but it can do nothing but idle. This fast agent has to wait its slow peers. To solve this problem, concurrent task coordination strategy is developed to replace the sequential one. The key point of concurrent coordination is that, when the fast agent finishes its current task, it will not wait its slow peers, but begin processing the next task. By using this method, multiple tasks can be executed concurrently.

This concurrent task coordination strategy can utilize the multiple CPU hardware and therefore accelerate the multiple task calculation to a maximum extent. It is particularly suitable for real-time application in EPCC. According to the double-dimensional architecture of WLGrid, concurrent coordination should be applied on both "Multi-Functions Dimension" and "Single-Function Dimension".

4.2.1 Concurrent Coordination among Multiple Function Modules

As described previously, online security assessment and control is composed by several function modules. If using conventional development methods, such as "Procedure-Oriented" or "Object-Oriented" paradigm, these function modules have to be implemented as subroutines or objects, and are executed in sequence.

Taking online DSA as an example, the whole process consists of "Contingency Filtering", "Detailed Simulation on Severe Contingencies" and "TTC Calculation". All these three modules should be executed in both "Real-time Mode" to analyze the security status of the current power grid, and "Forecast Mode" to predicate the security status 15 minutes later. If sequential strategy is used to coordinate these modules, only one of them can be executed at a specific time. And "Forecast Mode" analysis cannot be started unless all three function modules have finished analysis in "Real-Time Mode".

Things get changed in MAS. In "Agent-Oriented" paradigm, each agent has its own thread; different agents run in parallel on different threads and communicate with each other by passing messages. The basic feature of message passing is asynchronous, which means after sending out the message, the sender is not blocked and it waits until the receiver finishes dealing with that message.

By using this method in online DSA, when "Contingency Filtering" module finishes its real-time analysis, it sends a message to its subsequent modules to notify that they can start. Other than waiting for "Detailed Simulation on Severe Contingencies" and "TTC Calculation" to finish, "Contingency Filtering" module can begin its "Forecast Mode" job at once. As a result, real-time job and forecasting job can be run concurrently. In this way, computing resource can be utilized to a maximum extent and the total time cost can be cut.

As described in Section 2, WLGrid applies layered federation architecture in "Multi-Function Dimension". According to this layered architecture as shown in Fig. 6, each module in basic "Function Federation" consists of client agents, master agent and slave agents. Master agent is the manager, and the client agents and slave agents all act according to Master agent's command. So how to run master agents in different federations concurrently is the key problem to run multiple function modules concurrently. WLGrid solves the problem by using "Active Object" pattern.

"Active Object" is viewed as the best method to implement agent [Guessoum et al 1999]. Fig.15 shows structure of Active Object.

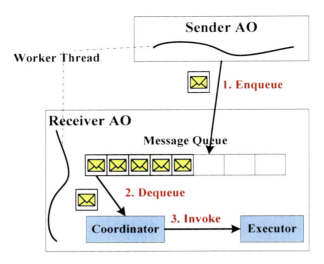

Fig. 15. Structure of Active Object

"Active Object" consists of following elements:

- **Worker thread.** Every active object runs on its own thread, which executes the "Coordinator" and "Executor".
- **Message Queue.** Each active object has a message queue to store the received messages. When ActiveObject-A wants to communicate with ActiveObject-B, the A will insert a message into B's message queue, and then A can do other jobs without being blocked for waiting B to deal with that message.
- **Coordinator.** Coordinator is the place to implement agent's "Autonomy" feature. After the coordinator takes out a message from the message queue, it has a fully control to the message: it can forward the message to "Executor" to process, or it can also discard the message, or it can prioritize the message and decide whether to process the message right now or later. If the message queue is empty, coordinator will be blocked and waits for appearance of new messages.
- **Executor.** It is the element that actually processes the message.

In WLGrid, a class, named as "MasterAO", is developed to implement this "Active Object" pattern. Each MasterAO has a master agent to work as the "Executor", which coordinates the client agents and slave agents in the federation. Each MasterAO runs on its own thread, and cooperates with other MasterAOs by writing messages onto the blackboard. The controller of the blackboard will insert notification about changes on the blackboard. Active Object based coordination among multiple function modules is shown in Fig. 16.

The structure in Fig. 16 is still not the final one. It will be extended in the next Subsection to import concurrency inside a single function module.

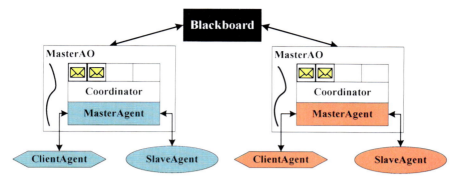

Fig. 16. Active Object based coordination among multiple function modules

4.2.2 Concurrent Coordination Inside Single Function Module

In function federation in Fig. 6, in order to avoid "Short-Board Effect", we import concurrency inside each single function module.

In this Subsection, an "Agent Pool", which contains multiple master agents, is built inside each MasterAO, as shown in Fig. 17, to implement concurrent coordination strategy inside single function module.

When the Coordinator of that active object takes a new task out from the message queue, if current Executor, a master agent, is idle, then the Coordinator just forwards the task to the current master agent. But if current master agent is busy (for example, it is blocked and waiting for results from slave agents), the Coordinator will fetch another idle master agent from the pool and forward the task to it immediately, other than waiting. As a result, multiple tasks can run concurrently within a single function module.

If no idle master agents are available in the pool, there are several options to be adopted: a new master agent can be created and inserted into the pool, or the task is re-inserted into the message queue and waits for executing later, or the task is

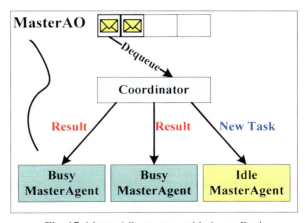

Fig. 17. MasterAO structure with Agent Pool

discarded directly. After a master agent finishes and becomes idle, it will return back to the pool and wait to be re-used.

The Coordinator is the key component in MasterAO whose working procedure is show in Fig. 18 and described as follows:

(1) Coordinator takes a message from the message queue, and checks its type. If the message is the result sent back from slave agent, then go to step (3). If the message is a new task, go to step (2).
(2) Coordinator fetches an idle master agent from the "Agent Pool" for the handling of the new task. After all subtasks have been sent out, go to step (4).
(3) The coordinator figures out the owner of the result, and forwards the result to that master agent. Go to step (4).
(4) Coordinator returns back to the outlet of the message queue, retrieves and processes the next message.

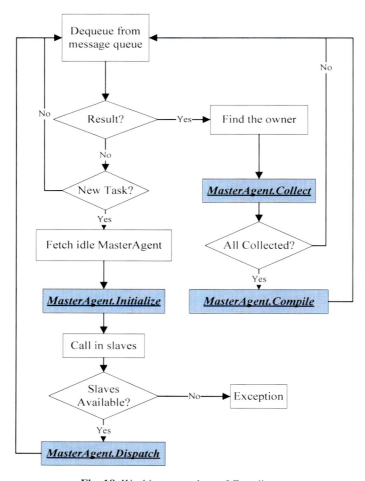

Fig. 18. Working procedure of Coordinator

4.2.3 Assembly Line Effect

To make the concurrent task coordination strategy more understandable, this Subsection compares the process of Online Security Assessment and Preventive Control with the production process in manufacturer.

Sequential coordination strategy is just like the case that there is only one worker in the factory. The worker should know all the steps to finish a product, and the worker has to move around different positions. That worker can only pick up the next job when current one is finished, so only one product is being made at a specific time.

When the factory grows larger, more workers can be employed and an assembly line can be built. The assembly line splits the whole production process into several stages, and each stage has its own worker. When a worker in a stage finishes his job, he just puts the component on the conveyor, and picks up another component from the conveyer immediately. By using assembly line, multiple products can be made concurrently on the line.

The same effect can be achieved by using concurrency mechanism among different function modules. Fig. 19 shows three task coordination strategies. By using "Active Object" pattern, each function modules ("stages" in the factory) has its own thread (like "worker" in the factory), so security assessment (like "product" in the factory) for different mode (e.g. real-time or forecast) can be performed at the same time, which accelerates the whole system.

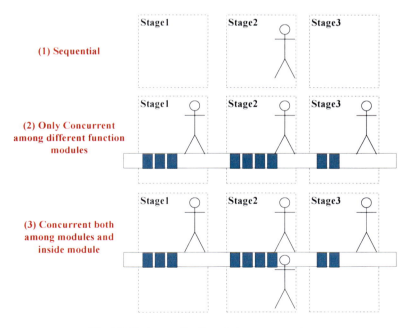

Fig. 19. Task coordination strategy comparsion

One problem in assembly line above is that the speed difference among stages may cause bottleneck that slows down the whole assembly line. For example, suppose speed of stage 2 in Fig. 19 is slower than that of stage 1 and stage 3, then stage 2 becomes a bottleneck. This bottleneck will cause that components finished from its previous stages are piled up and block the conveyor, while its subsequent stages are idle during most of the time. To solve this problem, more workers should be deployed to the slower stage (Stage 2) to share the workload. This is the idea of concurrent task coordination strategy both among modules and inside a module. By improving concurrency of a single stage, the slower stage can be sped up, then bottleneck problem can be alleviated.

Modern EMS has the same bottleneck problem. By using "Agent Pool" and adding more agents (like "worker" in the factory) to slower function modules (like "slower stage" in the factory), multiple tasks can be executed in a single function module concurrently, which can eliminate the bottleneck and accelerate the whole system.

From above description, it can be seen that, by importing concurrency both among multiple function modules and inside a single function module, modern EMS becomes an assembly line that produces security reports and control strategy for operators.

4.3 Dynamic Load Balancing

A common problem in distributed computation is load unbalance, which is, workload is unevenly distributed among computers, making some computers very busy while others are idle. One negative effect caused by load unbalance is "Short-Board Effect", which means the time cost of whole the system subjects to the slowest slave. Previous Subsection has discussed how to reduce the loss of "Short-Board Effect" by concurrent coordination strategy, and this Subsection covers how to reduce "Short-Board Effect" itself by a load balancing strategy.

Load balancing strategies can be categorized as "Static Load Balancing" and "Dynamic Load Balancing".

In Static Load Balancing, workload distribution strategy is predefined before the system starts. The advantage of this strategy is its simplicity, no new strategy is made during runtime. But the biggest disadvantage of Static Load Balancing is that it cannot adapt to changes. If workload of some task is increased, or some computers join or exit the cluster, static balancing strategy becomes obsolete, and the whole system has to be restarted. Therefore, static load balancing is not suitable for online security assessment and control system.

Dynamic Load Balancing strategy distributes workload dynamically on the task size and computer resources. It contains two categories, "Centralized" and "Distributed".

- **Centralized Dynamic Load Balancing Strategy.** In this strategy, each slave must report its current workload to the master first. After collecting the workload from all slaves, master makes a distribution plan and pushes subtasks to slaves, however, each slave has no pro-activeness but do whatever the master

sends to it. The advantage of this centralized method is that master has the global view and takes all slaves into consideration in making the distribution plan, while its disadvantage is that the workload on a slave may change after it reports, then the task the slave receives may not best suit its current conditions.
- **Distributed Dynamic Load Balancing Strategy.** In this strategy, master is not responsible for making load distribution plan to "push" subtasks, but each slave "pulls" appropriate subtasks from the master. The advantage of this strategy is that each slave is pro-active, and the subtask the slave gets always best suits its current condition. However, this strategy requires more frequent communication, which may cause overhead.

Different function modules in modern EMS have different computing requirements, so different dynamic load balancing strategies are needed. In our work, Contract-Net based centralized strategy and Auction-based distributed strategy are used for different cases.

4.3.1 Contract-Net Based Centralized Dynamic Load Balancing

Contract Net protocol is a well-known protocol used for task allocation in MAS [Contract Net]. Two kinds of agents exist in the protocol, "Manager" and "Contactor", and a task is allocated by following steps:

(1) Manager announces a bid.
(2) When receiving the announcement, a Contractor decides whether it should bid or not according to its ability and current state.
(3) Manager assesses all the bids and awards the job to the most appropriate Contractor.
(4) The winner contractor fulfills the job and sends result back to the Manager.
(5) Manger receives the result and the work is finished.

Based on "Contract Net Protocol", WLGrid develops a centralized dynamic load balancing strategy for online stability assessment.

(1) When the master agent wants to do security assessment on a huge number of predefined contingencies, it broadcasts an announcement in the cluster. The announcement includes the security aspect it wants to check (static, transient or voltage, etc) and the algorithm it wants to use (time-domain simulation, or EEAC, etc).
(2) The slave manager receives the announcement, and then searches for slave agent who is capable to undertake the task in the federation. If such an agent is found, the slave manager sends a message back to the master to bid for the task. The replying message here includes current workload of that computer.
(3) After a period of waiting time, the master agent splits the whole contingency set into several subsets for responded slaves to undertake. According to the workloads reported by the slaves, more contingencies will be allocated to the light-loaded slave, while fewer contingencies allocated to the heavy-loaded one.

(4) After a subset of contingencies is sent to a slave, a copy of those contingencies will be kept by the master agent. In case a slave fails to report result back within a specific period of time, the master agent will send that copy to another slave, to enhance the reliability of the whole system.

What should be stressed is that, in this centralized dynamic load balancing strategy, the master is responsible to design a scheme to distribute the workload evenly. But the problem is that the workload is affected by many factors, so it is hard to estimate workload before runtime. Therefore, some substitution should be used to replace the real workload when making the task distribution scheme.

In online stability assessment, the number of contingencies dispatched to a slave is a good substitution for real workload of that slave. This always works well, because the number of contingencies is large, and computing cost for each contingency is nearly the same, so as long as each slave is dispatched similar number of contingencies, their workloads are generally balanced.

4.3.2 Auction-Based Distributed Dynamic Load Balancing

Another main function in modern EMS is to calculate TTC for major transmission interfaces. TTC calculation has totally different picture from online security assessment, for example,

- Small number of tasks. Thousands of predefined contingencies should be assessed for a large power system in online security assessment, while TTCs should be calculated just for fewer transmission interfaces.
- Time costs of tasks differ greatly. Time cost to calculate TTC for one transmission interface depends on many factors. We have to increase power flow on transmission interface and to check stability for a pre-specified fault. The way how to increase load and how to adjust generator output to balance the load will affect the time cost for TTC calculation.

Therefore, it is difficult to find a substituted index to the real workload for each TTC task, so centralized load balancing strategy mentioned above doesn't fit.

To solve this problem, WLGrid applies a distributed load balancing strategy to perform online TTC. Each agent in centralized load balancing strategy is passive and it executes the task the master "pushes" it to do no matter the task is heavy or not. There is no competition among different slaves. Different from the centralized way, in distributed load balancing strategy, each slave agent is active and the slave "pulls" the workload from the master to take the one that best suits its condition. And competition exists among different slaves, so fast slave agent always gets more tasks than its slow peers.

WLGrid develops a simplified "Auction Protocol" to implement the distributed load balancing strategy for online TTC calculation. During auction, each slave agent will only get one TTC task to do in the first round. When some agent finishes its task, it reports the result to the master and asks for another task. This process repeats until all the tasks on the master side have been dispatched.

Different from the master agents in Contract Net Protocol, master agents in Auction Protocol will not design distribution plan beforehand, it only sends out task according to the request from the slave agents.

For Auction Protocol, the slow slave agent, no matter due to poor-performance hardware or current heavy task on its shoulder, will focus on the existing task rather than getting more, which avoids overloading; while the fast slave agent, no matter due to good-performance hardware or previous light tasks on its shoulder, will ask more tasks from the master. So the fast slave agent can share the workload for its slow peers.

Auction Protocol is suitable for the function whose time cost is hard to estimate beforehand. But different from Contract Net Protocol that all tasks are dispatched out in a single shot, auction protocol dispatches tasks many times, which needs frequent network communication. However, this is not a big problem for online TTC calculation, because there are fewer transmission interfaces in the TTC calculation.

4.4 Priority-Based Resource Allocation

A system is said to be real-time if the total correctness of an operation depends not only upon its logical correctness, but also upon the time when it is completed [Realtime Computing]. Real-time system can be divided into two categories:

- **Hard real-time system.** In such system, the completion of an operation after its deadline is considered useless - ultimately, this may cause a critical failure of the complete system.
- **Soft real-time system.** It can tolerate some lateness, and may respond with decreased service quality.

Modern EMS consists of various function modules. According to their requirements on real-time property, these function modules can be assigned different priority as follows:

- Modern EMS also has some functions for operation planning applications. These functions are calculated offline, and have no strict time constraints, so they are categorized as "non-real-time", and have the lowest priority.
- There are three ways to start an online function in modern EMS: periodically running, reacting to operator's command, or being triggered automatically by event in power grid. Functions running under periodic mode analyze predefined contingencies, and complete the calculation in a specific time period, for example in 15 min. Because the predefined contingencies are anticipated instead of really happened, these functions can tolerate slight lateness, categorized as "soft real-time", which have a higher priority.
- Functions launched by operator's command or event in the power grid are "hard real-time", which have the highest priority. They must finish within a strict deadline, otherwise blackout may occur in the power grid.

Computing resources (such as slave computers, CPU, etc) are allocated to different functions according to their priority, which guarantees that functions with higher priority always have sufficient resource to use and can be finished in time. Two strategies are used in WLGrid for resource allocation:

- **Service Partition.** This method distinguishes the computing resources used for "non-real-time" functions and "real-time" functions. Two clusters can be deployed in EPCC. The larger one is used for online calculation, while the smaller one is used for offline calculation. Since WLGrid dynamically gathers slaves, a computer can be easily transferred from one cluster to another at runtime, without stopping the whole system.
- **Service Isolation.** When operator's command is received or event in the power grid is detected, "hard real-time" functions will be launched. They will first kill any periodically running functions, take the computing resources away from those "soft real-time" functions and use all of them for "hard real-time" calculation without any contention.

4.5 Performance Test

The task coordination and dispatching strategies introduced above have already been used to EMS in two provincial EPCCs in China successfully. The scale of a provincial power grid is listed in Table 7.

Table 7. Scale of a provincial power grid in the test (number)

Equipment	Scale
Generators	175
Transmission Lines	516
Loads	920
Transformers	479

Performance tests are performed on this grid to show the effectiveness of the coordination and dispatching strategies introduced in this Section.

4.5.1 Test on Concurrent Task Coordination Strategy

16 computers are used in this cluster and the following three steps are executed in this test:

(1) Screen the transient stability situation for all in-service transmission lines. Use EEAC algorithm to calculate the CCT for each predefined contingency.
(2) Pick the top 15 contingencies with the smallest CCT, and use Time-Domain Simulation algorithm to calculate the precise CCT for them.
(3) If no unstable contingency is found, calculate TTC for 30 transmission interfaces.

Above steps should be performed under both "real-time mode" and "forecasting mode", providing the operator with the security status of the current power grid and the status 15 minutes later.

We use three coordination strategies to run the above processes as explained in Fig. 19, and each coordination strategy is tested five times.

- **Sequential coordination.** Multiple function module and multiple tasks in a single module are all executed sequentially. Analysis under forecasting mode cannot start until real-time mode is totally finished.
- **Single-Concurrency coordination.** Multiple function modules run concurrently, while multiple tasks inside a single function module are still executed sequentially. After a real-time calculation is finished, a function module can start its forecasting job immediately, other than wait for its subsequent modules to finish their real-time jobs. A function module cannot start forecasting task unless all results for real-time mode are collected.
- **Double-Concurrency.** Multiple function modules and multiple tasks in a single function module are all run concurrently. Different from "single-concurrency" strategy, after the master agent in a function module sends out real-time subtasks to its slaves, it will not wait for real-time results, but it starts its forecasting job immediately.

Table 8 gives CPU time for three concurrency coordination strategies. It can be seen from Table 8 that after importing concurrency both among different function modules and inside a single function module, the time cost for online DSA has been reduced from 217 sec. down to 116 sec., a 47% reduction.

Table 8. CPU time for three concurrency coordination strategies (second)

Test Number	Sequence	Single-Concurrency	Double-Concurrency
1	209	189	114
2	221	189	115
3	224	188	123
4	215	189	118
5	216	188	110
Average Time Cost	**217**	**188.6**	**116**

4.5.2 Test on Contract-Net Based Dynamic Load Balancing

In this test, we use Time-Domain Simulation algorithm to assess the transient stability for 465 contingencies. Each contingency is simulated 5 seconds, and the step size in the simulation is 0.01 second. A cluster of 16 computers are used in this test.

First, all contingencies are calculated on a single computer, and the time cost is recorded as t0. Then we perform the experiment multiple times to check the relationship between time cost and the size of the cluster. When there are n computers in the cluster, the ideal time cost should be t0/n and the actual time cost is tn in the test, the difference between them represents a time cost on communication and coordination. The test results on Contract-Net based dynamic load balancing is given in Table 9 and the "Time cost vs. Cluster Size" curves of the Table 9 are plotted in Fig. 20.

It can be seen from Table 9 and Fig. 20 that with more computers involved in the calculation, actual time cost decreases dramatically, while the difference (tn – t0/n) keeps stable. Actual time cost is very close to the ideal time cost which proves that Contract-Net based dynamic load balancing strategy is suitable for online security assessment.

Table 9. Test results on Contract-Net based dynamic load balancing (second)

Cluster size	Actual Time Cost (tn)	Ideal Time Cost (t0/n)	Difference (tn – t0/n)
0	467.76	467.76	0
1	469.76	467.76	2
2	242.68	233.88	8.8
3	177.13	155.92	21.21
4	129.39	116.94	12.45
5	105.91	93.55	12.36
6	90.83	77.96	12.87
7	76.32	66.82	9.50
8	71.67	58.47	13.2
9	64.21	51.97	12.24
10	59.89	46.78	13.11
11	51.55	42.52	9.03
12	49.9	38.98	10.92
13	45.11	35.98	9.13
14	41.75	33.41	8.34
15	39.58	31.18	8.40
16	38.9	29.24	9.67

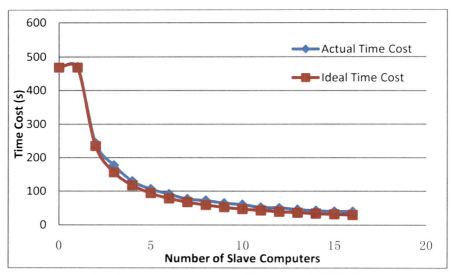

Fig. 20. "Time cost vs. Cluster Size" curves for Table 9

4.5.3 Test on Auction-Based Dynamic Load Balancing

This test uses 9 computers to calculate TTC for 30 transmission interfaces. Table 10 lists time costs of TTC calculations for 6 transmission interfaces. It can be seen from Table 10 that TTC time costs varies greatly for different interfaces. Therefore, centralized load balancing strategy doesn't fit TTC calculation.

Table 10. TTC time costs for several transmission interfaces

Transmission interface	Size of contingency set	Actual power flow (MW)	TTC (MW)	Time cost (second)
Interface 1	11	1266.05	2777.26	143
Interface 2	8	741.18	1203.15	137
Interface 3	6	264.17	504.33	25
Interface 4	2	653.29	1307.41	25
Interface 5	3	118.77	256.99	124
Interface 6	1	173.14	767.00	7

We use Contract-Net Protocol and Auction Protocol to calculate TTCs for 30 transmission interfaces separately by using 9 slave computers in the cluster. Workload and time cost on each slave computer is listed in Table 11 below.

Table 11. Performances for Contract-Net Protocol and Auction Protocol

Slave computer	Contract-Net Protocol		Auction Protocol	
	# of Tasks	Time cost (s)	# of Tasks	Time cost (s)
Slave 1	3	57	7	32
Slave 2	3	41	2	45
Slave 3	4	33	3	43
Slave 4	3	100	3	56
Slave 5	4	186	1	139
Slave 6	3	26	6	48
Slave 7	3	7	4	160
Slave 8	3	19	1	100
Slave 9	4	305	3	137
total time cost(s)	306		161	

It can be seen from Table 11 that for Contract-Net protocol, some slave agent(Slave 7 in Table 11) finishes its job in less than 10 seconds, while some others use more than 5 minutes(Slave 9 in Table 11). Total time cost depends on the time cost of the slowest Slave 9, thus "Short-Board Effect" is very serious.

For Auction protocol, some slave agents (Slave 5 and Slave 8 in Table 11) take only one heavy task, while some others (Slave 1 and Slave 6 in Table 11) undertake 7 or 6 light tasks respectively. The test results demonstrate that Auction Protocol allows each slave agent to get the workload that best suits its condition. Therefore, "Short-Board Effect" is highly alleviated.

Above process is repeated several times and average time costs for the two protocols are given in Table 12. It can be seen from Table 12 that average time cost by using Auction protocol is just 57% of that by using Contract-Net protocol, which confirms the advantage of distributed load balancing strategy.

Table 12. Average time costs for Contract-Net Protocol and Auction Protocol

Test Number	Contract-Net (s)	Auction (s)
1	275.37	160.8
2	306.6	175.27
3	308.57	160.82
4	308.39	175.35
5	275.42	160.79
Average time cost (s)	**294.87**	**166.61**

We do TTC calculations repeatedly by using both Contract-Net and Auction protocols, and add computer into the cluster one by one. The relationship between total time cost and the size of the cluster is listed in Table 13 and plotted in Fig. 21.

Table 13. Test on total time cost vs the size of the cluster

Cluster size	Contract-Net (s)	Auction (s)
1	767	765
2	436	459
3	359	325
4	338	277
5	327	242
6	321	211
7	321	200
8	306	179
9	275	161

It can be seen from Table 13 and Fig. 21 that, because of the frequent communication required, time cost of Auction Protocol is slightly larger than that of Contract-Net Protocol when two computers are used in the cluster. However, in other cases, Auction Protocol always costs far less time than Contract-Net Protocol. More importantly, as the cluster size grows larger, the time cost by Auction Protocol decreases much faster than Contract-Net Protocol, which proves that distributed load balancing strategy exploits the computing resource much more efficiently so that advantage in hardware improvement has been fully taken.

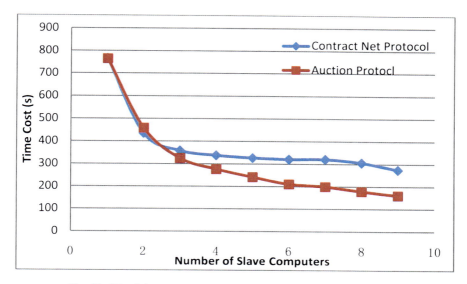

Fig. 21. "Total time cost vs the size of the cluster"-curve for Table 13

Another point should be mentioned here is that, the best dispatching scheme can be achieved when there are more slave computers than the number of TTC tasks. In this case, each task can run exclusively on a single computer without sharing computing resource with others. The total time cost of these tasks is the shortest we can achieve, which equals to the time cost of the slowest task.

In the test, if we have more than 30 computers, the shortest total time cost can be achieved, which is 143 seconds. However, by using Auction Protocol to properly dispatch the workload, we can complete all the tasks in 161 second with only 9 computers as shown in Table 11, which is very close to the shortest time cost (143 seconds) already. Therefore, Auction Protocol is proved to be suitable for online TTC calculation.

5 Concluding Remarks

Smart Grid has raised new requirement and challenges in high performance computing for modern EMS in EPCC. Taking advantage of "Reactivity", "Proactiveness" and "Social ability" features of Multi-Agent technology, a MAS based high performance computing platform is developed.

"Double-Dimensional Federal Architecture", which considers interactions inside single function module and among different function modules, is proposed to model modern EMS in detail. This architecture simplifies the implementation of the single function module and the communication and cooperation among different function modules.

A middleware named "WLGrid" is developed to implement the platform. During platform development, WLGrid works as a toolkit helping to reduce complexity in building autonomous, pro-active and distributed agent. During runtime, WLGrid works as a container which allows agents to be dynamically plugged in and out, and it can also provide services such as network communication, multi-processing and fault tolerance. With the help of WLGrid, computing resources in EPCC can be integrated into a "Super Computer", to address the challenges raised by Smart Grid.

Several methods, including concurrent dispatching strategy and dynamic load balancing strategy, are proposed to optimize system performance.

The proposed methods have been successfully applied to EMS platform in two provincial EPCCs in China. Onsite operation results prove that the MAS-based cluster computing platform developed is effective.

References

Azevedo, G.P., Feijo, B., Costa, M.: Control centers evolve with agent technology. IEEE Computer Applications in Power 13(3), 48–53 (2000)

Brooks, R.A.: Elephants don't play cheese. Robotics and Autonomous Systems 6, 3–15 (1999)

Buse, D.P., Sun, P., Wu, Q.H., Fitch, J.: Agent-based substation automation. IEEE Power Energy Mag 1(2), 50–55

Contract Net Protocol, http://en.wikipedia.org/wiki/Contract_Net_Protocol

Davidson, E.M., McArthur, S.D.J., McDonald, J.R., Cumming, T., Watt, I.: Applying multi-agent system technology in practice: automated management and analysis of SCADA and sigital fault recorder data. IEEE Trans. Power Syst. 21(2), 559–567 (2006)

Dimeas, A.L., Hatziargyriou, N.D.: Operation of a multi-agent system for microgrid control. IEEE Trans. Power Syst. 20(3), 1447–1455 (2005)

Dong, J., Chen, S.G., Jeng, J.J.: Event-based blackboard architecture for multi-agent systems. In: Int. Conf. Information Technology: Coding and Computing, Las Vegas, United States (2005)

Green II, R.C., Wang, L., Alam, M.: High performance computing for electric power systems: applications and trends. IEEE Power Eng. Soc. General Meeting, Detroit, MI, USA, July 24-28 (2011) PESGM2011-000865.pdf

Guessoum, Z., Briot, J.P.: From active objects to autonomous agents. IEEE Concurrency 14(3), 68–76 (1999)

Huang, Z., Nieplocha, J.: Transforming power grid operations via high performance computing, Panel of CAMS: Next-Generation Control Centers with High Performance Computing. IEEE Power Eng. Soc. General Meeting, Pittsburgh, PA, USA (2008)

Liu, C.C.: Strategic power infrastructure defense (SPID): a wide area protection and control system. In: IEEE/PES Trans. Distr. Conf. and Exhib., Asia Pacific, Yokahama, Japan (2002)

Liu, C.C., Bose, A., Cardell, J., et al.: Agent modeling for integrated power systems, Final Project Report. PSERC Publication 08-17 (2008), http://www.pserc.org/doc-sa/Liu_Agents_PSERC_Report_M13_Sep08_ExecSummary.pdf

McArthur, S.D.J., Strachan, S.M., Jahn, G.: The design of a multi-agent transformer condition monitoring system. IEEE Trans. Power Syst. 19(4), 1845–1852 (2004)

McArthur, S.D.J., Davidson, E.M., Catterson, V.M., et al.: Multi-agent systems for power engineering applications - Part I: concepts, approaches and technical challenges. IEEE Trans. Power Syst. 22(4), 1743–1752 (2007)

McArthur, S.D.J., Davidson, E.M., Catterson, V.M., et al.: Multi-agent systems for power engineering applications-Part II: technologies, standards and tools for building multi-agent systems. IEEE Trans. Power Syst. 22(4), 1753–1759 (2007)

Moslehi, K., Kumar, A.B.R., Dehdashti, E., et al.: Distributed autonomous real-time system for power system operations – a conceptual overview. IEEE PES Power Syst. Conf. and Exp., 27–34 (2004)

Real-time Computing, http://en.wikipedia.org/wiki/Real-time_computing

Schainker, R., Miller, P., Dubbelday, W., et al.: Real-time dynamic security assessment: fast simulation and modeling applied to emergency outage security of the electric grid. IEEE Power Energy Mag. 4(2), 51–58 (2006)

Sheng, G.H., Tu, G.Y., Luo, Y., et al.: Study on MAS-based secondary voltage control system. Automation of Electric Power Syst. 26(5), 20–25 (2002)

SQLite, http://sqlite.org/

Transmission Fast Simulation and Modeling (T-FSM), Architecture Requirements, http://www.epri-intelligrid.com/intelligrid/docs/000000000001011667.pdf

Widergren, S.E., Roop, J.M., Guttromson, R.T., Huang, Z.: Simulating the dynamic coupling of market and physical system operations. Proc. IEEE Power Eng. Soc. General Meeting, 748–753 (2004)

Wooldridge, M., Weiss, G. (eds.): Intelligent agents. Multi-Agent Systems, pp. 3–51. The MIT Press, Cambridge (1999)

Wu, W.C., Zhang, B.M., Sun, H.B., Zhang, Y.: Development and application of on-line dynamic security early warning and preventive control system in China. IEEE PES General Meeting, Mineapolis, MN, USA (2010) PESGM2010-000336.PDF

Zhang, B.M., Zhao, C.L., Wu, W.C.: A supporting platform for new generation of EMS based on PC cluster. IEEE PES General Meeting, Pittsburgh, PA, USA (2008)

Zhang, B.M., Zhao, C.L., Wu, W.C.: A multi-agent based distributed computing platform for new generation of EMS. IEEE PES Power Syst. Conf. Expo., Seattle, WA, USA (2009)

Zhang, B.M., Wu, W.C., Wang, H.Y., Zhao, C.L.: An efficient security assessment system based on PC cluster for power system daily operation planning validation. EEE PES General Meeting, Mineapolis, MN, USA (2010) PESGM2010-000705.PDF

Zhang, L., Hoyt, A., Gribik, P., Calson, B.: A Review of high performance technical computing in Mid-west ISO. In: IEEE Power Eng. Soc. General Meeting, Mineapolis, Minesota, USA, July 25-29 (2010) PESGM2010-001506.pdf

High-Performance Computing for Real-Time Grid Analysis and Operation

Zhenyu Huang, Yousu Chen, and Daniel Chavarría-Miranda

Pacific Northwest National Laboratory,
Richland, Washington, USA
{zhenyu.huang,yousu.chen,daniel.chavarria}@pnl.gov

Abstract. Power system computation software tools are traditionally designed as serial codes and optimized for single-processor computers. They are becoming inadequate in terms of computational efficiency for the ever increasing complexity of the power grid. The power grid has served us remarkably well but is likely to see more changes over the next decade than it has seen over the past century. In particular, the widespread deployment of renewable generation, smart-grid controls, energy storage, plug-in hybrids, and other emerging technologies will require fundamental changes in the operational concepts and principal components of the grid. The grid is in an unprecedented transition that poses significant challenges in power grid operation. Central to this transition, power system computation needs to evolve accordingly to provide fast results for power grid management.

On the other hand, power system computation should and has to take advantage of ubiquitous parallel computers. To bring HPC to power grid applications is not simply putting more computing units against the problem. It requires careful design and coding to match an application with computing hardware. Sometimes, alternative or new algorithms need to be used to maximize the benefit of HPC.

This chapter demonstrates the benefits of HPC for power grid applications with several examples such as state estimation, contingency analysis, and dynamic simulation. These examples represent the major categories of power grid applications. Each of the applications has its own problem structure and data dependency requirements. The approach to apply HPC to these problems has different challenges. The HPC-enhanced state estimation, contingency analysis, and dynamic simulation presented in this chapter are suitable for today's power grid operation.

1 Introduction

The term high-performance computing may evoke visions of biological and molecular systems, social systems, or environmental systems, as well as images of supercomputers analyzing human genes, massive data analysis revealing carbon evolution on the planet, and "Deep Blue" defeating world chess champion Garry

Kasparov or Watson's victory in Jeopardy. Such achievements are exhilarating—but limited thoughts have been given to the electricity that powers these brilliant computational feats and the power system that generates and delivers it.

Since the first commercial power plant was built in 1882 in New York City, the power grids in North America and elsewhere have evolved into giant systems comprising hundreds and thousands of components across many thousands of miles. These power grids has been called the most complex machine man has ever built. The reliable and relatively inexpensive electricity supplied by these power grids is the foundation of all other engineering advances and of our humankind's prosperity. This critical electric infrastructure has served us remarkably well but is likely to see more changes over the next decade than it has seen over the past century. In particular, the widespread deployment of renewable generation, smart-grid controls, energy storage, plug-in hybrids, and other emerging technologies will require fundamental changes in the operational concepts and principal components of the grid. The grid is in an unprecedented transition that poses significant challenges in grid operation, planning, and policy decisions. In this transition, power system computation needs to evolve accordingly to provide fast results for power grid management as well as to take advantage of ubiquitous parallel computers.

Power system computation can be generally classified as steady-state analysis and dynamic analysis depending on the models used. Steady-state analysis is based on a steady-state model, a.k.a. power flow model, which describes the grid topology and the associated impedance of transmission lines and transformers. Steady-state analysis determines a snapshot of the power grid, without considering the transition from a snapshot to the next. Dynamic analysis solves a set of differential-algebraic equations of dynamic models of generators, controllers and other dynamic devices in addition to the power flow model. The objective is to determine the evolving path of the power grid, giving a much finer details than steady-state analysis.

Today's power grid operation is based on steady-state analysis. Central to power grid operations is state estimation. State estimation typically receives telemetered data from the supervisory control and data acquisition (SCADA) system every few seconds (typically, four seconds) and extrapolates a full set of grid conditions based on the grid's current configuration and a theoretically based power flow model. State estimation provides the current power grid status and drives other key functions such as contingency analysis, optimal power flow (OPF), economic dispatch, and automatic generation control (AGC), as shown in Figure 1 [Nieplocha 2007].

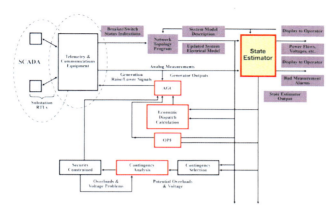

Fig. 1. Functional structure of real-time power system operations

Mathematically, power grid operation functions are built on complex mathematical algorithms and network theories in combination with optimization techniques (Figure 2) [Huang 2007-1]. Given the sheer size of a practical power grid, all these mathematical problems require significant time to solve. The computational efficiency in grid operations is very low and that leads to inability to respond to adverse situations.

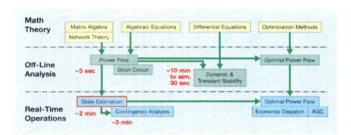

Fig. 2. Grid computational paradigm with typical computation time on personal computers

With today's computers and algorithms, the state estimation and contingency analysis can be updated only in an interval of minutes – much slower than the SCADA measurement cycle. This is not fast enough to predict system status because a power grid could become unstable and collapse within seconds [DOE 2004]. In addition, both of these analyses are conducted for areas within individual utility company's boundaries and examine an incomplete set of contingencies prioritized from experience as the most significant. The solution time would be longer if a larger footprint and more contingencies are considered. Other processes in Figure 1 take even longer time (>10 minutes) and are less frequently used in real time.

Dynamic analysis largely remains to be off-line applications due to computational intensity. Interconnection-scale dynamic simulation is at least 10 times slower than real time. It is not practical for on-line applications to determine the real-time dynamic view of the power grid. Grid operation margins determined by off-line analysis do not reflect real-time operating conditions. The consequence is that either the grid efficiency is low because of conservative off-line study scenarios and assumptions or the grid is at risk of blackouts when the critical real-time operating condition is not studied in the off-line analysis.

It is clear that computational issues of the analysis tools employed in today's power grid operations on serial computers significantly limits our understanding of real-time power grid behaviors and our responsiveness in the situation of emergencies such as the 2003 US-Canada Blackout [Chen 2011]. This problem calls for exploration of high performance computing (HPC) technologies to transform today's grid operations with improved computational efficiency and enhanced dynamic analysis capabilities. To illustrate the important role of HPC in power system applications, three major power system functions, i.e. state estimation, contingency analysis, and dynamic simulation, are discussed in Section 2, 3, and 4, respectively.

2 Parallelized State Estimation

The weighted-least-square (WLS) algorithm is the most widely used state estimation method. Real-time state estimation requires a fast and dependable implementation. The WLS estimator algorithm has traditionally incorporated direct methods based on the LDU, LU or Cholesky factorizations for solving large and sparse systems of linear equations in each iteration of the state estimation algorithm. Direct methods perform very well on sequential computers. However, they demonstrate somewhat limited speedups and performance on parallel computers due to the sequential character of the operations involved. The amount of parallelism in the direct solver may be improved by reordering or repartitioning of the gain matrix [Lau 1990][Abur 1988] because the fill-in elements generated in the factorization process render the ordering to be not optimal [Abur 1988][Tinney 1967]. However the reordering and repartitioning process is rather limited according to the precedence relationship graph in factorization and substitutions [Alvarado 1990]. Iterative methods are widely used for solving large, sparse systems of linear equations generated when finite differences methods are applied in the solution process of non-linear equations. Some iterative methods possess much higher levels of parallelism than the direct ones. In this section, we explore a preconditioned conjugate gradient for power grid state estimation.

2.1 The WLS State Estimation

The non-linear equations relating measurements to system state can be written as:

$$z = h(x) + e, \qquad (1)$$

where:
- z - Measurement vector of dimension m;
- x - State vector of dimension, n (=$2N - 1$);
- N- Number of buses;
- e - Vector of measurement errors;
- h - Vector of non-linear functions which relate states to measurements.

The truncated Taylor expansion of (1) yields:

$$z = h(x^\vartheta) + H(x^\vartheta)\Delta x + e, \quad (2)$$

where x^ϑ is the linearization point, and Jacobian matrix H is defined as:

$$H(x^\vartheta) = [\delta h(x_j^\vartheta)/\delta(x_j^\vartheta)]. \quad (3)$$

The WLS state estimation problem is formulated as the minimization of the following form:

$$J(x) = [z - h(x)]^T R^{-1} [z - h(x)], \quad (4)$$

where the covariance matrix of the noise is

$$R = E(ee^T) = \begin{cases} \sigma^2 & \text{if } i = j \\ 0 & \text{otherwise} \end{cases} \quad (5)$$

The state estimate can be found by iteratively solving the following equation:

$$A(x^\vartheta)(x^{\vartheta+1} - x^\vartheta) = H^T(x^\vartheta) R^{-1} [z - h(x^\vartheta)], \quad (6)$$

where A denotes the gain matrix,

$$A(x^\vartheta) = H^T(x^\vartheta) R^{-1} H(x^\vartheta). \quad (7)$$

2.2 Solution of Linear Equation Systems

In each cycle of WLS state estimation, a large and sparse system of linear equations is being solved:

$$A \Delta x = b, \quad (8)$$

where the matrix A is the symmetric positive-definite (SPD) gain matrix and (8) is another form of (6) with the variables defined as:

$$\Delta x = (x^{\vartheta+1} - x^\vartheta)$$
$$b = H^T(x^\vartheta) R^{-1} [z - h(x^\vartheta)]$$

2.2.1 SuperLU

Direct solution methods which have traditionally been employed in the WLS state estimation is based on Gaussian elimination, LDU [Turing 1948], LU, or Cholesky factorization, followed by the backward and forward substitutions. There are some direct solve packages available, such as SuperLU [SuperLU 2005], MUMPS [Amestoy 2000], and PARDISO [Schenk 2001]. While MUMPS and PARDISO are developed specified for solving symmetric matrices, in our study, SuperLU is selected to implement a direct solver for the WLS algorithm. It is because SuperLU can be used to solve symmetric and unsymmetrical matrices, which is much more compatible for different power system applications. One salient feature of SuperLU is the capability to keep track of dense submatrices in the L and U factors, hereby enabling the use of efficient, optimized Level 3 BLAS libraries [Dongarra 1990].

Solving for X in (8) is equivalent to an evaluation of following equation in SuperLU:

$$x = A^{-1}b = D_c \left(P_c \left(U^{-1} \left(L^{-1} \left(P_r \left(D_r b \right) \right) \right) \right) \right) \qquad (9)$$

where:

- L - Unit lower triangular matrix
- U - Upper triangular matrix
- r,c - Row and column subscripts;
- D - Diagonal matrices for system equilibration;
- P - Permutation matrices for sparsity, stability and parallelism enhancement.

The overall SuperLU algorithm computes a triangular factorization as:

$$P_r D_r A D_c P_c = LU \qquad (10)$$

by first equilibrating the matrix A with D_r and D_c, preordering with P_r and P_c and then performing the actual LU factorization. The system is then iteratively solved with Newton's method, guided by forward and backward error bounds computed in each step.

The stopping criterion for the iterative refinement method is determined by measuring:

$$BERR = \max_i \frac{(Ax - b)_i}{(|A||x| + |b|)_i}, \qquad (11)$$

where: i - row indices.

Once BERR exceeds the level of machine roundoff or ceases to decrease significantly the iterative refinement stops. The iteratively determined perturbed system of equations:

$$(A + E)x = b + f, \qquad (12)$$

will then satisfy $\forall_{ij}: |E_{ij}| \leq$ BERR $|A_{ij}|$ and $|f_i| \leq$ BERR $|b_i|$. SuperLU comes in a *sequential*, shared *memory multithreaded* – Pthreads [IEEE 1995] – and *distributed memory* – MPI [MPI] – version. More direct methods can be found at [Davis 2006].

2.2.2 Conjugate Gradient

Many iterative methods for sparse linear system can be found at [Saad 2003]. Among these methods, the conjugate gradient (CG) method is a widely used one. It was originally developed by Hestenes and Stiefel [Hestenes 1952] for solving symmetric positive definite (SPD) systems of linear equations. Although applying iterative solution methods, CG-type methods theoretically yield exact solutions within a finite number of steps. Discounting roundoff errors, the CG method would guarantee convergence to the exact solution in at most N iterations, where N is the dimension of the linear system (8). Despite its interesting properties, the method was largely ignored after its initial introduction because of the computational expense required for convergence. This handicap was eliminated after discovering that preconditioning the system of equations would lead to solution convergence in far fewer than N iterations. Despite the issue of robustness and consistency, the CG method has since become popular because of its capacity for efficiently solving sparse matrix problems, its suitability for implementation on vector and parallel computers, and developments which allow it to be generalized for use with nonsymmetric matrices. Several modifications to the Hestenes and Stiefel's original method [IEEE 1995] have been proposed which make the method suitable for nonsymmetrizable matrix problems [Jea 1983]. Practically, the convergence rate depends on the condition number, the ratio of the largest to smallest eigenvalue, of the matrix, A. When the matrix's condition number is minimized, the method usually converges much faster than in N iterations. The condition number can be minimized by premultiplying both sides of equation (8) by the inverse of a preconditioner matrix, Q,

$$Q^{-1} A \Delta x = Q^{-1} b \tag{13}$$

to yield a new system to be solved,

$$\hat{A} \Delta x = \hat{b}. \tag{14}$$

This preconditioned Conjugate Gradient process is shown below:

```
1. Input
   Initial guess of x i.e., u⁰
   Stopping number ζ
2. Compute
   δ⁰ = b - u⁰
   Solve linear system Qz⁰ = δ⁰ for z⁰
   Set p⁰ = z⁰ and a₀ = (z⁰,δ⁰)
3. For k = 1,2, ...
   Compute cᵏ⁻¹ = Apᵏ⁻¹
   Set λₖ₋₁ = aₖ₋₁/(pᵏ⁻¹,cᵏ⁻¹)
   Take uᵏ = uᵏ⁻¹ + λₖ₋₁pₖ₋₁
   Compute δᵏ = δᵏ⁻¹ -λₖ₋₁cₖ₋₁
   Solve for zᵏ the linear system Qzᵏ = δᵏ
   Check for convergence.
   If stopping number <ζ then STOP else
```

$$\text{Set } a_k = (z^k, \delta^k)$$
$$\text{Set } \alpha_k = a_k/a_{k-1}$$
$$\text{Compute } p^k = z^k + \alpha_k p^{k-1}$$

Our implementation of the CG algorithm uses the Global Arrays toolkit [Nieplocha 1996][Zhang 2005] as a container for the arrays and vectors. Using the Global Arrays toolkit to create the arrays and vectors makes it possible to place them in shared memory with little programming efforts. The matrix and all the vectors used are in shared memory so any process is able to access it without moving or copying data. The workload is divided by reordering vectors x and b such that each processor has almost the same amount of work.

2.3 Numerical Results

The numerical results focus on speed-up performance only. Two computers were used in our experiments: SGI Altix and an NWICEB cluster machine. The SGI Altix 3000 [SGI] located at Pacific Northwest National Laboratory has 128 1.5GHz Itanium-2 processors, 256GB memory and runs a Linux 2.4.21 kernel. The SGI Altix supports software multithreading. The NWICEB contains 192 nodes and two 2.33 GHz quad-core Xeon E5345 processor per node. The size of memory per node is 16GB and the interconnection is Gigabit Ethernet and Infiniband.

We evaluated effectiveness of our parallel implementation of the CG algorithm and compared with a multithreaded implementation of SuperLU version 3.0. The SuperLU solvers were compiled with the Intel C++ and Fortran Itanium compiler version 7.1. On NWICEB, preconditioners are implemented and evaluated for the state estimation problem.

2.3.1 Comparison of Direct and Iterative Methods

A 1177-bus system with 1770 lines and a data set with 6144 measurements was used to evaluate the performance of the WLS state estimator using SuperLU direct method and the parallel iterative CG method. In particular, the focus is on the solution of the sparse linear system as it represents the predominant component of computations in the state estimation algorithm. The initial sparse input matrix A of the state estimator has a size of 2353x2353 with 39521 non-zeros.

All the reordering schemes available in the SuperLU package were tested: *Multiple Minimum Degree (MMD)* [Liu 1985] permutations applied to $A'A$ and $A'+A$ as well as *Approximate Minimum Degree (AMD)* [Davis 2000] column permutations that do not require the extra step of A^TA formation. Figure 3 shows the execution time for the three permutation schemes as well as the non-permuted scheme labeled as "*Natural Ordering*". The scheme "*MMD A'A*" produces overall the fastest result up to 16 processors. Beyond 16 processors, "*Natural Ordering*" gains over the "*MMD A'A*" scheme, suggesting that the LU factorization speedups cannot compensate for the additional overhead introduced in the permutation phase as the work per processor is reduced. The performance/scalability tradeoffs for choosing the ordering are consistent with results reported in prior work [Lau 1990][Abur 1988][Tinney 1967][Alvarado 1990].

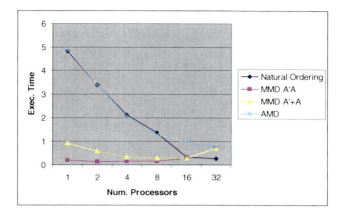

Fig. 3. Multi-threaded SuperLU on the SGI Altix 3000: execution time of one iteration in seconds in relationship to the number of processors

The execution time of the parallel CG method and the fastest version "*MD A'A*" of the SuperLU algorithm is compared in Figure 4. A Jacobian preconditioning matrix (i.e. diagonal scaling matrix) was used in the method. The parallel CG method performs very well in terms of both scalability and absolute run-time performance, while execution time increases with SuperLU running on more than two processors [Nieplocha 2006]. There might have some direct solver packages that have better performance than SuperLU, but in general, iterative methods will have better scalability.

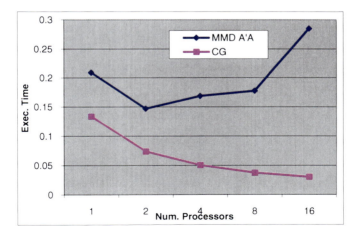

Fig. 4. Parallel CG compared to MT-SuperLU on the Altix 3000: execution time of one iteration in seconds in relationship to the number of processors

The parallel CG method scales better but the scaling is not linear and the curve flattens at higher numbers of processors. This is due to the smaller problem size. For example, on 16 processors the sparse-matrix-vector multiplication loop executes over as few as 2400 elements per processor. Reducing the loop size impacts the latency hiding made possible by instruction pipelining and hence the time for executing these loops does not scale linearly. Further evaluation was performed with a larger-size state estimation problem.

The larger problem is derived from the Western Electricity Coordinating Council (WECC) transmission grid, whose footprint encompasses a geographical area including western Canada, western United States, and the northern tip of Mexico. It has over 14,000 buses and over 2700 generators. The largest size of today's power system model is around 50,000 buses. The WECC represents today's large-scale matrices in the power system domain. The performance for the data corresponding to the first iteration is shown in Figure 5. It indicates that on 16 processors a speedup of 14.8 was achieved – a fairly linear scalability [Nieplocha 2007]. The solution time for one step is about five seconds. State estimation usually needs to take a few steps to solve. That would put the total solution time to be in the order of 10 of seconds. The goal is to bring the total solution time down to be comparable to SCADA cycles, that are typically 4-5 seconds. Next section continues to explore preconditioning methods for improving state estimation performance.

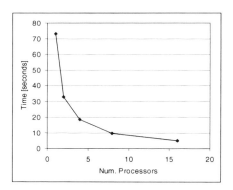

Fig. 5. Performance of CG on SGI Altix of the WLS algorithm for the WECC 14000-bus system

2.3.2 Performance of Preconditioning Methods

In practice, the CG method without preconditioning can require a significant number of iterations to converge. This is due to the large condition number of the gain matrix in the WLS state estimation problem. Table 1 shows condition numbers of example power systems. They are all in the order of 10^5. To reduce the condition number and thus the number of iterations, preconditioners are often considered in the CG method.

Table 1. Condition number of the gain matrix

WLS cycle	1	2	3	4
30-bus system	2.705 10^5	3.919 10^5	3.859 10^5	3.859 10^5
118-bus system	7.712 10^5	8.803 10^5	9.035 10^5	9.042 10^5

Several methods exist for preconditioning a linear equation system: a Jacobi matrix for diagonal scaling as used in the previous evaluation studies; a matrix approximating the inverse of matrix A; an incomplete Cholesky or LDL^T decomposition, and as a splitting matrix from a basic iterative algorithm. The Jacobi preconditioner is very cheap while the Incomplete Cholesky preconditioner is expensive in their CPU time consumption. In the Hypre library package [Hypre], several preconditioners are available for solving very large sparse linear systems with good scalability. These preconditioners were evaluated, and the Euclid preconditioner [Hysom 1999][Hysom 2001] shows the best performance for the state estimation problem running on the NWICEB computer. The Euclid preconditioner is based on a parallel incomplete LU decomposition algorithm. Our test shows the Euclid preconditioner can reduce the number of iterations in solving the linear system of Equation (8) by more than 40 times for a WECC-size power grid. The actual execution time as well as the scalability is shown in Figure 6. It shows the parallel CG method with the Euclid preconditioner can solve the linear equation 10 times faster than with the simple Jacobi preconditioner. On 16 processors, the Euclid preconditioner needs less than one second to solve one step of the WLS state estimation problem. Euclid preconditioner also demonstrates excellent scalability in solving the problem as shown in Figure 6. A full state estimation package has been implemented using the parallel CG method with the Euclid preconditioner. It achieves five-second solution time for the full state estimation problem of the 14,000-bus WECC-size system, a speed comparable with SCADA cycles. As indicated by the scalability in Figure 6, the solution time can be further reduced by the Euclid preconditioner based CG method if more processors are used, though it may not be necessary from a practical perspective.

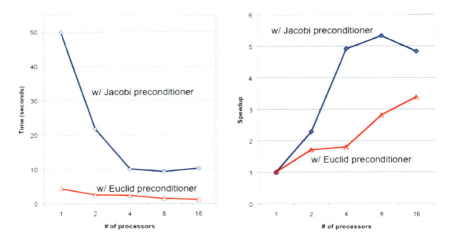

Fig. 6. Performance of preconditioners for parallel CG on the NWICEB machine in solving one step of state estimation with the WECC 14000-bus system

2.4 GPU Implementation

Graphic Processing Unit (GPU) is an emerging architecture. It addresses the issue of memory bottlenecks by leveraging concurrent memory access, which leads to increased bandwidth, and thus, improved performance. An evaluation of throughput computing on CPU and GPU can be found at [Lee 2010]. In [Bell 2010], the authors explored several efficient implementation techniques for sparse matrix-vector multiplication that is well-suited for GPU architectures. Averaging 16 GFLOP/s and 10 GFLOP/s in double precision for structured grid and unstructured mesh matrices are reported in this paper, which is more than 10 times that of a quad-core Intel Clovertown system.

Realizing the advantages of GPU, power system researchers are starting to apply GPU into power system applications. In [Zhao 2011], the authors reported a CG-based state estimation for a 14-bus system using GPU. The results show that the GPU version has a factor of 7-10 improvement over the CPU version. While the test system size is very small, the GPU could potentially have a big impact on power system engineering. To further improve performance with GPU implementation for large-scale power system is an important short-term future task.

2.5 Summary

In adapting software to HPC architectures, data and computations need to be partitioned so all available processors are deployed to perform the required computations. The challenge lies in redeveloping certain algorithms to take advantage of parallel computing architectures. SuperLU was designed for solving linear systems of equations in other computational areas such as computational fluid dynamics. This section illustrates that a specific algorithm may perform well for some computational problems, but a good fit of the algorithm, the characteristics of the problem domain data, and high-performance computing hardware is needed to achieve the desired results.

Iterative methods can perform better in solving the state estimation problem than directive methods as shown in this section. The Euclid preconditioner-based parallel CG method has been implemented and tested with a 14,000-bus WECC-size problem. The results very well demonstrate the improved computational performance with good scalability as well as a solution time that now is comparable with SCADA cycles.

3 Parallel Contingency Analysis

State estimation is used to estimate system status based on measurements. Its outputs provide inputs for many power system operation functions, such as contingency analysis. Contingency analysis is used to assess the ability of the power grid to sustain various combinations of component failures based on state estimates. The outputs of contingency analysis, together with other EMS functions, provide the basis for preventive and corrective operation actions. Contingency analysis is also extensively used in power market operation for feasibility test of market solutions.

Due to heavy computation involved, contingency analysis is currently limited to be selected "N-1" cases within a balancing authority's boundary. A typical example is to solve 500 contingency cases in a time interval of five minutes. Power grid operators manage the system in a way that ensures *any single credible contingency will not propagate into a cascading blackout*, which approximately summarizes the "N-1" contingency standard established by the North American Electric Reliability Corporation (NERC) [NERC].

Though it has been a common industry practice, analysis based on limited "N-1" cases may not be adequate to assess the vulnerability of today's power grids due to new development in power grid, variable grid status, and market operations. Power grids are operated closer to their capacity, and high penetration of intermittent renewable energy demands faster analysis of massive contingency cases to safely and reliably operate today's power grids. One consequence of operating power grids closer to the edge is massive blackouts resulting in significant electricity supply disruptions and economic losses [Kosterev 1999][DOE 2004]. Power grid blackouts often involve failure of multiple elements as revealed in recent examples. Preventing and mitigating blackouts requires "N-x" contingency analysis. The North American Electricity Reliability Corporation (NERC) moves to mandate contingency analysis from "N-1" to "N-x" in its grid operation standards [NERC]. All this calls for a massive number of contingency cases to be analyzed. As an example, the Western Electricity Coordinating Council (WECC) system has about 20,000 elements. Full "N-1" contingency analysis constitutes 20,000 cases, "N-2" is roughly 10^8 cases, and the number increase exponentially with "N-x".

On the power market side, one example is the introduction of Financial Transmission Rights (FTR) [PJM 2007][MISO 2008]. FTR provides market participants a means to hedge risks due to power transmission congestions. It is operated as an auction market. When clearing the FTR auction market, the feasibility of the FTR solution has to be evaluated by contingency analysis. There exist multiple FTR categories such as annual FTRs, seasonal FTRs and monthly FTRs. Each category requires contingency analysis of a full system-size model. Multiple categories couple the contingency problem and multiply the size of the model. The resulting issue is that the number of contingency cases is multiplied. With regular personal computers, it takes hours and even days to clear the FTR auction market.

It is obvious that the computational workload is beyond what a single personal computer can achieve within a reasonable time frame for real-time operation. Parallel computers or multi-core computers as emerging in the high performance computing (HPC) industry hold the promise of accelerating power grid contingency analysis. Contingency cases are relatively independent of one another, so contingency analysis is inherently a parallelizable process. Mathematically, there is a relatively straightforward parallelization path, but the issue with parallelization schemes remains due to the unevenness in computation time of individual cases.

Previous work in parallel computing for contingency analysis has been focused on "N-1" analysis with a small set of cases [Chen 1996][Morante 2006]. Scalability remains to be an issue when more processors are used and more cases are analyzed. The performance of parallel contingency analysis heavily relies on computational load balancing. A well-designed computational load balancing scheme considering the CPU speed, network bandwidth and data exchange latency is key to the success. Parallelization schemes for computational load balancing of massive contingency analysis need to be investigated.

3.1 Computational Load Balancing Schemes for Massive Contingency Analysis

Contingency analysis is naturally a parallel process because multiple contingency cases can be easily divided onto multiple processors and communication between different processors is very minimal. Therefore, cluster-based parallel machines are well suited for contingency analysis. For the same reason, the challenge in parallel contingency analysis is not on the low-level algorithm parallelization but on the computational load balancing (task partitioning) to achieve the evenness of execution time among the processors.

The framework of parallel contingency analysis is shown in Figure 7 [Chen 2010]. Each contingency case is essentially a power flow run. In our investigation, full Newton-Raphson power flow solution is implemented. Given a solved base case, each contingency updates its admittance matrix with an incremental change from the base case. One processor is designated as the master process (Proc 0 in Figure 7) to manage case allocation and load balancing, in addition to running contingency cases.

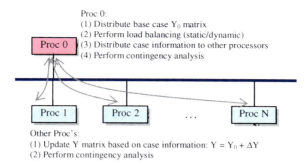

Fig. 7. Framework of parallel contingency analysis

3.1.1 Static Load Balancing

A straightforward load balancing of parallel contingency analysis is to pre-allocate equal number of cases to each processor, i.e. static load balancing. The master processor only needs to allocate the cases once at the beginning. Due to different execution time for different cases, each power flow run may require different number of iterations and thus take different time to finish. The extreme case would be non-converged cases which iterate until the maximum number of iterations is reached. The variations in execution time result in unevenness, and the overall computational effort is determined by the longest execution time of any of the individual processors. Computational power is not fully utilized as many processors are idle while waiting for the last one to finish.

3.1.2 Dynamic Load Balancing

Another load balancing scheme is to allocate tasks to processors based on the availability of a processor, i.e. dynamic load balancing. In other words, the contingency cases are *dynamically* allocated to the individual processors on-demand so that the cases are more evenly distributed in terms of execution time by significantly reducing processor idle time. Our implementation of the scheme is based on a shared variable (task counter) updated by an atomic fetch-and-add operation. The master processor (Proc 0) does not distribute all the cases at the beginning. Instead, it maintains a single task counter. Whenever a processor finishes its assigned case, it requests another task from the master processor and the task counter is updated by one. This process is illustrated in Figure 8. Different from the evenly-distributed number of cases on each processor with the static scheme, the number of cases on each processor with the dynamic scheme may not be equal, but the computation time on each processor is optimally equalized.

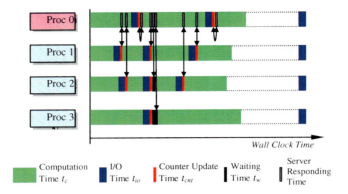

Fig. 8. Task-counter-based dynamic computational load balancing scheme

3.1.3 Multi-counter Dynamic Load Balancing

In the dynamic load balancing scheme, counter congestion will occur when multiple counter requests arrive at the counter server at the same time, a situation shown in Figure 8 where the waiting time t_w is resulted. The possibility of counter congestion increases when more processors are used for massive contingency analysis. As a result, the scalability of dynamic load balancing schemes will be limited by counter congestion.

In order to better manage counter congestion, we propose a multi-counter based dynamic load balancing scheme with task stealing. The framework of this new scheme is illustrated in Figure 9 with two counters. The equal number of cases is pre-allocated to the two counter groups. Each group has its own counter (Proc 0 in Figure 9) [Chen 2010]. Inside each group, the dynamic load balancing scheme is applied based on the availability of processors. When the pre-allocated tasks are finished in one group, the counter in this group "steals" tasks from the other group to continue the computation until all tasks are done. By implementing the multi-counter dynamic load balancing scheme, counter congestions can be reduced, and further speedup is expected.

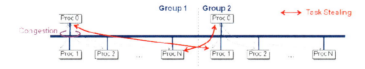

Fig. 9. Framework of multi-counter based dynamic load balancing scheme with task stealing

The cost to minimize counter congestion using multi-counter schemes is the overhead with managing multiple counters. Even though additional counters can reduce counter congestion, it is possible that the overhead would compromise the benefit gained by reducing counter congestion. Therefore, it is important that we evaluation the performance of these load balancing schemes and determine under what conditions the multi-counter scheme had superior performance over the single-counter scheme.

3.2 Computational Performance Analysis

Conceptually, the dynamic computational load balancing scheme balances execution time among processors better than the static scheme. But the cost is the overhead of managing the task counter. As shown in Figure 8, the execution time of each case consists of four parts: t_c – the computation time spent on solving one contingency case, t_{io} – the I/O time used to write the results to disks, t_{cnt} – the time to update the task counter, and t_w – the time to wait for the master processor to respond with a new case assignment when counter congestion occurs. The server responding time is very minimal and thus neglected in the following analysis.

Running all the cases on only one processor would take a total time as estimated in (1):

$$t_{total} = \sum_{i=1}^{N_C} \left(t_c^{(i)} + t_{io}^{(i)} \right) = N_C \left(\bar{t}_c + \bar{t}_{io} \right) \tag{1}$$

where N_C is the total number of cases, and $\bar{t}_c$ and $\bar{t}_{io}$ are the average computation time and I/O time, respectively. On one processor, there is no counter management needed, so no t_{cnt} and t_w should be included in (1).

Running the cases on multiple processors with dynamic load balancing scheme would evenly distribute the total time in (1), but involves counter management. If the total number of processor is N_P, the worst-case scenario with counter congestion is that all the N_P counter update requests arrive at the same time at the master processor. Then the first processor has not waiting time, the second waits for time t_w, and the last one has the longest waiting time $(N_P - 1)t_w$. The average waiting time of a processor can be estimated as:

$$t_{w,N_P} \approx \frac{\frac{N_C}{N_P} \sum_{i=1}^{N_P} (i-1) t_w}{N_P} = \frac{N_C}{N_P} \cdot \frac{(N_P - 1) t_w}{2} \tag{2}$$

Therefore, the total wall clock time required to run all the contingency cases can be estimated as (3):

$$\begin{aligned} t_{total,N_P} &\approx \frac{\sum_{i=1}^{N_C} \left(t_c^{(i)} + t_{io}^{(i)} \right)}{N_P} + \frac{N_C}{N_P} t_{cnt} + \frac{N_C}{N_P} \cdot \frac{(N_P - 1) t_w}{2} \\ &= \frac{N_C}{N_P} \left(\bar{t}_c + \bar{t}_{io} + t_{cnt} + \frac{(N_P - 1) t_w}{2} \right) \end{aligned} \tag{3}$$

The speedup of the dynamic load balancing scheme can be expressed as the following estimate:

$$S_{N_P} = \frac{t_{total}}{t_{total,N_P}}$$

$$= \frac{N_C(\bar{t}_c + \bar{t}_{io})}{\frac{N_C}{N_P}\left(\bar{t}_c + \bar{t}_{io} + t_{cnt} + \frac{(N_P - 1)t_w}{2}\right)} \quad (4)$$

$$= N_P \frac{(\bar{t}_c + \bar{t}_{io})}{\bar{t}_c + \bar{t}_{io} + t_{cnt} + \frac{(N_P - 1)t_w}{2}}$$

Several observations can be drawn from (4) [Huang 2009]:

1) It is clearly shown that the dynamic load balancing scheme is scalable with the number of cases as the speedup performance is irrelevant to the number of cases, N_C.
2) If the counter update is instantaneous and no counter congestion would occur, i.e. $t_{cnt} = 0$ and $t_w = 0$, then the ideal speedup performance would be N_P, equal to the number of processors.
3) For practical implementation, improving speedup performance would require to minimize the overhead t_{cnt} and t_w.
4) Counter update time t_{cnt} is mainly determined by the network latency and speed. Minimizing t_{cnt} usually means choosing high-performance network connection between processors.
5) Waiting time t_w is due to counter congestion. Though more processors would improve the speedup, but they also increase the possibility of counter congestion as shown in (4).

3.2.1 Network Environment Comparison

As stated earlier, minimizing the counter update time t_{cnt} usually means choosing a high-performance network connection among processors. However, due to the cost issues, PC-based Ethernet network dominates the current utility control center environment. In order to better understand the effect of high-performance networks, it is important to study the performance of multi-counter dynamic load balancing schemes under different network environments.

The main network properties are latency and bandwidth. The latency and bandwidth in two common networks are of interest here: 1GB/Sec Ethernet and InfiniBand network. The typical values of latency and bandwidth in these two networks are listed in Table 2. The latency of InfiniBand network is approximately 1/20 of that of Ethernet.

Table 2. The typical values of latency in 1GB/Sec Ethernet, and InfiniBand networks

	1GB/Sec Ethernet	InfiniBand
Latency (μSec)	~30	~1-2
Bandwidth	100~200 MB/Sec	10GB/Sec*

* InfiniBand is a type of communications links between processors and I/O devices that offers throughput of up to 2.5 GB/Sec, and it can achieve 10GB/Sec or higher bandwidth through double-rate and quad-rate techniques.

3.2.2 Performance Simulation

Before the actual case studies are conducted, the contingency analysis process is modeled and simulated to predict the performance under different computing platforms including different network environments. The advantage of using simulated environments is the flexibility of studying different configurations and reducing the time implementing the schemes on actual parallel computers.

The main factors that affect speedup include: (a) the number of processors, N_P; (b) the number of cases, N; (c) the latency of the network communication; (d) the counter updating time; and (e) the bandwidth of the network. Since the effects of factor (c), (d), and (e) are equivalent in terms of the contribution of total time, these factors can be treated as one term, t_{cnt}, for the purpose of performance simulation. The computation time of each contingency case is selected based on the histogram derived from the actual computation time of WECC full N-1 contingency analysis.

Figure 10 shows the speedup performance with different numbers of cases. The t_{cnt} is fixed to 0.0001 for this set of simulations. The horizontal axis is the number of processors in base-2 exponentials, and the vertical axis is the speedup. It is clear that better speedup can be achieved when the number of cases increases.

The sensitivity of t_{cnt} is shown in Figure 11. In this simulation, the number of contingency cases is fixed to be 20,000. Same as in Figure 10, the horizontal axis shows the number of processors in base-2 exponentials, and the vertical axis shows the speedup. There are three observations from Figure 11 [Chen 2010]:

1) For certain number of processors, the larger t_{cnt}, which represents the low speed communication network (e.g. Ethernet network), the less speedup archived for both single-counter and two-counter schemes;
2) With larger time of t_{cnt}, the two-counter dynamic load balancing scheme shows better performance than the single-counter scheme when the number of processors increases. The larger the number of processors, the better performance for two-counter scheme appears; and
3) When the number of processors is relatively small, e.g. less than 128, and with t_{cnt}=0.0001s, which represents the high speed communication network, the performance of single-counter scheme is better than that of the two-counter scheme. However the two-counter scheme has better performance than single-counter scheme when t_{cnt} is larger for the number of processors less than 128.

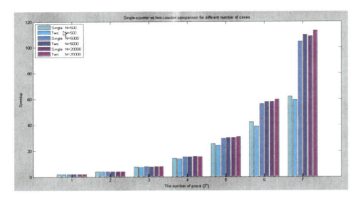

Fig. 10. Single-counter vs. two-counter comparison for different number of cases with respect to different number of processors

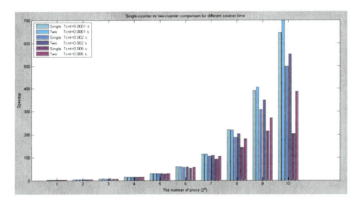

Fig. 11. Single-counter vs. two-counter comparison for different counter time with respect to different number of processors

The third observation is important for most utility/control center users. Currently, most utilities/control centers do not own a cluster computer with a large number of processors. Mostly available are networked computers using Ethernet. Therefore, the two-counter dynamic load balancing scheme would be useful for contingency analysis applications in current control centers.

3.3 *Case Studies of Parallel Contingency Analysis*

The 14,000-bus WECC power grid model is used to test the performance of parallel contingency analysis.

Four scenarios of cases are selected for the studies:

1) **Scenario 1:** 512 "N-1" contingency cases, representing a typical scenario today's contingency analysis practice.

2) **Scenario 2:** 20,094 full "N-1" cases, which consist of 2748 generator outage cases and 17346 line outage cases. Generator outages do not need to update the admittance matrix, but line outages do by adding an incremental change to the admittance matrix.
3) **Scenario 3:** 150,000 "N-2" cases, which randomly choose 50,000 cases from each of the three combinations: double-generator outages, double-line outages, and generator-line outages.
4) **Scenario 4:** 300,000 "N-2" cases, including 100,000 cases from each of the three combinations mentioned above.

The "N-2" scenarios have many more divergence cases as power flow would be more difficult to solve with double elements outaged.

Parallel contingency analysis is implemented with Colony2A, MPP2, and NWICEB cluster machines, all located at Pacific Northwest National Laboratory. Colony2A is a relatively older machine with 24 dual-core 1-GHz Itanium-2 processors. Each processor has 6 GB memory interconnected with other processors through Infiniband connection. MPP2 has 940 nodes with dual 1.5-GHz Itanium processors. Each node has 10 GB memory connected through Quadrics ELAN4. NWICEB has 192 nodes with dual sockets. each socket has a 2.33 GHz quad-core Xeon E5345 processor. Each node has 16 GB memory connected through Gigabit Ethernet and Infiniband.

3.3.1 Comparison of Static and Dynamic Load Balancing Schemes

Both computational load balancing schemes are tested with 512 "N-1" contingency cases of the 14,000-bus WECC system (Scenario 1) on the Colony2A machine. The results are shown in Figure 12. For the static scheme, though computational efficiency continuously increases when more processors are used, the performance is not scalable and exhibits the tendency of saturation. The performance of dynamic load balancing, in comparison with its static counterpart, shows much better linear scalability. The dynamic scheme achieves eight times more speedup with 32 processors as shown in Figure 12, and the difference is expected to be greater with more processors.

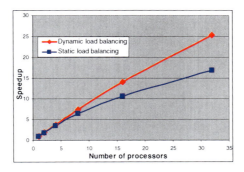

Fig. 12. Performance comparison of static and dynamic computation load balancing schemes with 512 WECC contingency cases

Figure 13 further compares the processor execution time for the case with 32 processors. With dynamic load balancing, the execution time for all the processors is within a small variation of the average 23.4 seconds, while static load balancing has variations as large as 20 seconds or 86%. The dynamic load balancing scheme successfully improves speedups. It is also worth pointing out that the contingency analysis process of 512 WECC cases with full Newton-Raphson power flow solutions can be finished within about 25 seconds. It is a significant improvement compared to several minutes in current industry practice.

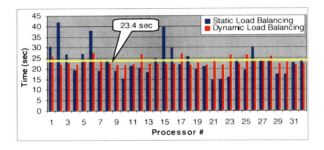

Fig. 13. Evenness of execution time with different computational load balancing schemes

Figure 13 shows the performance comparison of the Scenario 2 analysis on Colony2A with static and dynamic load balancing schemes. The dynamic load balancing scheme again exhibits superior speedup performance over the static one because the dynamic scheme perfectly balances execution time among all the processors as shown in Figure 13(b). Compared with Figure 12, the performance curve of the static scheme considerably approaches that of the dynamic scheme as shown in Figure 13(a). This attributes to the increased number of cases on each processor: 16 (=512/32) in Figure 12 vs. 628 (=20,094/32). With more cases on each processor, the randomness effect with the static scheme tends to smooth out the unevenness in execution time. With more processors used and less number of cases on each processor, the two curves are expected to further depart, and the dynamic scheme will have much better speedup performance than the static one.

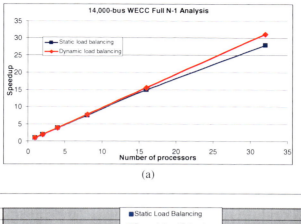

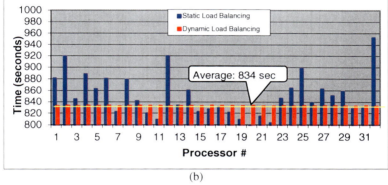

Fig. 14. Full WECC "N-1" contingency analysis on Colony2A (a) speedup performance and (b) execution time

3.3.2 Massive Contingency Analysis with Dynamic Load Balancing

The massive contingency analysis framework with the dynamic balancing scheme is further tested on the larger MPP2 machine with all the scenarios excluding Scenario 1. The results are summarized in Table 3. 512 MPP2 processors are used, and excellent speedup performance is achieved: about 500 times with the "N-2" scenarios and slightly less with the "N-1" scenario. For all the scenarios, the counter time is very insignificant compared to the computation time. It indicates that the implementation of the dynamic balancing scheme using a counter adds very little overhead to the overall process and good scalability can be ensured. The speedup increases as the number of cases increases. This result is consistent with the performance simulation results as shown in Figure 10, i.e. in the case of fast counter update, better speedup can be achieved when the number of cases increases.

Table 3. Summary Results of the Massive Contingency Analysis on the MPP2 Machine

512 Processors used	Wall Clock Time (seconds)	Total Computation Time (seconds)	Total I/O Time (seconds)	Total Counter Time* (seconds)	Speedup
Scenario 2 (20,094 N-1 cases)	31.0	14235.2	82.7	0.899	462
Scenario 3 (150,000 N-2 cases)	187.5	93115.5	489.1	5.550	503
Scenario 4 (300,000 N-2 cases)	447.9	226089.8	1087.1	9.984	507

* Includes waiting time.

As stated earlier, dynamic computational load balancing aims to balance execution time on each processor by dynamically distributing cases based on the availability of processors. This is clearly confirmed by Figure 14. It shows that the number of cases processed by individual processors varies from 20 to 50, but the computation time stays flat across the processors. A noticeable larger time on Processor #51 is due to the last case on this processor being a diverged one and taking longer to solve.

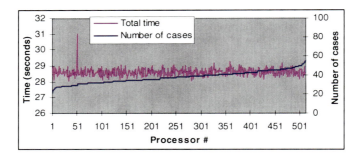

Fig. 15. Total time vs. the number of cases on each processor of Scenario 2

Figure 15 shows the increase in counter time with the increase in the number of cases on each processor for Scenario 2. This is understandable as more cases would need more counter update requests. For the MPP2 machine, the communication between processors is done with very high speed networking. One counter update takes about 10^{-5} seconds. The counter time is minimal compared to the computation time. But in other situations such as slow Ethernet-connected computers, the communication between processors is much slower and the counter time would be more costly. This would more likely cause counter congestion and should be taken into consideration when tuning the load balancing schemes.

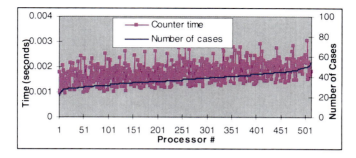

Fig. 16. Counter time vs. the number of cases on each processor of Scenario 2

3.3.3 Massive Contingency Analysis With Multi-counter-Based Scheme

The massive contingency analysis framework with the single-counter and two-counter dynamic computational load balancing schemes is implemented on NWICEB cluster machine. NWICEB machine uses InifiniBand network. In order to emulate the Ethernet environment on the NWICEB machine, the operation of counter update is executed 20 times more to mimic the low speed Ethernet communication. The reason for using the number of 20 is because that the latency of an Ethernet network is approximately 20 times of that of InfiniBand.

The 14,000-bus WECC power grid model is used as the study model. Three different scenarios with different number of contingency cases, $N = 500, 5000$, and 17346, are tested on NWICEB using both the single-counter and two-counter schemes. Two different environments (InfiniBand and emulated Ethernet) are compared. The execution time of all scenarios, excluding disk I/O time for the purpose of eliminating side-effects, with InfiniBand network is listed in Table 3, while Table 5 shows the execution time with the emulated Ethernet network.

Table 4. Execution time* in seconds of massive contingency analysis on the NWICEB machine under InfiniBand network

# of cases	500		5000		full N-1 (17346)	
# of procs	1_cnt	2_cnt	1_cnt	2_cnt	1_cnt	2_cnt
1	260.7	260.7	2285	2285	9038	9038
2	147.39	154.40	1327.9	1348.5	5095.6	5177.2
4	85.487	91.243	748.69	751.41	2880.3	2887.5
8	50.1	53.044	406.3	417.26	1488	1546.6
16	28.494	27.609	205.95	210.78	765	767.6
32	14.8325	14.949	108.46	110.35	397.81	386.2
64	8.1825	9.4521	62.339	57.92	187	201.9
128	5.7331	6.2403	30.01	33.251	96.63	97.87

* Disk I/O time excluded

Table 5. Execution time* in seconds of massive contingency analysis on the NWICEB machine under emulated Ethernet network

# of cases	500		5000		full N-1 (17346)	
# of procs	1_cnt	2_cnt	1_cnt	2_cnt	1_cnt	2_cnt
1	261.12	261.12	2359.9	2359.9	8994.1	8994.1
2	147.51	144.16	1341.5	1316.4	5098.4	5015.2
4	83.546	85.655	754.4	764.7	2869.9	2804.6
8	49.191	48.145	402.45	419.18	1569.5	1532.7
16	26.236	26.007	207.39	210.18	804.6	778.95
32	14.12	16.903	109.03	112.7	408.36	399.34
64	8.1423	8.7142	58.26	60.537	211.03	202.66
128	5.8304	6.869	29.959	32.098	112.17	108.17

* Disk I/O time excluded

In Table 3, the execution times with the two-counter scheme are larger than those with the single-counter scheme, which indicates that the two-counter dynamic load balancing scheme does not show better speedup performance than the single-counter scheme under InfiniBand environment. Corresponding to Table 3, the speedup results with both single-counter vs. two-counter schemes under InfiniBand environment are shown in Figure 16. This result matches the performance simulation results in Section 5.2.2. The main reason for this phenomenon is that the network latency in InfiniBand is low and the communication speed is fast. As such, the counter congestion is less likely to happen with a relatively small number of processors. Therefore, the overhead introduced by an additional counter degrades the performance under this testing environment.

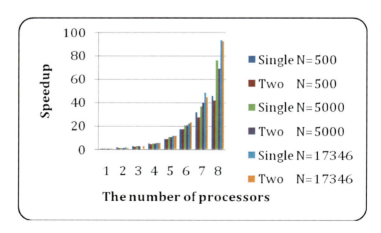

Fig. 17. Single-counter vs. two-counter comparison with respect to different number of under InfiniBand environment

When the communication speed is relatively low, counter congestion is more likely to happen. As shown in Table 5, when the number of contingency cases is large, (N=17,346), the two-counter scheme can improve the overall performance under the emulated Ethernet environment. Very importantly, this is true when the number of processors is relatively small and the number of contingency cases is large enough. For example, for full N-1 contingency analysis cases and with 16 processors, the execution time with single counter is 804.6 seconds, while the time with two counters is 778.95 seconds, which is 26 seconds less. The speedup with single-counter vs. two-counter schemes under the emulated Ethernet environment is shown in Figure 17. It shows that when N is large (=17,364), the performance of the two-counter scheme is better than that of the single-counter scheme; while the performance of the two-counter scheme is worse when the number of cases is small. These results again match the performance simulation results in Section 5.2.2.

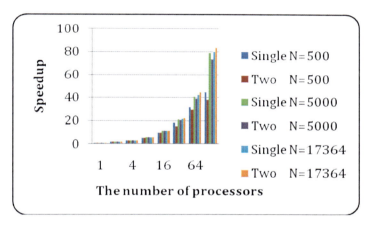

Fig. 18. Single-counter vs. two-counter comparison with respect to different number of under Emulated Ethernet environment

3.4 Summary

Contingency analysis computes solutions for a set of contingency cases. The contingency cases are computationally decoupled with little data exchange among the cases. Therefore, contingency analysis is relatively straightforward to be implemented on parallel computers. The challenge lies at computational load balancing, not the parallelization of algorithms and codes such as those with state estimation.

In this chapter, a dynamic computational load balancing scheme is implemented using a shared task counter updated by atomic fetch-and-add operations to facilitate the work load management among processors for massive contingency analysis. The computational performance of the dynamic load balancing scheme is analyzed, and the results provide guidance in using high-performance computing machines for large number of relatively independent computational jobs. An "N-x" massive contingency analysis framework with the dynamic balancing scheme is tested on two different high performance cluster machines. The test results indicate excellent scalability of the dynamic load balancing scheme. On 512 processors, massive contingency analysis can achieve over 500 times speedup compared with a single processor, and full "N-1" WECC contingency analysis can be completed with half a minute.

The computational performance analysis through case studies, as well as MATLAB simulation results, provides insights regarding the performance of load balancing schemes. These insights can serve as guidance for utilities to select suitable counter schemes to implement massive contingency analysis under their computing environments. Considerations for the implementation include:

(a) In the case of a computing environment with an Ethernet network or equivalent for a large number of processors (e.g. a networked computer farm), the two-counter dynamic scheme is expected to have better performance for the application of massive contingency analysis;

(b) In the case of a computing environment with a high-speed communication network but only a small number of processors, the single-counter dynamic scheme is suggested;

(c) In the case of a computing environment with a high-speed communication network and a dedicated cluster computer with a large number of processors, the performance of the two-counter dynamic scheme will be better than that of the single-counter scheme.

4 Parallelized Dynamic Simulation

Power system dynamic simulation aims to determine the time-series trajectory when the system is subject to disturbances such as a short-circuit fault, generator tripping, or line switching. It solves a set of differential-algebraic equations that describes the electro-mechanical interaction of generators as well as power electronic devices and their controllers. Dynamic simulation is widely used for off-line planning purposes, but rarely used as an on-line function due to its computational complexity. Solving the set of differential algebraic equations in the time domain using numerical integration techniques is far more time consuming than solving static power flow equations. The computational algorithms shown in Figure 2 have been optimized to the maximum performance of single-processor computers. As an example, on a regular personal computer with a 3.8GHz processor and 1-GB memory, it takes about 5 minutes to simulate 30-second dynamics of the 14,000-bus western U.S. power grid, i.e. 10 times slower than real time. Unfortunately, computer chip manufacturers are unable to increase single processor speed and have turned to parallel processors to continue the increase of computing power. Further improving the computational performance of dynamic simulation will require parallel computing implementation of the solution methods.

4.1 Power System Dynamic Modeling and Solution Methods

A simple system with one machine supplying power to an infinite bus through two parallel lines (Figure 19) is used to illustrate power system dynamic modeling.

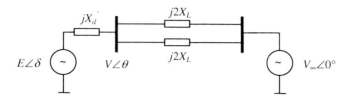

Fig. 19. A simple system for illustrating power system dynamic modeling

For the purpose of this example, a classical model is used for the generator, composed of a voltage source $|E|\angle\delta$ behind an impedance jX'_d; the magnitude of the voltage source $|E|$ is constant. The lines are modeled as constant impedance elements each with a value of $j2X_L$. The infinite bus is modeled as a magnitude, V_∞, of 1 per unit and an angle of zero degrees. The electro-mechanical dynamics can be described using the following differential algebraic equations [Kundur 1994]:

$$\begin{cases} \dfrac{d\delta}{dt} = \omega_B(\omega-\omega_0) \\ \dfrac{d\omega}{dt} = \dfrac{\omega_0}{2H}\left(P_m - P_e - D(\omega-\omega_0)\right) \\ P_e = \dfrac{|E||V_\infty|}{X'_d + X_L}\sin\delta \\ Q_e = \dfrac{|E|^2}{X'_d + X_L} - \dfrac{|E||V_\infty|}{X'_d + X_L}\cos\delta \end{cases} \qquad (8)$$

where the state variables, δ and ω, are the generator rotor angle and speed, respectively, ω_0 is the synchronous speed in per unit with ω_B being the speed base, also in per unit. H is the machine inertial constant, and D is the damping coefficient. P_m and P_e are the per unit mechanical input and the electrical output power of the generator. Q_e is the reactive power from the generator internal bus to the infinite bus.

There are multiple numerical integration methods for solving these differential algebraic equations. A simple method is the second-order Euler method. Starting with a known initial power flow condition, the process of dynamic simulation can be described as follows ($t_1, t_2, t_3, \ldots$ are time steps):

```
At t=0, V⁽⁰⁾∠θ⁽⁰⁾ is known
Initializaiton:
    Assume ω⁽⁰⁾=1, i.e. starting from a steady state
```

```
        Compute Pₘ=Pₑ⁽⁰⁾ and Qₑ⁽⁰⁾ using V⁽⁰⁾∠θ⁽⁰⁾ and V∞⁽⁰⁾∠0°
        Compute E'⁽⁰⁾∠δ⁽⁰⁾ using V⁽⁰⁾∠θ⁽⁰⁾ adnd jXd'
At t=t₁
    Solving differential equations:
            Compute δ⁽¹⁾ and ω⁽¹⁾ using the second-order Euler method
    Solving algebraic equations:
            Computer Pₑ⁽¹⁾ and Qₑ⁽¹⁾
At t=t₂
    Solving differential equations:
            Compute δ⁽²⁾ and ω⁽²⁾ using the second-order Euler method
    Solving algebraic equations:
            Computer Pₑ⁽²⁾ and Qₑ⁽²⁾
At t=t₃, t₄, …
      Repeat the process until the simulation is terminated
```

In a general multiple-machine system, the equations follow the same structure but become much more complicated. The general form is as follows:

$$\begin{cases} \dfrac{dx}{dt} = f(x,y) \\ 0 = g(x,y) \end{cases} \quad \text{or} \quad \begin{cases} \dfrac{dx_1}{dt} = f(x_1,y) \\ \dfrac{dx_2}{dt} = f(x_2,y) \\ \dots \\ 0 = g(x_1, x_2, \dots, y) \end{cases} \tag{9}$$

where x is the dynamic state variables, y is the algebraic variables. x_i is the state variables for the ith device. $x = (x_1, x_2, \dots)$.

The differential equations include controller models such as governors and excitation systems and models of other dynamic devices such as dc transmission and Flexible AC Transmission System devices. The equations of each dynamic device do not directly relate to other devices but they all relate to the network. The algebraic equations are no longer a simple power calculation. They become essentially the power flow equations that couple the generator variables.

4.2 Dynamic Contingency Analysis (On-Line Dynamic Security Assessment)

The ability to sustain the potential loss of one or several components is important to power grid reliability. Static contingency analysis, i.e. power-flow-based contingency analysis, only studies the post-disturbance condition of the power grid. The transition from the pre-disturbance condition to the post-disturbance condition is ignored. However, this transition may not be stable, or the trajectory may violate some limits or trigger protection devices. That is, the power grid may not sustain the disturbance even of the post-disturbance condition is a feasible operation point. To full understand the grid sustainability needs real-time dynamic contingency analysis, i.e. on-line dynamic security assessment. Each dynamic contingency is a dynamic simulation case with initial conditions determined by the current operating point.

Similar to the traditional power-flow-based contingency analysis, dynamic contingency cases are relatively independent of one another, so dynamic contingency analysis is naturally a parallel process. The total computation time is reduced by allocating cases to multiple computers or processors. But each case is still a regular dynamic simulation process and takes the same amount of time to finish. Similarly, the challenge for dynamic contingency analysis is not on algorithmic parallelization or code-level parallelization, but on computational load management. Contingency load management is to balance the workload among processors since not all cases require the same amount of time. Some cases can be terminated early if it can be determined to be stable or unstable without finishing the full time-period simulation. In this case, the load balancing schemes presented in Section 5 about parallel static contingency analysis apply to dynamic contingency analysis. And cluster machines or networked computers are suitable for this kind of analysis. Discussion about computer environments in Section 5.2 applies here, too.

4.3 Look-Ahead Dynamic Simulation

If dynamic simulation can be run faster than real time, the simulation would then be able to predict future status of a power grid. A skilled operator or planner can "look ahead" and anticipate developing problems in the grid. Different from dynamic contingency analysis, look-ahead dynamic simulation requires shortening computation time for a single dynamic simulation process. Algorithmic and code parallelization is needed.

As mentioned in Section 6.1, dynamic simulation mainly includes two major steps – numerical integration for the differential equations and nonlinear system solution for the algebraic equations. Other steps include one-time initiation and output step (Figure 20).

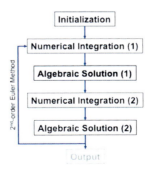

Fig. 20. Dynamic simulation process using the second-order Euler method

Benchmarking studies using General Electric's PSLF software package [GE] have been performed to determine the most time consuming steps in dynamic simulation. Several test systems were used to perform dynamic simulation, and the time spent on each step was recorded. Table 6 clearly shows the algebraic solution step (i.e. the power flow solution step that couples all the state variables) is the dominant time-consuming step and the numerical integration step is second. The algebraic solution step involves significant matrix operations and thus takes time to solve, while the numerical integration step is mainly scalar operations involving simple addition, division, multiplication, and so on.

Table 6. Benchmarking Studies of Dynamic Simulation with Different Power Systems

	Time of each step in % of Total time			
	16-Machine-68-Bus System	50-Machine-145-Bus System	100-Machine-300-Bus System	14,000-Bus WECC System
Initialization	Negligible	Negligible	Negligible	Negligible
Numerical Integration (1)	7.0%	3.7%	3.0%	11.9%
Algebraic Solution (1)	40.8%	43.6%	46.4%	38.1%
Numerical Integration (2)	10.0%	6.6%	3.8%	11.9%
Algebraic Solution (2)	42.4%	45.5%	46.8%	38.1%

High performance computing techniques can improve the computational performance of both of these two mathematical aspects to a great extent. Numerical integration is inherently parallel because each equation is integrated separately. Methods for linear equation solution are similar to that tested in the parallelized state estimation.

A version of parallel dynamic simulation is implemented on HP Superdome 128-core machine using OpenMP Application Program Interface (API) was used [OpenMP]. The OpenMP API allows for the dynamic equations, and some matrix operations, to be implemented on multiple processors without changing the core algorithms. HP Superdome belongs to the shared-memory family of high performance computers, which has a large block of memory accessible from multiple processing cores. A shared memory system is relatively easy to program since all processors share a single view of data and the communication between processors can be as fast as memory accesses to a same location. It is worth pointing out that large special-purpose machines are *not* required to speed up the dynamic simulation process. Similar performance was observed with implementation on a computer that has two quad-core Xeon E5345 nodes with 8 available cores running at 2.33 GHz and using 16 GB of shared memory [Schneider 2009].

Several test systems of size small to medium have been used to test the parallel dynamic simulation implementation. Figure 21 shows results in speeding up the transient simulation in a research setting. For small systems, it quickly runs out of work for the processors. The overhead required to coordinate the computation among processors quickly dominates the simulation. Thus the speed-up curve saturates at a relative small number of processors. However, with the system size increasing, better speed-up performance is observed, and saturation occurs at a larger number of processors. For example, the 400-machine test system achieved 14x speed-up with 32 processors. Speed-up performance is expected to be even better with larger system sizes (e.g. WECC has 3000 generators). As mentioned in the beginning of this section, dynamic simulation of a WECC-size system on a single PC is 10x slower than real-time. To make dynamic simulation to run in real-time, at least 10x speed-up is required. To look ahead, dynamic simulation needs to run faster than real-time, say 10-20x faster. Figure 21 shows it is very promising to run the simulation much faster and bring transient analysis into on-line applications to enable more transparency for better reliability and asset utilization.

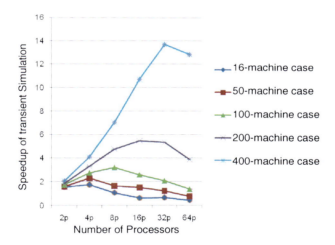

Fig. 21. Speedup of transient simulation with respect to the number of processors

If implicit numerical integration methods such as the trapezoidal method or Runge-Kuta method are used, the solution process for differential equations and algebraic equations can no longer be decoupled. It effectively becomes solving a set of nonlinear equations at each time step in dynamic simulation. The parallelization techniques for solving linear equations can be applied to speed up the computation.

4.4 Summary

Dynamic simulation remains to be off-line applications due to heavy computation. This chapter presents parallelized implementation of the solution process of differential-algebraic equations for power grid dynamic simulation. Test results demonstrate that HPC can significant improve the computational efficiency, especially for large-scale power grids. On a medium-scale power grid, more than 10 times of speedup has been achieved. For a larger power grid, the overhead introduced by the parallelization process is expected to outweigh the benefit of using multiple processors, and thus the speedup performance is expected to be better.

Parallel dynamic simulation can have multiple important applications in improving power grid reliability and efficiency. Faster-than-real-time dynamic simulation provides the capability to look ahead and predict grid behaviors. This look-ahead capability enables predictive management and control compared to today's reactive management and control. Fast dynamic simulation also enables real-time path rating that will be able to take actual operating conditions into consideration instead of computing transmission path limits based on assumed off-line conditions. The path rating can be more accuracy and thus improve asset utilization and confidence in operation.

5 Conclusion and Future Work

The power grid is evolving into the future to be a more dynamic, probabilistic, and complex system. Traditional computational resources and applications are not meeting the emerging requirements due to this grid evolution. Information revolution, on the other hand, provides unprecedented amounts of data and presents opportunities in managing the grid of the future. The resulting large amount of data further demands adoption of advanced computational resources. High-performance computing has been demonstrated in other domains with great success. It holds the promise to enable faster and more comprehensive analysis and transform grid applications for better reliability and efficiency.

To bring HPC to power grid applications is not simply putting more computing units against the problem. It requires careful design and coding to match an application with computing hardware. Sometimes, alternative or new algorithms need to be used to maximize the benefit of HPC.

This chapter demonstrates the benefits of HPC for power grid applications with several examples of state estimation, contingency analysis, and dynamic simulation. They represent the major categories of power grid applications. Each of the applications has its own problem structure and data dependency requirements. The approach to apply HPC to these problems has different challenges. State estimation equations are tightly coupled and require frequent data exchange between computing units. For solving these equations, LU decomposition algorithms, well suited for a single-processor environment, are not optimal for the state estimation problem. Instead, iterative methods exhibit obvious advantages. Contingency analysis represents another end of the problem spectrum: it is an inherently parallelizable process with each of the computation task relatively independent of others. Data exchange between tasks is very minimal. The challenge for parallel contingency analysis is not about parallelizing the algorithm and codes, but about balanced task allocation among the processors. A dynamic computational load balancing scheme with one or more task counters is developed and evaluated with both theoretical analysis and computational experiments. With this dynamic load balancing scheme, tasks can be evenly distributed, and utilization of computing resources is maximized, suitable for fast massive "N-x" contingency analysis. Dynamic simulation is a category of applications with mixed features of state estimation and contingency analysis. Each dynamic simulation case involves both decoupled differential equations and highly coupled algebraic equations. Parallel dynamic simulation involves both task allocation and algorithm parallelization. Built on parallel dynamic simulation, other applications such as on-line dynamic security assessment and real-time path rating require analysis of a large number of dynamic simulation cases. These applications require allocating dynamic simulation cases to multiple processors as well as parallelizing individual dynamic cases. The test results show a promising path forward to achieve faster-than-real-time performance for interconnection-scale dynamic simulation.

The HPC-enhanced state estimation, contingency analysis, and dynamic simulation presented in this chapter are suitable for today's power grid operation and planning. With the ever-decreasing hardware cost, it is also becoming practical and cost-effective.

Power grid HPC applications are still at its early stage. Expanding from these applications presented in this chapter, future HPC applications need to consider a more dynamic and probabilistic environment, that introduces new complexities due to increased modeling size, larger amount of data, and significant uncertainties. HPC techniques and hardware are essential to tackle the complexities.. This chapter focuses on the traditional shared-memory platform and cluster-machine platform. Future power grid applications will also push the envelope of HPC technologies in many aspects such as real-time computing, distributed computing, and hybrid computing that are important for power grid applications with task parallelization and data parallelization.

Acknowledgement. The following individuals are acknowledged for being the sources of significant elements to this chapter: Dr. Gilbert Bindewald with the U.S. Department of Energy, and Dr. Ning Zhou, Robert Pratt, Dr. Shuangshuang Jin, Dr. Kevin Schneider, Dr. Barry Lee, Dr. Steve Elbert, Ruchi Singh, Dr. Ruisheng Diao, and Dr. Marcelo Elizondo, all with Pacific Northwest National Laboratory. The authors would like to dedicate this chapter to the memory of Dr. Jarek Nieplocha. Dr. Nieplocha's vision has guided us in the endeavor of pursuing the interdisciplinary adventure to apply high-performance computing to power grid applications.

References

[Abur 1988] Abur, A.: A parallel scheme for the forward/backward substitutions in solving sparse linear systems. IEEE Trans. Power Systems 3 (November 1988)

[Ajjarapu 1992] Ajjarapu, V., Christy, C.: The continuation power flow: a tool for steady state voltage stability analysis. IEEE Trans. Power Systems 7(1), 416–423 (1992)

[Alvarado 1990] Alvarado, F.L., Yu, D.C., Betancourt, R.: Partitioned A-1 methods. IEEE Trans. Power Systems 5 (May 1990)

[Amestoy 2000] Amestoy, P.R., Duff, I.S., L'Excellent, J.-Y.: Multifrontal parallel distributed symmetric and unsymmetric solvers. in Comput. Methods in Appl. Mech. Eng. 184, 501–520 (2000)

[Bell 2009] Bell, Garland, M.: Implementing sparse matrix-vector multiplication on throughput-oriented processors. In: Proc. Supercomputing 2009 (November 2009)

[BPA 2010] Bonneville Power Administration, Fact Sheet: BPA's wind power efforts surge forward (March 2010), http://www.bpa.gov/corporate/pubs/fact_sheets/10fs/BPA_Wind_Power_Efforts_March_2010.pdf (accessed May 30, 2010)

[Chen 1996] Chen, R.H., Gao, J., Malik, O.P., Wang, S.-Y., Xiang, N.-D.: Automatic contingency analysis and classification. In: The Fourth International Conference on Power System Control and Management, April 16-18 (1996)

[Chen 2010] Chen, Y., Huang, Z., Chavarria, D.: Performance Evaluation of Counter-Based Dynamic Load Balancing Schemes for Massive Contingency Analysis with Different Computing Environments. In: Proceedings of the IEEE Power and Energy Society General Meeting 2010, Minneapolis, MN, July 25-29 (2010)

[Chen 2011] Chen, Y., Huang, Z., Zhou, N.: An Advanced Framework for Improving Situational Awareness in Electric Power Grid Operation. In: Proceeding of the 18th World Congress of the International Federation of Automatic Control (IFAC), Milano, Italy, August 28-September 2 (2011)

[Davis 2000] Davis, T.A., Gilbert, J.R., Larimore, S.I., Ng, E.: A column approximate minimum degree ordering algorithm, Technical Report TR-00-005, CISD, University of Florida (2000)

[Davis 2006] Davis, T.A.: Direct methods for sparse linear systems. Society for Industrial Mathematics (2006)

[DOE 2004] U.S. Canada Power System Outage Task Force, Final Report on the August 14, Blackout in the United State and Canada: Causes and Recommendations (August 14, 2003), https://reports.energy.gov/ (April 2004)

[Dongarra 1990] Dongarra, J., Croz, J., Duff, I., Hammarling, S.: A Set of Level 3 Basic Linear Algebra Subprograms. ACM Trans. Math. Soft. 16, 1–17 (1990)

[Ellis 2006] Ellis, A., Kosterev, D., Meklin, A.: Dynamic Load Models: Where Are We? IEEE PES Transmission and Distribution Conference and Exhibition 2005/2006, May 21-24 (2006)

[GE] PSLF Software Manual, General Electric

[Grainger 1994] Grainger, J.J., Stevenson, W.D.: Power System Analysis. McGraw-Hill Co. (1994)

[Hestenes 1952] Hestenes, M.R., Stiefel, E.L.: Methods of Conjugate Gradient for Solving Linear Systems. Nat. Bur. Std. 49, 409–436 (1952)

[Huang 2007-1] Huang, Z., Guttromson, R., Nieplocha, J., Pratt, R.: Transforming Power Grid Operations via High-Performance Computing. Scientific Computing (April 2007)

[Huang 2007-2] Huang, Z., Schneider, K., Nieplocha, J.: Feasibility Studies of Applying Kalman Filter Techniques to Power System Dynamic State Estimation. In: Proceedings of IPEC 2007 – The 8th International Power Engineering Conference, Singapore, December 3-6 (2007)

[Huang 2009] Huang, Z., Chen, Y., Nieplocha, J.: Massive Contingency Analysis with High Performance Computing. In: Proceedings of PES-GM 2009 – The IEEE Power and Energy Society General Meeting 2009, Calgary, Canada, July 26-30 (2009)

[Hypre] High Performance Preconditioner Library (hypre), Libarary Lawrence Livermore National Laboratory, http://acts.nersc.gov/hypre/

[Hysom 1999] Hysom, D., Pothen, A.: Efficient Parallel Computation of ILU(k) Preconditioners, SC, published on CDROOM 1999, ISBN 1-58113-091-0, ACM Order #415990. IEEE Computer Society Press Order #RS00197

[Hysom 2001] Hysom, D., Pothen, A.: A Scalable Parallel Algorithm for Incomplete Factor Preconditioning. SIAM J. Sci. Comput. 22(6), 2194–2215 (2001)

[IEEE 1995] IEEE Standards, POSIX System Application Program Interface: Threads extension 1003.1c (1995)

[Jea 1983] Jea, K.C., Young, D.M.: On the Simplification of Generalized Conjugate-Gradient Methods for Nonsymmetrizable Linear Systems. Linear Algebra Appl. 52/53, 399–417 (1983)

Kosterev, D.N., Taylor, C.W., Mittelstadt, W.A.: Model Validation for the August 10, 1996 WSCC System Outage. IEEE Trans. Power Syst. 14(3), 967–979 (1999)

[Kundur 1994] Kundur, P.: Power System Stability and Control. McGraw-Hill Professional (1994) ISBN-10 007035958X

[Lau 1990] Lau, K., Tylavsky, D.J., Bose, A.: Coarse grain scheduling in parallel triangular factorization and solution of power system matrices. Presented at IEEE/PES Summer Meeting, Minneapolis (July 1990)

[Lee 2010] Lee, V.W., Kim, C., Chhugani, J., Deisher, M., Kim, D., Nguyen, A.D., Satish, N., Smelyanskiy, M., Chennupaty, S., Hammarlund, P., Singhal, R., Dubey, P.: Debunking the 100X GPU vs. CPU myth: an evaluation of throughput computing on CPU and GPU. Computer Architecture News 38, 141–153 (2010)

[Liu 1985] Liu, J.: Modification of the minimum degree algorithm by multiple elimination. ACM Trans. Math. Soft. 11, 141–153 (1985)

[Makarov 2010] Makarov, Y.V., Etingov, P.V., Ma, J., Huang, Z., Subbarao, K.: Incorporating Uncertainty of Wind Power Generation Forecast into Power System Operation, Dispatch, and Unit Commitment Procedures. submitted to IEEE Transactions on Sustainable Energy (2010)

[MISO 2008] Midwest ISO, Financial Transmission Rights (FTR) and Auction Revenue Rights (ARR) (March 13, 2008)

[Moore 1965] Moore, G.E.: Cramming more components onto integrated circuits. Electronics Magazine. Electronics 38(8) (April 19, 1965)

[Morante 2006] Morante, Q., Ranaldo, N., Vaccaro, A., Zimeo, E.: Pervasive Grid for Large-Scale Power Systems Contingency Analysis. IEEE Transactions on Industrial Informatics 2(3) (2006)

[MPI] Message Passing Interface forum, http://www.mpi-forum.org

[NERC] NERC Transmission System Standards – Normal and Emergency Conditions, North American Electricity Reliability Corporation, http://www.nerc.com

[Niagara] Rijk, C.: Niagara: A Torrent of Threads, http://www.aceshardware.com/read.jsp?id=65000292

[Nieplocha 1996] Nieplocha, J., Harrison, R.J., Littlefield, R.J.: Global Arrays: A nonuniform memory access programming model for high-performance computers. The Journal of Supercomputing 10, 197–220 (1996)
[Nieplocha 2006] Nieplocha, J., Marquez, A., Tipparaju, V., Chavarría-Miranda, D., Guttromson, R., Huang, Z.: Towards Efficient Power System State Estimators on Shared Memory Computers. In: Proceedings of PES-GM 2006 – The IEEE Power Engineering Society General Meeting, Montreal, Canada, June 18-22 (2006)
[Nieplocha 2007] Nieplocha, J., Chavarría-Miranda, D., Tipparaju, V., Huang, Z., Marquez, A.: Parallel WLS State Estimator on Shared Memory Computers. In: Proceedings of IPEC 2007 – The 8th International Power Engineering Conference, Singapore, December 3-6 (2007)
[Obama 2011] Obama, B.: State of the Union 2011: Winning the Future (January 2011), http://www.whitehouse.gov/state-of-the-union-2011 (accessed March 15, 2011)
[OpenMP] OpenMP, http://openmp.org
[Pereira 2002] Pereira, L., Kosterev, D., Mackin, P., Davies, D., Undrill, J., Zhu, W.: An Interim Dynamic Induction Motor Model for Stability Studies in the WSCC. IEEE Transactions on Power Systems 17(4) (November 2002)
[PJM 2007] PJM, PJM eFTR Users Guide, PJM (2007)
[Saad 2003] Saad, Y.: Iterative methods for sparse linear systems, Society for Industrial Mathematics (2003)
[Schenk 2001] Schenk, O., Gärtner, K., Fichtner, W., Stricker, A.: PARDISO: A High-Performance Serial and Parallel Sparse Linear Solver in Semiconductor Device Simulation. Journal of Future Generation Computers Systems 18 (2001)
[Schneider 2009] Schneider, K.P., Huang, Z., Yang, B., Hauer, M., Nieplocha, J.: Dynamic State Estimation Utilizing High Performance Computing Methods. In: Proceedings of PSCE 2009 – The IEEE Power and Energy Society Power System Conference and Exposition 2009, Seattle, WA, March 15-18 (2009)
[SGI] SGI Inc., http://www.sgi.com/products/servers/altix/3000/
[SuperLU 2005] An Overview of SuperLU: Algorithms, Implementation and User Interface. ACM Transactions on Mathematical Software 31, 302–325 (2005)
[Tinney 1967] Tinney, W.R., Walker, J.W.: Direct Solutions of sparse network equations by optimally ordered triangular factorization. Proc. IEEE 55, 1801–1809 (1967)
[TOP500 2010] TOP500, TOP500 list of the world's most powerful supercomputers (2010), http://www.top500.org/
[Turing 1948] Turing, A.M.: Rounding-off errors in matrix processes. Quart. J. Mech. Appl. Math. 1, 287–308 (1948)
[Ungerer 2003] Ungerer, T., Robi, B., Jurij, A.: A survey of processors with explicit multithreading. ACM Comput. Surv. 35, 29–63 (2003)
[Vaahedi 1988] Vaahedi, E., El-Din, H., Price, W.: Dynamic Load Modeling in Large Scale Stability Studies. IEEE Transactions on Power Systems 3(3) (August 1988)
[Zhang 2005] Zhang, Y., Tipparaju, V., Nieplocha, J., Hariri, S.: Parallelization of the NAS Conjugate Gradient Benchmark Using the Global Arrays Shared Memory Programming Model. In: Proceedings of IPDPS 2005 (2005)
[Zhao 2011] Zhao, L., Donde, V.D., Tournier, J.-C., Fang, Y.: On limitations of traditional multi-core and potential of many-core processing architectures for sparse linear solvers used in large-scale power system applications. In: 2011 IEEE Power and Energy Society General Meeting (July 2011)

Dynamic Load Balancing and Scheduling for Parallel Power System Dynamic Contingency Analysis

Siddhartha Kumar Khaitan and James D. McCalley

Electrical and Computer Engineering,
Iowa State University, Iowa, USA
{skhaitan,jdm}@iastate.edu

Abstract. Power system simulations involving solution of thousands of stiff differential and algebraic equations (DAE) are extremely computationally intensive and yet crucial for grid security and reliability. Online simulation of minutes to hours for a large number of contingencies requires computational efficiency several orders of magnitude greater than what is todays state-of-the-art. We have developed an optimized simulator for single contingency analysis using efficient numerical algorithms implementation for solving DAE, and scaled it up for large-scale con-tingency analysis using MPI. A prototype parallel high speed extended term simu-lator (HSET) on in-house high performance computing (HPC) resources at Iowa State University (ISU) (namely Cystorm Supercomputer) is being developed. Since the simulation times across contingencies vary considerably, we have focused our efforts towards development of efficient scheduling algorithms through work stealing for maximal resource utilization and minimum overhead to perform faster than real time analysis. This chapter introduces a novel implementation of dynamic load balancing algorithm for dynamic contingency analysis. Results indicate potential for significant improvements over the state-of-the-art methods especially master-slave based load balancing typically used in power system community. Simulations of thousands of contingencies on a large real system were conducted and computational savings and scalability results are reported.

1 Introduction

Despite the availability of computation resources, full exploration of solution space remains a challenge due to ineffective scheduling techniques. The conventional scheduling methods are generally based on static allocation methods and thus fail to scale to the case where a large number of contingencies need to be tested which may arrive arbitrarily and have considerably variable simulation times. Moreover, inefficient techniques prohibit the operators from full exploration of solution search-space and hence conclusions derived from such simulations/experimentation, may be inaccurate and/or partial. In this work, we integrate the HSET optimized power simulator with efficient dynamic load scheduling techniques to maximize the utilization of the computation -resources available using task-stealing algorithm and taking into account the constraints present on the task-set. Acceleration of individual task is accomplished by efficient numerical algorithms for integration, solution of nonlinear and linear system of equations. A brief description of the numerical algorithms implemented in the optimized power simulator is presented below. The experiments performed show the effectiveness of our proposed technique.

1.1 Design of the Power System Time Domain Simulator

We have developed a research-grade power simulator for performing steady state and time domain power system simulations. The simulator is developed in C++ which facilitates object oriented programming. This is very convenient for a modular design and interfacing different components of the software. It also allows easy addition and extensibility and adaptability to modern high performance computing architecture. The developed simulator is a cross platform simulator with industry standard input format.. It has extensive power system component library for modeling generators, excitoers, governors and so on. It also provides an interface to a number of numerical algorithms for integration, solution of nonlinear and linear system of equations. We have been working on primarily developing a high speed extended term time domain simulator (HSET) as the future functionality for control center security assessment [26–29]. It is intended to be a decision support tool for the operational personnel in the event of a low probability high consequence events involving multiple elements and leading to cascading events spanning a time period ranging from minutes to hours. Such events are normally deemed to be unlikely and therefore undeserving of preventive actions [9, 16]. However such events do occur in reality at frequency more than ever before due to highly stressed power system and increased penetration of renewable energy sources and highly interconnected infrastructure. HSET is intended to serve as a decision support tool in identifying appropriate corrective actions and in tracing the evolution of the system trajectory under such abnormal power system operation conditions. In addition, it also has the ability to perform short term transient simulation studies under normal conditions. The chapter is mainly devoted to transient simulation based contingency analysis. HSET is intended to be a part of the energy management system (EMS) and would have the following attributes:

1. Contingency selection based on the topological processing of the power grid which is represented by the node breaker model rather than bus branch model.
2. Extended-term: Cascading events can last for long periods of time up to several hours and HSET is intended to have the capability to simulate for this time frame.
3. Fast and Slow Dynamics: HSET is intended to be able to capture both the fast and the slow transients corresponding to the fast and the slow progression sequences in the evolution of cascading events.
4. Failure detection: HSET should be able to detect unacceptable system performances within the simulation based on the state variables.
5. Corrective action: After the successful failure detection, HSET must identify within the simulation, the best possible corrective action strategy to arrest the evolution of cascading events. This would involve optimization.
6. Computational Efficiency: Time domain simulation involving the solution of thousands of stiff system of differential and algebraic equations (DAE) are computationally intensive and extremely challenging even for a short amount of simulated time such as 10 seconds, even for few limited number of contingencies. On-line simulation of minutes to hours for a very large number of contingencies requires computational efficiency several orders of magnitude greater than what is todays state-of-the-art.
7. Anticipatory Computing: HSET-TDS may be helpful in a responsive mode where it is run following initiation of a severe disturbance. Alternatively, we envision that it will play a heavy role in an anticipatory mode, continuously computing responses to many contingencies and storing preparatory corrective actions that would be accessed by operational personnel should one of the contingencies occur. Results can be archived and re-used when similar conditions are met.

1.2 Numerical Algorithms

Power system is represented by thousands of DAEs. These are highly stiff in nature due to multi-scale dynamics with components with widely varying time constants. These DAEs are discretized using an integration algorithm. The resulting set of nonlinear equations together with the algebraic equations from the network and the different power system components are solved through Newton iteration. The core of the solution of the nonlinear system of equations is the solution of the linear system of equations. The software is designed to allow interfacing with any numerical algorithm at these three different levels namely the integration algorithm, nonlinear solver and the linear solver.

Integrator: The simulator is interfaced with a number of implicit and explicit integrators including trapezoidal method, Euler method, BDF method and so on. The simulator is also interfaced with the public domain variable order and variable time step BDF integrator IDAS from LLNL [36]. Our results have shown that IDAS outperforms other integrators significantly and is able to efficiently and accurately handle the stiff power system dynamics.

Non Linear Solver: Newton methods are the ubiquitous choice for the solution of nonlinear equations. We have developed several Newton-based nonlinear solvers, including a conventional Newton, a Newton algorithm with line search, and an algorithm to identify an initial solution based on a least square minimization (which gives very good starting point for fast convergence of the Newton method). We have also implemented the relaxed Newton method to avoid repeated jacobian calculations. This is commonly called very dishonest Newton method (VHDN).

Linear Solver: The simulator is fully modular to interface any linear solver. A number of linear solvers have been interfaced and tested. The IDAS integration solver does not support direct sparse linear equation solvers. There are only inbuilt linear equation solvers for dense matrices (BLAS routines and LAPACK routines). These are not suitable for power system applications, which have high dimension sparse matrices. IDAS also supports iterative solvers like GMRES with user supplied preconditioner routines. However the power system dynamics jacobian is highly ill conditioned and iterative solvers convergence is very poor. This is a potential research field which would require design of innovative preconditioners for power system specific problems. We will explore this later. Therefore, to use IDAS, we have spent significant effort to make modifications of the code to facilitate the use of sparse solvers with IDAS. We have restructured the IDAS code in such a modular way that now any linear solver can be easily interfaced with it. Of all the linear equation solvers interfaced with IDAS (KLU, UMFPACK, SuperLU, PARDISO etc.), [12,13,35] KLU is found to be most computationally efficient and fastest. KLU is a serial platform based linear equation solver.

1.3 High Performance Computing

High performance computing (HPC) refers to parallel processing at a large scale by efficient deployment of multiple processing units in parallel using suitable programming interfaces to solve large problems. Thus, HPC presents challenges and opportunities both for computer architecture designers and software developers, since architecture must be designed to facilitate high performance, and appropriate software interfaces must be developed to tap the potential of the underlying architecture. The advances in HPC over the last few decades has made it extremely popular in both research and industry as it provides the ability to solve larger problems in lesser time, thereby resulting in increased throughput and potential cost savings. There are many parallel programming models, most popular of which are the shared memory model using threads (OpenMP or POSIX threads), distributed memory model (MPI), and their hybrids. General purpose graphical processing units (GPGPUs) have also become increasingly common in the last few years. All these HPC techniques have been successfully applied in many fields like aerospace [24, 33], climate science [30, 32], sequence analysis in bioinformatics [1, 2, 22, 40–44], nuclear physics [6, 11, 37], power systems [18, 19, 25], and so on. Developing an HPC implementation involves several approaches and challenges in terms of data partitioning, functional decomposition, data dependencies, communication overheads, synchronization, load balancing, I/O, etc., of which we would particularly discuss load balancing in detail next; as in this work, we focus on optimizing the load distribution across multiple processors for the application of parallel power system contingency analysis.

1.4 Load Balancing

The scheduling techniques can be broadly divided into two categories, namely static and dynamic scheduling techniques.

Static Load Balancing. A static-load balancing approach pre-computes the schedule for each process in advance and thus requires little runtime overhead of scheduling and monitoring. However, use of this technique requires that all the jobs (contingencies) should be available (as in the case of power system) and their characteristics (e.g. precise run-time) be known at the beginning of the work to achieve fairly balanced scheduling. However, if the simulation times of the individual contingency vary too much, it would result in significant load imbalance. Except for few processors, all the rest would remain idle, waiting for other processors to finish. Thus the finish time of the schedule is the time for the last job/contingency to finish.

This idea of the static allocation is presented through figures (Fig. 1 Fig. 4). Fig. 1 shows the loading on the different processors in the beginning. All the processors are equally loaded or almost equally loaded in terms of number of tasks/contingencies to be analyzed. Since different contingencies

Fig. 1. Static Scheduling: Initial contingency allocation

Fig. 2. Static Scheduling: Load on individual processor as time progresses

could have different simulation times the work load on different processors can change with time as shown in Fig. 2. With further evolution of time, different processors finish their tasks and sit idle waiting for the last job to finish as shown in Fig. 3 and Fig. 4.

Fig. 3. Static Scheduling: One Processor finishes all its tasks and sits idle

Fig. 4. Static Scheduling: Multiple processors sitting idle waiting for last task to finish

Thus static scheduling could be very inefficient and have high load imbalance. Thus it is not preferred in most applications. However most of the parallel contingency analysis studies reported in the literature implement static load allocation strategy.

Static load balancing has the advantage that it does not required online scheduling and has low overhead. Also, it is useful when the task lengths are almost equal. However, if task lengths are quite different, then it leads to poor load balancing. This leads to poor computation efficiency.

Dynamic Load Balancing. A dynamic load scheduler generates schedules in an online manner by dynamically allocating the contingency cases to the individual processors on demand. Dynamic scheduling generates evenly-distributed schedules by significantly reducing processor idle time. A simple dynamic scheduling technique is known as the master-slave scheduling, where one node is used as a master and other nodes are used as slaves, e.g. [17,23]. The master schedules the tasks on slave-nodes and on finishing a task, the slaves request the master for allocation of new tasks. This technique has the disadvantage that the master-processor becomes occupied with scheduling and hence cannot be used for useful work.

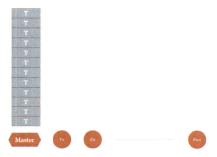

Fig. 5. Master-Slave Scheduling:Initial contingency allocation

Fig. 6. Master-Slave Scheduling: Contingency allocation by the master to the slave nodes

To begin with, the master node has all the contingencies queued at its node as shown in Fig. 5. It then allocates tasks to each of the slave nodes as in Fig. 6. When the individual processor finishes the assigned contingency analysis it requests a task from the master node, if available (Fig. 7) , and the master node assigns a task to the processor if it has pending contingencies to be simulated and reduces its queue by one (Fig. 8).

However a situation may arise where multiple processors finish their assigned tasks and send request to the master node for additional tasks simultaneously. This would lead to contention and need for proper synchronization resulting in potential extra overhead as shown in Fig. 9. This problem increases dramatically with large number of processors when the chances for contention increase with number of processors.

Fig. 7. Master-Slave Scheduling: Slave sends a request to the master upon job completion

Fig. 8. Master-Slave Scheduling: Master responds to the request of the slave

Fig. 9. Master-Slave Scheduling: Multiple slave nodes request the master node leading to contention

Master-slave scheduling overcomes the limitations of static scheduling by providing better load balancing. Also, compared to work stealing, it does not require communication between every pair of processors (slaves) and hence it has low overhead. However, master-slave technique also has the disadvantage that the master processor does not perform any useful work and hence gets wasted. Also, if multiple slaves simultaneously make a request to the master, contention may arise.

To reduce the contention at a single master, a variant of the master-slave is to use multiple masters with multi counters [10] to avoid contention on the master. However this is not efficient as multiple master nodes do not do useful work and thus get wasted. And if one master exhausts work it needs to request work from the other masters which results in high latency. Work-stealing [3, 8] is a well-known dynamic scheduling method, which works on the intuition that scheduling with load-balancing can be easily achieved, if a process with no pending job (called thief) is allowed to steal a pending

Fig. 10. Work-Stealing: Initial contingency allocation

job from the job-queue of other running processes (called victim). The initial allocation of different contingencies could be similar as in the case of static allocation. Fig. 10 depicts the initial contingency allocation in the case of work stealing based dynamic load balancing. As can be seen from the figure, it looks similar as in the case of static allocation shown in Fig. 1. In the same way, Fig. 11 shows the progression of the load on the different processors as time progresses. This is similar to Fig. 2. However, after this stage static scheduling and work stealing completely diverges in the sense that after a processor finishes its assigned contingencies, it does not remain idle unless none of the other processors has any pending contingencies to be analyzed.

Fig. 11. Work-Stealing: Load on individual processor as time progresses

Fig. 12. Work-Stealing: The thief P3 tries to steal a task from the victim P2

Fig. 12, 13 and 14 show the progression of the work stealing based load balancing. When a processor finishes all its allocated tasks, it sends a request (as in MPI) or checks the queues (as in multithreaded implementations) of the different processors, to see if they have additional tasks to steal from and achieve fine grained load balancing. While the processor (victim) responds to the request of the thief, it does not suspend its own operation. The detail of such an implementation is discussed below.

Fig. 13. Work-Stealing: Successful stealing by P3 from P2

Fig. 14. Work-Stealing: P1 trying to steal from Pn

Work-stealing has the advantage of providing fine load-balancing and it is provably efficient. It does not lead to wastage of the master node. Also, due to distribution of the stealing requests, there is no contention on a single node. However, master-slave has the disadvantage that it requires special hardware, where every node should be able to communicate with every other node. This facility may be unavailable in some situations. Also, in worst case its communication overhead can become very high.

Blumofe and Leiserson [8], show that using their work-stealing based scheduler, the expected time of executing a fully-strict computation with P processors is given by:

$$T = T_1/P + O(T_\infty) \qquad (1)$$

Here T1 is the minimum serial execution time of the computation and T shows the minimum execution-time with infinite number of processors. Further, the space required by the algorithm is S1P, where S1 is the minimum serial space required. Also, the expected total communication-cost of the algorithm is PT(1 + nd)* Smax, where Smax is the size of largest activation record of any thread, nd is the maximum number of times, a thread synchronizes with its parent [8]. Thus, work-stealing algorithm is efficient in terms of time, space and communication . Work-stealing is also known as task-stealing and random-stealing [39]. Work-stealing has been studied in context of various application contexts such as wide-area networks, distributed memory machines [39] and several variations of work-stealing have been proposed in the literature (e.g. [20, 21, 31, 38]). Further, work-stealing has also been used in achieving load-balancing in GPUs [45] and several programming languages designed for multithreaded parallel computing, such as Cilk [7].

Pezzi et al. propose a hierarchical work-stealing (HWS) algorithm for MPI platforms [34], which is an extension of single-master work-stealing algorithm.

HWS uses a hierarchical structure of managers (i.e. master) and workers (i.e. slaves), arranged in a binary tree structure. In this tree, the inner-nodes are all managers and the leaf nodes represent the worker nodes. In MPI platforms, processes can only communicate if they share an inter communicator and hence using N-processes will require collective synchronization between them. HWS aims to reduce this synchronization-cost by using managers which mediate and facilitate the task of communication between processes. This technique, however, has the limitation of requiring several extra manager nodes. Further, each work-stealing request is slowed-down, since it must go through manager-nodes. A variant of this is cluster aware hierarchical work-stealing [4, 5], which provisions that the processors be arranged in a specific topology (e.g.tree topology) to minimize the communication-cost. Dinan, Olivier et al implement work-stealing for MPI platforms [15]. To reduce the communication-overhead due to stealing (polling) requests, they provision a fixed time interval (called polling interval) which must be elapsed before a process examines and answers incoming stealing requests. They have observed that the length of the polling interval has a significant impact on the algorithm performance. For implementing distributed termination detection, a modified version of Dijkstra termination-detection algorithm is used.

Dinan, Larkins et al implement work-stealing for 8,192 processors with a distributed memory systems [14].The authors use Partitioned Global Address Space (PGAS) programming model, which provides a global view of physically distributed data and efficient one-sided access. The distributed task queues are stored in the global address space and thus, steal-operations proceed without interrupting the victim process. Each process maintains a local task-queue, which is split into public and private portions to enable non-blocking stealing. Regarding the decision of a suitable number of jobs which a thief should steal from victim process, the authors suggest that stealing half the number of jobs available in the public-queue of the victim gives better performance and scalability, than stealing only 1 or 2 jobs. For our work, we implement non-blocking version of work stealing algorithm [3], which is provably efficient [8]. This technique alleviates the disadvantages of previous techniques by allowing work stealing by other processes without the need to wait for other processes to steal work from their queue and hence this technique leads to minimum synchronization overhead. During work-stealing, a process with no remaining work (called free-process), requests other processes to find an unfinished work. When using a large number of processes, the order of stealing request becomes important. A naive method of polling might always poll in a fixed order, for example, polling process 0,1,2,.... This, however, is likely to create contention on the processes 0,1 etc., since multiple free-processes will poll processes 0 and 1 etc. To address this, we provision random polling.. Statistically, random-polling distributes the stealing-requests to all the processes equally.

2 Implementation

Even an initially balanced distribution of contingencies across the processors could become highly unbalanced with time. Therefore dynamic load balancing algorithms are employed to ensure minimum idle time for the processors and maximal resource utilization of each processor. We now discuss the algorithms for static and dynamic load balancing approaches.

2.1 Static Load Balancing

Algorithm 1 shows the pseudo code for the implementation of the static load balancing based parallel contingency analysis algorithm.

```
Input: A task-list T and a processor-list P
Output: A static allocation of tasks to the processors
while True do
    foreach processor p in P do
        if T is empty then
            break;
        end
        Remove a task t from T;
        Allocate t on p;
    end
end
```

Algorithm 1. The Static Scheduling Algorithm

2.2 Dynamic Load Balancing

Algorithm 2 and 3 below show the pseudo code for the dynamic load balancing based on the master-slave and the work stealing algorithm.

3 High Performance Computing Resources at Iowa State University (ISU)

ISU recently installed a supercomputer called Cystorm which consists of 400 dual quad core nodes with AMD processors distributed through 12 racks. Thus there are a total of 3200 processors with a peak performance of 15.7 TF and data storage capacity of 44TB. All the results reported here are obtained by simulation on Cystorm supercomputer.

```
Input: A task-list T and a slave processor-list S and a Master-node m
Output: A load-balanced (best-effort) allocation of tasks to processors
//Initialization;
foreach Slave-Processor s in S do
    if T is empty then
    |   break;
    end
    Remove a task t from T;
    Assign t to s;
end
For Master-node m ;                         For any slave-node s ;
while T is not empty do                     while there is a task to be run do
    Wait for task-request from a slave;         Finish the task;
    if a task-request comes from slave s        Request a task from m;
    then                                        if no task is available then
    |   Remove a task t from T ;                |   break;
    |   Allocate t to s;                        end
    end                                     end
end
```

Algorithm 2. The Master-slave Scheduling Algorithm

4 Test System

The test system is a large size test case with more than 13000 buses and 400 generators. The generators are represented by GENROU model, excitors are presented by the IEEET1 model and the governors are represented by TGOV1 model. The different contingencies are generated through bus faults on each of the bus and through a combination of bus fault followed by branch faults on connected branches and/or other branches. Other contingencies could be generated through generator tripping, bus fault followed by line fault and line tripping and so on. The simulation time of the individual contingency varies from 15 seconds to 25 seconds depending on the number of events (bus fault, bus fault followed by line fault, duration of fault) and the time to settle to steady state.

5 Results and Discussion

5.1 Validation and Comparison with PSS/e

The developed simulator is validated and compared against different commercial power simulator such as PSSE, DSA tools, PowerWorld as discussed in the previous chapter. This establishes the accuracy of the developed simulator and also gives insight into the computational gain that can be derived

```
Input: A task-list T and a processor-list P
Output: A load-balanced (best-effort) allocation of tasks to processors
//Initialization
while True do
    foreach processor p in P do
        if T is empty then
        |   break;
        end
        Remove a task t from T ;
        Allocate the task t to p;
    end
end
//Each Processor has two threads: Worker-thread and Polling-thread
For Worker-thread on processor p ;    For Polling-thread on processor p ;
while p has unfinished tasks do        while True do
    foreach unfinished task t at p do      if p' sends stealing request then
    |   Finish the task t ;                    if p has an unstarted task then
    end                                            Let t be such task ;
    foreach  p' in P-{p} do                        remove t from tasks of p ;
        try stealing a task from p' ;              return t to p' ;
        if stealing was successful then        else
            assign stolen task to p ;          |   return None to p' ;
            break;                             end
        end                                end
    end                                end
end
wait on a barrier for all processors;
terminate all threads and processors;
```

Algorithm 3. The Work-Stealing Scheduling Algorithm

with the choice of efficient numerical algorithms at all levels. Variable time strategy combined with efficient sparse solvers results in significant computational gain compared with PSSE which has explicit integrator with fixed time step.

5.2 Parallel Contingency Analysis

Simulations are performed on thousands of contingencies ranging from 10000 to 50000 contingencies on a large test system. As discussed earlier, even an initially balanced distribution of contingencies across the processors could become highly unbalanced with time. Therefore dynamic load balancing algorithms are employed to ensure minimum idle time for the processors and maximal resource utilization of each processor. Below the results are presented for the master-slave based dynamic load balancing and for work

Table 1. Master-Slave Scheduling: Simulation Time (seconds)

No. of Contingencies	P=8	P=16	P=32	P=64
10000	18408	8595	3954	1951
20000	35183	16369.5	7970.4	3919
30000	52789	24508	12082	5895
40000	69483	32138	15588.2	7663
50000	87631.3	41047	19856	9658

Table 2. Master-Slave Scheduling: Speed-up values Relative to $P=8$

No. of Contingencies	P=8	P=16	P=32	P=64
10000	1	2.15	4.65	9.43
20000	1	2.15	4.41	8.98
30000	1	2.15	4.37	8.95
40000	1	2.16	4.46	9.07
50000	1	2.13	4.41	9.07

stealing based dynamic load balancing for massively parallel contingency analysis on Cystorm.

Simulation results on master-slave based contingency analysis are presented in Table 1 and Table 2. Table 1 shows the simulation time in seconds taken to simulate contingencies (10000-50000) for different number of processors (from 8 to 64 processors). It can be noted from the simulation times that if all the contingencies were to be simulated on one processors, it would take days to do that. Thus use of HPC resources are highly recommended for future grid operations. Using HPC the total simulation time could be reduced from day to merely couple of hours as seen in Table 1 and Table 3.

Table 3. Work-Stealing Scheduling: Simulation Time (seconds)

No. of Contingencies	P=8	P=16	P=32	P=64
10000	15983	7989.3	3801	1903
20000	30594	15221.4	7657.3	3819
30000	45846.3	22832	11578	5723
40000	60052	29859	14971	7504
50000	75627	37691	18928	9454

From the simulation times in Table 1 and the scalability results in Table 2 as the number of processors increases it is obvious, that dynamic load balancing achieves excellent scalability with processors. However an important observation in Table 2 is that as the number of processors are multiplied by a

Table 4. Work-Stealing Scheduling: Speed-up values Relative to $P=8$

No. of Contingencies	$P=8$	$P=16$	$P=32$	$P=64$
10000	1	2	4.2	8.4
20000	1	2	4	8.01
30000	1	2.01	3.96	8.01
40000	1	2.01	4.01	8
50000	1	2.01	4	8

factor of 2, the speedup is more than multiplication factor. This is to be expected since in master-slave approach, as the number of processors increase, the effective efficiency of the working processors (number of slaves/total number of processors) increases.

Similarly Table 3 and Table 4 show the simulation times for the different number of contingencies and by different number of processors and the scalability with increase in number of processors, for work stealing based dynamic load balancing. The trend is similar as in the case of master-slave for the simulation times. However it is much more efficient computationally than master-slave based dynamic load balancing.

Comparing Table 1 and Table 3, we observe that for the same number of contingencies and processors, it takes much lesser time to simulate when work stealing based scheduling is used than with master-slave strategy. Moreover the amount of time savings increases linearly with increase in the number of contingencies for the same number of processors. This is a significant result.

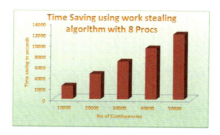

Fig. 15. Time-Saving for $P=8$

Figure 15 through Figure 18 show the amount of time savings for a certain number of processors as the number of contingencies increase. The amount of saving is more than 3 hours just with 50000 contingencies as shown in Fig. 18.

As can be seen from Figures 16, 17, 18 the percentage saving in time reduces by a factor of two as the number of processors increase by a factor of two.

Fig. 16. Time-Saving for P=16

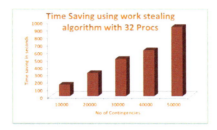

Fig. 17. Time-Saving for P=32

Fig. 18. Time-Saving for P=64

Thus as the number of processors double, same savings could be achieved by increasing the task list. Additionally, as the unbalance across the processors increase, more computational gain could be expected. Similarly with increase in individual tasks simulation time further imbalance and gain is expected as in the case of extended-term simulations which includes slow dynamics also.

6 Summary

A novel work stealing based dynamic load balancing implementation is proposed for massively parallel dynamic contingency analysis. The proposed algorithm is compared with master-slave based dynamic load balancing algorithm. A large number of contingencies (up to 50000) of a large realistic system are simulated. The results show excellent scalability of the proposed algorithm and huge amount of time saving compared with master-slave approach. The proposed algorithm combines the advantages of the static scheduling and master-slave approach and overcomes their defects. The huge computational saving could allow further exploration of the solution search space, and provides further aid to the operator to take appropriate action when required.

References

1. Agrawal, A., Misra, S., Honbo, D., Choudhary, A.: Mpipairwisestatsig: Parallel pairwise statistical significance estimation of local sequence alignment. In: Proceedings of the 19th ACM International Symposium on High Performance Distributed Computing, pp. 470–476. ACM (2010)
2. Agrawal, A., et al.: Parallel pairwise statistical significance estimation of local sequence alignment using message passing interface library. Concurrency and Computation: Practice and Experience (2011)
3. Arora, N., Blumofe, R., Plaxton, C.: Thread scheduling for multiprogrammed multiprocessors. In: Proceedings of the tenth Annual ACM Symposium on Parallel Algorithms and Architectures, pp. 119–129. ACM (1998)
4. Backschat, M., Pfaffinger, A., Zenger, C.: Economic-based dynamic load distribution in large workstation networks. In: Fraigniaud, P., Mignotte, A., Robert, Y., Bougé, L. (eds.) Euro-Par 1996. LNCS, vol. 1124, pp. 631–634. Springer, Heidelberg (1996)
5. Baldeschwieler, J., Blumofe, R., Brewer, E.: A tlas: an infrastructure for global computing. In: Proceedings of the 7th Workshop on ACM SIGOPS European Workshop: Systems Support for Worldwide Applications, pp. 165–172. ACM (1996)
6. Berger, N.: GPUs in experimental particle physics. Bulletin of the American Physical Society 57 (2012)
7. Blumofe, R., Joerg, C., Kuszmaul, B., Leiserson, C., Randall, K., Zhou, Y.: Cilk: An efficient multithreaded runtime system, vol. 30. ACM (1995)
8. Blumofe, R., Leiserson, C.: Scheduling multithreaded computations by work stealing. In: Proceedings. 35th Annual Symposium on Foundations of Computer Science, pp. 356–368. IEEE (1994)
9. Chen, Q., McCalley, J.: Operational defence of cascading sequences. In: 2011 IEEE Power and Energy Society General Meeting, pp. 1–8 (July 2011)
10. Chen, Y., Huang, Z., Chavarría-Miranda, D.: Performance evaluation of counter-based dynamic load balancing schemes for massive contingency analysis with different computing environments. In: 2010 IEEE Power and Energy Society General Meeting, pp. 1–6. IEEE (2010)

11. Collazuol, G., Lamanna, G., Pinzino, J., Sozzi, M.: Fast online triggering in high-energy physics experiments using GPUs. Nuclear Instruments and Methods in Physics Research Section A: Accelerators, Spectrometers, Detectors and Associated Equipment (2011)
12. Davis, T.: Algorithm 832: Umfpack v4. 3—an unsymmetric-pattern multifrontal method. ACM Transactions on Mathematical Software (TOMS) 30(2), 196–199 (2004)
13. Davis, T., Stanley, K.: Klu: a" clark kent" sparse lu factorization algorithm for circuit matrices. In: SIAM Conference on Parallel Processing for Scientific Computing, PP 2004 (2004)
14. Dinan, J., Larkins, D., Sadayappan, P., Krishnamoorthy, S., Nieplocha, J.: Scalable work stealing. In: Proceedings of the Conference on High Performance Computing Networking, Storage and Analysis, p. 53. ACM (2009)
15. Dinan, J., Olivier, S., Sabin, G., Prins, J., Sadayappan, P., Tseng, C.: Dynamic load balancing of unbalanced computations using message passing. In: IEEE International Parallel and Distributed Processing Symposium, IPDPS 2007, pp. 1–8. IEEE (2007)
16. Dobson, I., McCalley, J., Liu, C.: Fast simulation, monitoring and mitigation of cascading failure. Tech. rep., Power System Engineering Research Center (PSERC) (October 2010)
17. Gorton, I., Huang, Z., Chen, Y., Kalahar, B., Jin, S., Chavarria-Miranda, D., Baxter, D., Feo, J.: A high-performance hybrid computing approach to massive contingency analysis in the power grid. In: Fifth IEEE International Conference on e-Science 2009, pp. 277–283 (December 2009)
18. Green, R., Wang, L., Alam, M.: High performance computing for electric power systems: Applications and trends. In: 2011 IEEE Power and Energy Society General Meeting, pp. 1–8. IEEE (2011)
19. Green, R., Wang, L., Alam, M., Singh, C.: Intelligent and parallel state space pruning for power system reliability analysis using MPI on a multicore platform. In: 2011 IEEE PES Innovative Smart Grid Technologies (ISGT), pp. 1–8. IEEE (2011)
20. Guo, Y., Barik, R., Raman, R., Sarkar, V.: Work-first and help-first scheduling policies for async-finish task parallelism. In: IEEE International Symposium on Parallel & Distributed Processing, IPDPS 2009, pp. 1–12. IEEE (2009)
21. Hiraishi, T., Yasugi, M., Umatani, S., Yuasa, T.: Backtracking-based load balancing. In: ACM Sigplan Notices, vol. 44, pp. 55–64. ACM (2009)
22. Honbo, D., Agrawal, A., Choudhary, A.: Efficient pairwise statistical significance estimation using fpgas. In: Proceedings of BIOCOMP 2010, pp. 571–577 (2010)
23. Huang, Z., Chen, Y., Nieplocha, J.: Massive contingency analysis with high performance computing. In: IEEE Power and Energy Society General Meeting 2009. IEEE (July 2009)
24. Jacobsen, D., Thibault, J., Senocak, I.: An MPI-CUDA implementation for massively parallel incompressible flow computations on multi-GPU clusters. Mechanical and Biomedical Engineering Faculty Publications and Presentations, p. 5 (2010)
25. Jalili-Marandi, V., Zhou, Z., Dinavahi, V.: Large-scale transient stability simulation of electrical power systems on parallel GPUs. IEEE Transactions on Parallel and Distributed Systems 99, 1 (2011)

26. Khaitan, S., Fu, C., McCalley, J.: Fast parallelized algorithms for on-line extended-term dynamic cascading analysis. In: Power Systems Conference and Exposition, PSCE 2009. IEEE/PES, pp. 1–7. IEEE (2009)
27. Khaitan, S., McCalley, J.: A class of new preconditioners for linear solvers used in power system time-domain simulation. IEEE Transactions on Power Systems 25(4), 1835–1844 (2010)
28. Khaitan, S., McCalley, J., Chen, Q.: Multifrontal solver for online power system time-domain simulation. IEEE Transactions on Power Systems 23(4), 1727–1737 (2008)
29. Khaitan, S., McCalley, J., Raju, M.: Numerical methods for on-line power system load flow analysis. Energy Systems 1(3), 273–289 (2010)
30. Kurowski, K., Kulczewski, M., Dobski, M.: Parallel and GPU based strategies for selected cfd and climate modeling models. Information Technologies in Environmental Engineering, 735–747 (2011)
31. Michael, M., Vechev, M., Saraswat, V.: Idempotent work stealing. In: ACM Sigplan Notices, vol. 44, pp. 45–54. ACM (2009)
32. Mielikainen, J., Huang, B., Huang, H., Goldberg, M.: GPU acceleration of the updated goddard shortwave radiation scheme in the weather research and forecasting (wrf) model. IEEE J. Sel. Topics Appl. Earth Observ. Remote Sens. 5(2), 555–562 (2012)
33. Milshteyn, A., Alegre, A., Estrada, J., Lam, A., Beltran, S., Adigwu, J., Boussalis, H., Liu, C.: High-performance parallel processing aerospace information server. Journal of Next Generation Information Technology 1(3) (2010)
34. Pezzi, G., Cera, M., Mathias, E., Maillard, N.: On-line scheduling of MPI-2 programs with hierarchical work stealing. In: 19th International Symposium on Computer Architecture and High Performance Computing, SBAC-PAD 2007, pp. 247–254. IEEE (2007)
35. Schenk, O., Gärtner, K.: Solving unsymmetric sparse systems of linear equations with pardiso. Future Generation Computer Systems 20(3), 475–487 (2004)
36. Serban, R., Petra, C., Hindmarsh, A.C.: User documentation for IDAS v1.0.0 (2009), https://computation.llnl.gov/casc/sundials/description/description.html
37. Srinivasa, A., Sosonkina, M., Maris, P., Vary, J.: Dynamic adaptations in abinitio nuclear physics calculations on multicore computer architectures. In: 2011 IEEE International Symposium on Parallel and Distributed Processing Workshops and Phd Forum (IPDPSW), pp. 1332–1339. IEEE (2011)
38. Tzannes, A., Caragea, G., Barua, R., Vishkin, U.: Lazy binary-splitting: a runtime adaptive work-stealing scheduler. In: ACM SIGPLAN Notices, vol. 45, pp. 179–190. ACM (2010)
39. Van Nieuwpoort, R., Kielmann, T., Bal, H.: Efficient load balancing for widearea divide-and-conquer applications. In: ACM SIGPLAN Notices, vol. 36, pp. 34–43. ACM (2001)
40. Zhang, Y., Misra, S., Agrawal, A., Patwary, M., Liao, W., Qin, Z., Choudhary, A.: Accelerating pairwise statistical significance estimation for local alignment by harvesting GPU's power. BMC Bioinformatics 13(suppl. 5), S3 (2012)
41. Zhang, Y., et al.: Efficient pairwise statistical significance estimation for local sequence alignment using GPU. In: 2011 IEEE 1st International Conference on Computational Advances in Bio and Medical Sciences (ICCABS), pp. 226–231. IEEE (2011)

42. Zhang, Y., et al.: Accelerating pairwise statistical significance estimation using NUMA machine. Journal of Computational Information Systems 8(9), 3887–3894 (2012)
43. Zhang, Y., Patwary, M.M.A., Misra, S., Agrawal, A., Liao, W.K., Choudhary, A.: Enhancing parallelism of pairwise statistical significance estimation for local sequence alignment. In: 2nd HiPC Workshop on Hybrid Multi-Core Computing, WHMC 2011, pp. 1–8 (2011)
44. Zhang, Y., et al.: Par-psse: Software for pairwise statistical significance estimation in parallel for local sequence alignment. International Journal of Digital Content Technology and its Applications (JDCTA) 6(5), 200–208 (2012)
45. Zhou, K., Hou, Q., Ren, Z., Gong, M., Sun, X., Guo, B.: Renderants: interactive reyes rendering on GPUs. ACM Transactions on Graphics (TOG) 28, 155 (2009)

Reconfigurable Hardware Accelerators for Power Transmission System Computation

Prawat Nagvajara[1], Chika Nwankpa[1], and Jeremy Johnson[2]

[1] Electrical and Computer Engineering, Drexel University, Philadelphia, PA 19104, USA
nagvajara@ece.drexel.edu, nwankpa@ece.drexel.edu
[2] Computer Science, Drexel University, Philadelphia, PA 19104, USA
jjohnson@cs.drexel.edu

Abstract. This chapter reviews designs and prototypes of reconfigurable hardware implemented on a Field Programmable Gate Array (FPGA) to speedup ubiquitous linear algebra subroutines used in system security analysis. The grid operators use Energy Management System (EMS) software to analyze system security to assure normal operating state. EMS consists of three main computations: 1) state estimation, 2) contingency analysis, and 3) optimal power flow. These computations involve sparse linear algebra algorithms such as matrix orthogonal (QR) decomposition, Lower-Upper (LU) decomposition and matrix multiplication. Currently, EMS computations are performed on a general-purpose processor system. A benchmark study of several state-of-the-art sparse linear solver packages running on these systems reveals inefficient utilization of the floating-point computational throughput. A custom hardware sparse linear solver that maximizes floating-point hardware utilization based on pipeline architecture and efficient data caching offers an alternative. A prototype on reconfigurable hardware demonstrated that despite more than an order of magnitude deficit in clock speed as compared to general purpose processor based systems, a specialized sparse LU hardware running on FPGA is capable of an order of magnitude speedup relative to these systems for power system Jacobian matrix sparse LU decomposition. Performance analysis of sparse QR decomposition hardware showed a similar potential speedup over general-purpose processors.

1 Introduction

Reliable operation of electrical power grids depends on real-time system security analysis performed on high-performance computers. The grid operators use Energy Management System (EMS) software to analyze system security to assure normal operating state. Normal operating state means the system can supply power to the loads with the available generators and within the system constraints. EMS consists of three main computations: 1) state estimation, 2) contingency analysis, and 3) optimal power flow (OPF) [Bergen and Vittal 2000, Albur and Gomez-Exposito 2004]. These computations involve sparse linear algebra algorithms such as matrix orthogonal (QR) decomposition and Lower-Upper (LU) decomposition [Strang 2006].

EMS involves sparse matrix computation where the number of state variables is large, e.g., 19,825 variables in an electric power grid with 10,279 nodes. Solutions to these large-scale systems can be computed within a reasonably amount of time because the matrices involved have few non-zero elements relative to matrix size, e.g., the computation time with today's technology is 0.16 second for the 19,895×19,895 matrix having 0.035% of its elements as non-zero. Sparse matrix computation arises in operation, planning and market analysis for electrical power infrastructure where the solutions of nonlinear equations and optimization problems are computed by direct method linear solvers based on LU or QR decomposition.

Currently, EMS computations are performed on a general-purpose processor system. A benchmark study of several state-of-the-art sparse linear solver packages running on these systems reveals inefficient utilization of the floating-point computational throughput. This is due to the irregularity of the sparse matrix data access pattern. Thus, the processor's floating-point unit – pipeline architecture, is not efficiently processing data in all of its stages.

A custom hardware sparse linear solver, implemented on Field Programmable Gate Array (FPGA), which maximizes floating-point hardware utilization based on pipeline architecture and efficient data caching offers an alternative. A prototype on reconfigurable hardware demonstrated that despite more than an order of magnitude deficit in clock speed as compared to general purpose processor based systems, a specialized sparse LU hardware running on FPGA is capable of an order of magnitude speedup relative to these systems, for power system Jacobian matrix sparse LU decomposition. Performance analysis of sparse QR decomposition hardware showed a similar potential speedup over general-purpose processors.

The relevance of using FPGA custom hardware to accelerate sparse linear algebra routines is due to high percentages of the computing time spent on sparse linear solver routines within the EMS computations. In state estimation computation, QR decomposition makes up approximately 90% of the computing time. In contingency analysis, where the power flow is performed for each contingency, LU decomposition makes up approximately 88% of the computing time. In optimal power flow computation, LU decomposition makes up 40% of the computing time and matrix multiplication makes up 53% of the computing time.

This chapter reviews results on reconfigurable hardware accelerators, implemented on FPGA, for LU and QR decompositions used in EMS computation [Johnson et. al. 2005, Murach et. al. 2005, Vachranukunkiet 2007, Nagvajara et. al. 2007, Chagnon et. al. 2008, Nwankpa et. al 2008, Nwankpa et. al. 2009]. Section 2 describes sparse matrix computation and efficiency of general-purpose processor floating-point units when computing sparse LU and QR decompositions. The section also includes a discussion on the sparsity of the benchmark Jacobian matrices Section 3 reviews the EMS computation; state estimation, load flow and optimal power flow problems. Section 4 reviews the custom design of the sparse LU and QR hardware and summarizes the performances of the hardware. Section 5 concludes the chapter.

2 Sparse Matrix Computation in EMS

In this section the characteristic of sparse matrices involved in the power grid computations is first reviewed. Efficiency of the sparse solver software packages is discussed as the motivation in developing alternative custom reconfigurable hardware. The section closes with an overview of matrix preordering for reducing the computation complexity.

2.1 Sparse Matrices

For electrical power transmission grids, the state variables are voltages (phase angles and magnitudes) at the nodes (generators and substations). For practical grids, such as the mid-Atlantic grid, the number of nodes (or busses) in 2006 was on the order of three thousand busses. The sparse nature of the system equations (matrix) follows from the fact that the grid is a graph with most of the nodes having degree two or three. Table 1 [Johnson et. al. 2005] shows the statistics of the Jacobian matrices in the Newton-Raphson power flow of the benchmark grids. (Sources: PSS/E, Siemens Power Systems Simulator for Engineering, the 1648-bus and 7917-bus; and PJM Interconnect, a regional transmission authority, 10279-bus and 26829-bus.) The sparsity is the percentage of the number of non-zeros divided by the square of the matrix size.

Table 1. Benchmark Grids Sparse Matrices

System	Matrix Size	# of non zeros	Sparsity
1648 Bus	2,982	21,196	0.24%
7917 Bus	14,508	105,522	0.050%
10279 Bus	19,285	134,621	0.036%
26829 Bus	50,092	351,200	0.014%

The efficiency of a general-purpose processor floating-point unit when computing the LU decomposition of the benchmark Jacobian matrices, can be presented in terms of the relative FLoating-point OPeration per Second (FLOPS), i.e., the obtained FLOPS×100/peak-FLOP. LU decomposition software packages used in the efficiency study included UMFPACK (Un-Symmetric Multi-Frontal Package) [Davis and Duff 1997], SuperLU [Demmel et. al. 1999] and WSMP (Watson Sparse Matrix Package) [Gupta and Joshi 2001]. These packages use Level 2 and Level 3 BLAS (Basic Linear Algebra Subprograms) operations, which are optimized for the processor (Pentium 4, 2.6 GHz) floating-point unit and data cache. UMFPACK provided the best performance for the power grid matrices given in Table 1. Table 2 [Vachranukunkiet et. al. 2005] shows the UMFPACK Mega-FLOPS and the efficiency. The low floating-point hardware utilization is apparent, e.g., when solving the largest Jacobian matrix with 50,092 variables, the UMFPACK sparse LU solver utilized only 3.45% of the Pentium 4 peak FLOPS.

Table 2. General-purpose Processor Floating-point Hardware Utilization

System	UMFPACK MFLOPS	Efficiency
1648-Bus	27.42	1.05%
7917-Bus	34.69	1.33%
10278-Bus	24.9	0.96%
26829-Bus	89.74	3.45%

2.2 Sparse Matrix Ordering

Sparse system analysis almost always requires solutions of $Ax = b$ where A is an M×N sparse matrix, x is an N×1 state-variables vector and b is an M×1 constant vector. A direct solver based on LU decomposition is used when the number of equations M is equal to the number of variables N. For an over-determined system when M > N the least-square solutions are computed from the normal equations or by using orthogonal decomposition (Givens rotations or Householder reflections) [Strang 2006].

To reduce the number of floating-point operations and storage size, sparse matrices are pre-permuted (preordered) by column and/or rows prior to decomposition. Ordering techniques have been studied extensively [Yannakakis 1981, George and Liu 1989, Amestoy et. al. 1996, 14, 16, 24]. One of the commonly used ordering is the approximate minimal degree [George and Liu 1989]. The objective for optimal preordering is to reduce the number of new non-zero elements called "fill-ins" generated as the result of elimination in the decomposition algorithm, e.g., Gauss elimination in LU decomposition and Givens rotations in orthogonal decomposition.

Another type of analysis of sparse matrices is blocking, which is performed by permuting the matrix into loosely coupled blocks (denser blocks). The blocking allows parallelism where hardware units or processors in the cluster can compute different blocks in parallel and combine the partial solutions into the final solution.

Researchers and engineers have been using Bordered Block Diagonal (BBD) techniques [Koester et. al. 1995, 23] for triangular decomposition on cluster computers. The difficulty associated with BBD technique on clusters is communication latency overhead associated with the standard protocol such as the Message Passing Interface (MPI). Tu and Flueck [Tu and Flueck (2002)] applied BBD technique in LU decomposition for power flow calculation on cluster computers. The communication latency was reported to limit the parallelism scalability (e.g., at most 6 computers were used before the performance gain saturated). Wang and Ziavras [Wang and Ziavras (2003)] applied BBD for multiple embedded processors in Field Programmable Gate Array (FPGA) for power flow calculation. However, FPGA embedded processors are not high-performance architecture and do not have floating-point units and thus are not suited for cluster computing.

Figure 1 below shows non-zero element structures of a series of linear transformed 2,982×2,982 Jacobian matrices originally having 21,196 non-zero elements (nz) from power flow problem. The top left shows the non-zero structure of the original matrix. The top right shows the non-zero elements permuted with the BBD technique. The bottom left shows the BBD of the Approximate Minimum Degree (AMD) ordered matrix. The bottom right shows the addition (superimpose) of the Lower and Upper triangular matrices of the BBD AMD-ordered matrix. The number of fill-ins is approximately equal to the number of non-zero elements (nz) of the original matrix, however, without preordering the number of fill-ins can be an order of magnitude higher.

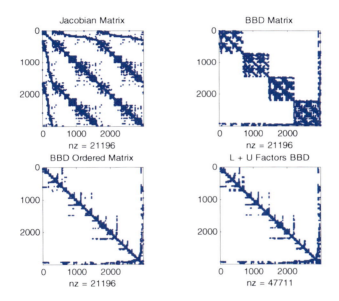

Fig. 1. Non-zero Structures of Sparse Matrix and its Transformation

3 Energy Management System Computation

System security relies on energy management system computations that includes i) the power flow and contingency analysis, ii) state estimation and (iii) optimal power flow. This section reviews these computations.

3.1 Power Flow Calculation and Contingency Analysis

Calculation of electric power flow between nodes in the electrical power grid involves solutions of sparse systems derived from the Newton-Raphson method for solving non-linear power equations. The method uses a direct solver based on LU factorization and forward/backward substitution. High-performance computing is utilized to compute the power flow calculation- the fundamental computation in the operation of the grid. Power flow analysis is a crucial computation to prevent voltage collapse and to maintain continuous power distribution. This requires performing a contingency analysis involving computations of the network equations for all possible cases of load or generator outages. Performing single contingency analysis on such a system requires evaluating the load flow for each of the possible contingency cases.

Power flow solution via Newton-Raphson method involves iterating the following equation:

$$-J \cdot \Delta x = f(x), \qquad (1)$$

until $f(x)$ is sufficiently small. The Jacobian, J of the power system is a sparse matrix (see Table 1), which while not symmetric has a symmetric pattern of non-zero elements. Δx is a vector of the change in the voltage magnitude and phase angle for the current iteration. And $f(x)$ is a vector representing the real and reactive power mismatch.

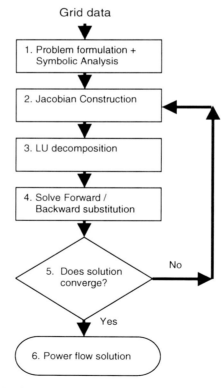

Fig. 2. Power Flow Calculation Flowchart

The above set of equations are of the form Ax = b, which can be solved using a direct linear-solver. Direct linear-solvers perform Gaussian elimination, which is performed by decomposing the matrix A into Lower (L) and Upper (U) triangular factors, followed by forward and backward substitution to solve for the unknowns. Fig. 2 shows the process used in computing power flow. This process generally repeats 4 to 10 times until a tolerance (usually 10^{-4} pu) is satisfied.

A breakdown of the computation time for the stages in the power flow calculation showed that on an average the LU decomposition (Stage 3) takes up 85% of the total time [Vachranukunkiet et. al. 2005] for the systems given in Table 1. The forward/backward substitution (Stage 4) and the Jacobian construction (Stage 2) take on the order of 10% and 5%, respectively. The problem formulation and symbolic analysis (sparse matrix ordering), Stage 1 takes relatively negligible percentage.

3.2 State Estimation

State estimation is an essential technique for monitoring, control and optimization of power system to achieve reliable power system operation. It determines the system state, which is the voltage magnitude and relative phase angle at each bus. State estimation uses redundant real-time measurements provided by the Supervisory Control and Data Acquisition (SCADA) system. These measurements typically are the power flow on the transmission lines, the real power and voltage on the generators, and the current magnitudes. Since the data provided by the SCADA system may not always be reliable due to errors in the measurement devices and telemetry failures, state estimator can filter out measurement errors and provides optimal statistical estimation of the system states. As the size of power systems increases numerical stability and computational efficiency become critical issues.

J. Widles first introduced power system static state estimation in 1968 [Wildes 1968)]. Different methods for solving power system state estimation had been proposed. Among these the normal equations, orthogonal transformation and hybrid method have been studied and compared [Holten et. al. 1988] where the fast decoupled version of normal equations was shown to have best performance (computing time) but less numerical stability. The orthogonal transformation method can overcome the numerical stability problem [Simoes-Costa and Quintana 1981, Gu et. al. 1983]. However, the calculation for the orthogonal matrix Q requires costly memory space and computational load. The hybrid method is a combination of the normal equations and orthogonal transformation and provides less numerical stability than orthogonal transformation method [Monticelli et. al. 1985] however does not require computing Q. It computes only the upper triangular matrix R and requires efficient computational load and storage space.

One of the most commonly used criteria for power system state estimation is the maximum likelihood [Wood and Wollenberg 1996]. The estimation is based on the measurement z with additive errors commonly assumed a Gaussian normal distribution having mean μ and variance σ^2 [Albur and Gomez-Exposito 2004].

The mean value μ_i is the expected value $E(z_i)$, which can be expressed as $h_i(x)$, a non-linear measurement function that relates the system state vector x to the ith measurement; the error variance σ_i^2 is called weighted factor W_{ii}. For m statistically independent measurements, the maximum likelihood estimation can be obtained by minimizing:

$$\sum_{i=1}^{m} \frac{[z_i - h_i(x)]^2}{W_{ii}} \qquad (2)$$

The solution x which minimizes (2) is called the weighted least-squares estimator for the state vector x.

The measurements h(x) commonly include the line power flows, bus power injections, bus voltage magnitudes. These measurements can be expressed as non-linear functions of voltage magnitudes and phase angels (the state variable vector x). If there are N buses (nodes) in the system, the number of state variables will be 2N-1, N bus voltage magnitudes and N-1 voltage phase angles (one bus is a reference bus with 0 radian phase angle

3.2.1 Normal Equation Formulation

Consider the set of measurements z, an m×1 vector; h(x), an m×1 expected measurement vector that is related to x, an n×1 state vector; and recall the objective function J(x) of the weighted least squares estimator (2):

$$J(x) = \sum_{i=1}^{m} (z_i - h_i(x))^2 / W_{ii} = [z - h(x)]^T W^{-1}[z - h(x)] \qquad (3)$$

where W is diagonal m by m matrix whose elements are the measurement error variances $\{\sigma_1^2, \sigma_2^2, ..., \sigma_m^2\}$.

To minimize the weighted least squares objective function, the first-order optimality conditions have to be satisfied:

$$g(x) = \frac{\partial J(x)}{\partial x} = -H^T(x)W^{-1}[z - h(x)] = 0 \qquad (4)$$

$$\text{where } H(x) = [\frac{\partial h(x)}{\partial x}]$$

The estimation is solved by an iterative method, which computes the correction Δx by solving normal equations:

$$[G(x^k)]\Delta x^{k+1} = H^T(x^k) \cdot W^{-1} \cdot [z - h(x^k)] \qquad (5)$$

$$\text{where } \Delta x^{k+1} = x^{k+1} - x^k$$

3.2.2 Orthogonal Transformation Formulation

The normal equations state estimation requires significantly more computations than those required by the power flow solution for the same network [Holten et. al. 1988] and it could cause numerical instabilities and ill-conditioning problem, and such situations will prohibit the state estimation solution from reaching an acceptable solution or even will cause divergence. To overcome ill-condition and obtain the numerical stable solution orthogonal transformation is applied.

Consider the normal equations:

$$H^T W^{-1} H \Delta x = H^T W^{-1} \Delta z \qquad (6)$$

where $\Delta z = z - h(x)$, weighting factor matrix $W^{-1} = W^{-1/2} W^{-1/2}$. Since W is the diagonal matrix, then $W^{-1/2} = [W^{-1/2}]^T$, thus the normal equations can be written as:

$$\hat{H}^T \hat{H} \Delta x = \hat{H}^T \Delta \hat{z} \qquad (7)$$

where $\hat{H} = W^{-1/2} H \quad \Delta \hat{z} = W^{-1/2} \Delta z$

The m×n matrix $\hat{H}$ can be decomposed into orthogonal matrix Q and upper triangular matrix R:

$$\hat{H} = QR \qquad (8)$$

where Q is an m×m orthogonal matrix, which has the property $Q^T Q = I$ (identical matrix), and R is an m×n matrix with first n rows are upper triangular matrix and m-n rows are null. By partitioning Q and R the reduced form of factorization is:

$$\hat{H} = [Q_n \quad Q] \begin{bmatrix} U \\ 0 \end{bmatrix} = Q_n U \qquad (9)$$

Thus, the equation (7) can be transformed as follows:

$$U \Delta x = Q_n^T \Delta \hat{z} \qquad (10)$$

where U is an n×n upper triangular matrix, Q_n^T is the first n rows of Q^T. State variable mismatch Δx can be solved by backward substitution.

3.2.3 Hybrid Method Formulation

Hybrid method [Albur and Gomez-Exposito 2004, Holten et. al. 1988, Tinney et. al. 1991] is a combination of normal equations method and orthogonal transformation. It performs orthogonal transformation on the left hand side of normal equations without changing its right hand side. Recalling Equation (7), instead of computing the Gain Matrix $\hat{H}^T\hat{H}$, the left hand side of the equation can be transformed into:

$$\hat{H}^T Q Q^T \hat{H} \Delta x = \hat{H}^T \Delta \hat{z} \tag{11}$$

Since $Q^T\hat{H} = R$. (11) becomes:

$$U^T U \Delta x = \hat{H}^T \Delta \hat{z} \tag{12}$$

Using forward and backward substitutions one solves for the state variable mismatch Δx.

In the hybrid method, it is not necessary to store orthogonal matrix Q, thus eases memory requirement, and the orthogonal transformation by obtaining only the upper triangular matrix requires less computation (faster).

In the hybrid method, solving the state variables consists of three steps: (i) Jacobian matrix $\hat{H}$ construction, (ii) orthogonal decomposition for upper triangular matrix U, and (iii) backward/forward substitutions for the solutions. A breakdown of the computational load showed that the orthogonal decomposition takes up approximately 90% of the entire execution time consequently improvement in computation performance for the upper triangular matrix U will most impact the overall performance thus orthogonal decomposition hardware can accelerate the computation [Nagvajara et. al. 2007].

3.3 *Optimal Power Flow*

To review, grid security analysis involves state estimation, contingency analysis and optimal power flow operation. To assure reliable operation of the power grid, the state estimation is the front-end real-time processing of the measurement data being transmitted from the SCADA system via network and telemetry. Based on the state estimation computation, if the state of the grid (system) is normal; a contingency analysis is performed to check the system status against a set of predetermined contingencies. If the system is found to be unstable (insecure) under certain contingency, a security constraint optimal power flow is performed to determine a correct dispatch of the generators. The changes in the dispatch also affect the electricity prices, which are different for the different locations on the grid. The computation for setting the prices is the security constraint economic dispatch, which is similar to the optimal power flow computation. Linear programming optimization is used to find an optimal dispatch that satisfies the security constraint at minimal cost. The interior-point method of solution for linear programming problem has been used. The calculation requires sparse matrix multiplication and triangular decomposition.

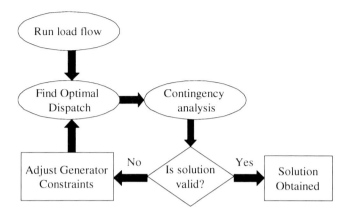

Fig. 3. Optimal Power Flow Computation

Optimal Power Flow (OPF) can be broken into two stages, solving the base case and running contingency analysis. The base solution determines the best generator dispatch for normal system operation. Contingency analysis ensures that the system will continue to operate under all single outage conditions. This is often referred to as Security Constrained OPF (SCOPF). A flow chart of the DC OPF is shown below in Fig. 3. If an optimal dispatch is invalid, a contingency causes additional outages; the generation constraints must be modified to ensure that additional outages will not occur. The base case is then recomputed and the process is repeated until the contingencies of interest are satisfied.

Solving the base case in DC optimal power flow (OPF) is generally accomplished by using linear programming (LP) techniques. The primal-dual interior point method (PDIPM) [Astfalk et. al. 1992] was chosen since it almost exclusively utilizes matrix operations. The base case problem for OPF can be constructed as follows:

List of terms for equations 13 through 16
c_i: Cost of power per MW at bus i
P_{gi}: Power dispatched by generator i;
S_{ki}: Shift factor relating the flow in branch k to generator i
T_k: Active load through branch k
P_{k_s}: Flow in branch k due to load profile;
L_i: Load at bus i
M: Number of busses

Objective:
$$\min \sum_{i=1}^{M} \left(c_i * P_{gi}\right)$$
such that

$$\sum_{i=1}^{M}\left(S_{gki} * P_{gi}\right) + T_k = P_k \qquad (13)$$

$$T_k^{\min} \le T_k \le T_k^{\max} \qquad (14)$$

$$P_{gi}^{\min} \le P_{gi} \le P_{gi}^{\max} \qquad (15)$$

$$\sum_{i=1}^{M} P_{gi} - \sum_{i=1}^{N} L_i - losses = 0 \qquad (16)$$

The objective function is to minimize the cost of running the system. The constraints for this problem are given by (13)-(16). The first constraint represents the line flow constraints, which arise from the admittances of the transmission lines. The second set of constraints represents transmission limitations on lines due to physical limitations such as thermal capacity. Thermal capacity is defined as the maximum allowable load that a line can deliver without undergoing permanent plastic deformation. The third set of constraints defines limits at which each generator can operate. The last constraint is based on the conservation of power in the system. The net flow injected must equal the sum of the power withdrawn and net losses in the transmission system.

In the PDIPM a sparse linear system given by the matrix ADA^T, where A is a rectangular matrix constructed from (13)-(16) and D is a diagonal matrix comprised of the current solution must be repeatedly solved. A breakdown of the OPF computation showed that 46% of the program's runtime is spent on LU calculations in which reconfigurable hardware accelerator can provide speedup [Murach et. al. 2005].

4 Reconfigurable Hardware for Sparse Linear Algebra

Energy management system involves sparse linear solvers based on LU and QR decomposition. Reconfigurable hardware, i.e., Field Programmable Gate Array (FPGA), can provide a cost-effective alternative architecture. This section reviews the architecture for the LU and QR hardware and the performance analysis.

4.1 Hardware Accelerator Architecture

Field Programmable Gate Array (FPGA) is a semiconductor device which contains programmable - logic components, interconnects, embedded static RAM and input/output blocks. FPGAs have become more suitable for scientific computation due to their unique features such as embedded multiplier, memory and processor. These features allow implementation of hardware for high-performance computing with floating-point operations. Unlike Application Specific Integrated Circuit (ASIC) which is a semi-custom process involving foundry and mass production, FPGA has advantages, such as shorter time to market, ability to reprogram to perform a different function and much lower cost. As their size, capabilities, and speed are increasing, FPGAs are increasingly being used for high performance computing applications. It is notable that, the embedded SRAM (Static RAM) blocks in FPGA allow the designer to create a specific memory hierarchy (data cache).

Sparse matrix decomposition FPGA hardware is connected to its host processor (computer) via interconnection data bus such as the PCI-Express bus. The host processor transfers data to and from hardware on FPGA. The data transfers include the transfer of Jacobian matrix to memory, e.g., DDR RAM (double data RAM), connected to the FPGA, and the transfer of the result decomposition matrices back to the host processor. Using forward and backward substitutions, the host processor software solves for the state variable mismatch Δx in (1) or (12). The amount of time spent in transferring the Jacobian matrices (see Table 1) and the result decomposition matrices between the host processor and FPGA hardware was relatively negligible compare to the total computation time.

The architecture of the hardware for sparse decomposition consists of memory unit, cache and processing unit [Johnson et. al. 2005, Nagvajara et. al. 2007, Nwankpa et. al. 2008, Chagnon 2008]. Fig. 4 shows a block diagram of the Givens-rotation QR decomposition. The host processor has access to the external memory SDRAM (Synchronous Dynamic RAM) via a data bus (not shown in the diagram). Sparse matrix data are transferred to SDRAM and the results are written back to the SDRAM.

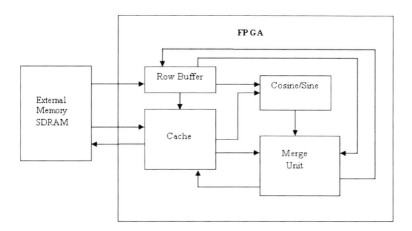

Fig. 4. Sparse Matrix QR Decomposition Hardware Architecture

The QR hardware on the FPGA contains cache and row buffer, which are responsible for caching data to be processed by the floating-point arithmetical units in a stream fashion. The cosine/sine and merge units are pipeline architecture such that the data processing rate is equal or close to the hardware clock frequency. Floating-point hardware is a pipeline architecture where the number of stages is on the order of 10 clock cycles. The cosine/sine and merge unit produces result every clock cycle.

Generally, high-performance computing relies on cache in keeping the pipeline hardware close to 100% utilization. Both sparse LU and QR decomposition hardware on FPGA include optimized cache design based on the access patterns of the sparse matrix rows.

The QR decomposition arithmetical hardware includes the cosine/sine unit and the merge unit. The hardware blocks (cores) consist of floating-point adder, multiplier and square root. To annihilate a_{ji} (the element (j, i) in Row j), the merge unit computes

$$\text{Row i} \leftarrow \cos\alpha * \text{Row i} + \sin\alpha * \text{Row j} \qquad (17)$$

$$\text{Row j} \leftarrow -\sin\alpha * \text{Row i} + \cos\alpha * \text{Row j} \qquad (18)$$

where

$$\cos\alpha = \frac{a_{ii}}{\sqrt{a_{ii}^2 + a_{ji}^2}} \qquad \sin\alpha = \frac{a_{ji}}{\sqrt{a_{ii}^2 + a_{ji}^2}} \qquad (19)$$

The assignments (17) and (18) are equivalent to left multiplying the matrix under transformation by an n×n Givens rotation matrix, consists of sinα and -sinα at row-column locations (i, j) and (j, i); cosα at locations (i, i) and (j, j); and all other diagonal elements are one; and the rest of the elements are zero.

Sparse matrices are stored in a compressed format. With the row-compressed format the locations of row data are stored as an array (array of row pointers). Each row is a variable-length array of pairs. A pair comprises the value of non-zero elements and its column position (number). The pairs are sorted in an increasing column numbers. This computation is a serial process in which pipeline hardware is efficient.

4.2 *Performance Analysis*

This subsection reviews the performance of the FPGA-based hardware for LU and QR decompositions applied to power flow, state estimation and optimal power flow [Johnson et. al. 2005, Nagvajara et. al. 2007, Nwankpa et. al. 2008, Chagnon 2008].

A prototype of LU hardware was developed on FPGA board. Figure 5 shows a performance of the LU hardware prototype for 1 and 2 merge units and the projected performance based on a model for hardware with more than 2 merge units. The system Jacobian matrices were those given in Table 1. Performance models are software programs simulating the hardware to determine the computation times (clock cycle counts). The clock cycle counts obtained from the model of the LU hardware were compared against the actual counts obtained from the prototype. The model accurately predicted the clock cycle when compared to the prototype. The performance gains shown in Fig. 5 were compared against those obtained from the UMFPACK sparse LU solver running on Pentium 4. An order of magnitude performance gains indicates a practical alternative when using 250 MHz clock-frequency reconfigurable hardware comprising 4 merge units.

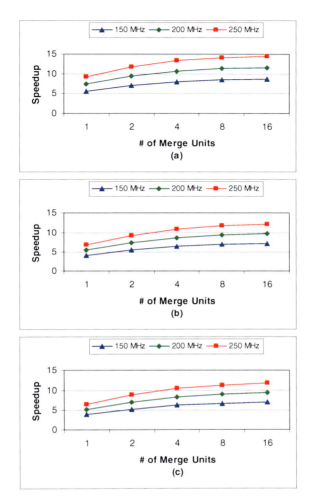

Fig. 5. Performance Gain over General-purpose Processor on LU Decomposition on (a) 1,648-Bus (b) 7,917-Bus (c) 10,278-Bus System

For the DC Optimal Power Flow (OPF) problem, Fig. 6 shows the performance gains of the LU hardware (200 MHz) over UMFPACK sparse LU solver running on Pentium 4 (P4), with ADA^T matrices arising in DC OPF where A is a rectangular matrix constructed from (13)-(16) and D is a diagonal matrix comprised of the current solution. Software program for OPF calculation was developed to obtain the matrix data. Correctness of the linear solver portion was verified using lp_solve [Berkelaar 1995]. Note that the speedup decreases for the larger system (7,917-bus) due to an increase in the number of non-zero elements of the matrix ADA^T, which the UMFPACK performs more efficiently.

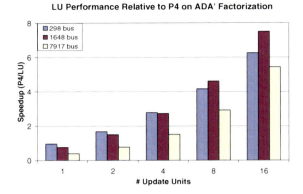

Fig. 6. LU Decomposition Performance Gain in OPF Calculation

Based on the performance model, a speedup of 2.5 is achievable with 4 update (merge) units and a speedup of 6 is possible with 16 units. The 298-bus system is an IEEE benchmark. The results indicate that the LU hardware is useful for smaller systems in linear programming, since these matrices are considerably less sparse than the Jacobian matrices used in power flow.

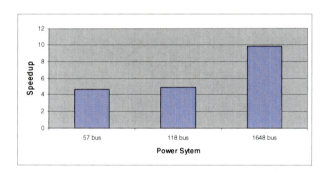

Fig. 7. QR Hardware Performance Gain in State Estimation Calculation

In the case of state estimation, a performance model for Givens rotation QR decomposition hardware was developed. Figure 7 shows the performance gain for the row-oriented Givens rotation FPGA hardware over sparse QR software running on Pentium 4. The software and hardware implemented the same algorithm. For the 57-bus and 118-bus systems (IEEE benchmark systems), the row-oriented hardware can achieve approximately 5 times speedup over the software design; for a larger system, such as 1648 bus system, the speedup is approximately 10 fold.

5 Concluding Remarks

Reconfigurable hardware provides a cost-effective platform for practical applications such as power transmission systems computation. The main advantage is that it can be customized for specific algorithms. Custom hardware is particularly useful for sparse linear algebra routines, due to the limitations of general-purpose processors in the optimization of their floating-point unit's peak performance. By utilizing reconfigurable hardware such as FPGA, data caching and pipeline architecture arithmetical units can be optimized. More specifically, it has been found that by utilizing FPGA pipeline architecture, performance gain of at least an order of magnitude or more, in comparison to general-purpose processor (in terms of lower clock frequency) can be realized. Future research should focus on reducing the complexity of the custom design in order to make the approach more ubiquitous.

References

Astfalk, G., Lustig, I., Marsten, R., Shanno, D.: The Interior-Point Method for Linear Programming. IEEE Software 9(4), 61–68 (1992)

Albur, A., Gomez-Exposito, A.: Power System State Estimation: Theory and Implementation. Marcel Dekker, New York (2004)

Amestoy, P., Davis, T.A., Duff, I.S.: An approximate minimum degree ordering algorithm. SIAM J. Matrix Anal. Applic. 17, 886–905 (1996)

Bergen, A.R., Vittal, V.: Power Systems Analysis. Prentice Hall, Upper Saddle River (2000)

Berkelaar, M.: Lp_solve Mixed Integer Linear Program Solver (1995), ftp://ftp.es.ele.tue.nl/pub/lp_solve/lp_solve

Davis, T.A., Duff, I.S.: An Unsymmetric-Pattern Multifrontal Method for Sparse LU Factorization. SIAM Journal on Matrix Analysis and Applications 18(1), 140–158 (1997)

Demmel, J.W., Eisenstat, S.C., Gilbert, J.R., Li, X.S., Liu, J.W.H.: A Supernodal Approach to Sparse Partial Pivoting. SIAM J. Matrix Analysis and Applications 20(3), 720–755 (1999)

Chagnon, T., Johnson, J., Vachranukunkiet, P., Nagvajara, P., Nwankpa, C.: Sparse LU Decomposition Using FPGA. In: PARA 2008: International Workshop on State-of-the-Art in Scientific and Parallel Computing, Trondheim, Norway (2008)

George, J.A., Liu, J.W.H.: Evolution of the Minimum Degree Ordering Algorithm. SIAM Review 31(1), 1–19 (1989)

Gu, J.W., Clements, K.A., Krumpholz, G.R., Davis, P.W.: The solution of ill-conditioned power system state estimation problem via the method of Peters and Wilkinson. In: Pica Conference Proceedings, Houston, TX (1983)

Gupta, A., Joshi, M.: WSMP: A High-Performance Shared- and Distributed-Memory Parallel Sparse Linear Equations Solver. IBM Research Report RC 22038 (98932) (2001)

Holten, L., Andrers Gjelsvik, A., Sverre Aam, S., Felix, F., Wu, F.F., Liu, W.-H.E.: Comparison of Different Methods For State Estimation. IEEE Transactions on Power Systems 3(4) (1988)

Johnson, J., Vachranukunkiet, P., Nagvajara, P., Tiwari, S., Nwankpa, C.: Performance Analysis of Load Flow on FPGA. In: 15th Power Systems Computational Conference, Liege, Belgium (2005)

Koester, D.P., Ranka, S., Fox, G.C.: Parallel Direct Methods for Block-Diagonal-Bordered Sparse Matrices. NPAC Technical Report, SCCS 679 (1995)

Matstoms, P.: Sparse QR Factorization in MATLAB. AMC Trans. Math. Software 20, 136–159 (1994)

Monticelli, A., Murari, C.A.F., Wu, F.F.: A Hybrid State Estimator: Solving Normal Equations By Orthogonal Transformations. IEEE Transactions on Power Apparatus and Systems PAS-104(12) (1985)

Murach, M., Nagvajara, P., Johnson, J., Nwankpa, C.: Optimal Power Flow Utilizing FPGA Technology. North American Power Symposium, Ames, IA (2005)

Nagvajara, P., Lin, Z., Nwankpa, C., Johnson, J.: State Estimation Using Sparse Givens Rotation Field Programmable Gate Array. IEEE North America Power System Symposium, Las Cruces, NM (2007)

Nwankpa, C., Johnson, J., Nagvajara, P., Lin, Z., Murach, M., Vachranukunkiet, P.: Special purpose hardware for power system computation. Power System Engineering General Meeting 2008, Pittsburgh, PA (2008)

Nwankpa, C., Johnson, J., Nagvajara, P., Chagnon, T., Vachranukunkiet, P.: FPGA Hardware Results for Power System Computation. In: IEEE Power Systems Conf. and Exposition, Seattle, WA (2009)

Vachranukunkiet, P.: Power Flow Computation Using FPGA. Ph.D. Dissertation, Drexel University, Philadelphia PA (2007)

Simoes-Costa, A., Quintana, V.H.: A robust numerical technique for power system state estimation. IEEE Transactions on Power Apparatus and Systems 100, 691–698 (1981)

Soman, S.A., Pandian, A., Parthasarathy, K.: Towards Faster Givens Rotations Based Power System State Estimator. IEEE Transactions on Power Systems 14(3) (1999)

Strang, G.: Linear algebra and its applications. Thomson-Brooks/Cole Pacific Grove, CA (2006)

Tinney, W.F., Vempati, N., Slutsker, I.W.: Enhancements To Givens Rotations For Power System State Estimation. IEEE Transactions on Power Systems 6(2) (1991)

Tu, F., Flueck, A.J.: A message-passing distributed-memory parallel power flow algorithm. IEEE Power Engineering Society Winter Meeting, New York, NY (2002)

Wang, X., Ziavras, S.: Parallel Direct Solution of Linear Equations on FPGA-Based Machines. In: IPDPS 2003 Proceedings of the 17th International Symposium on Parallel and Distributed Processing, Nice, France (2003)

Wildes, J.: Static State Estimation for a Power System Network. Power System Engineering Group, Department of Electrical Engineering, Massachusetts Institute of Technology, Cambridge, Rept. 8 (1968)

Wood, J.A., Wollenberg, B.F.: Power Generation, Operation and Control. John Wiley & Sons, New York (1996)

Yannakakis, M.: Computing the minimum fill-in is NP-Complete. SIAM J. Alg. Disc. Meth. 2(1), 77–79 (1981)

Polynomial Preconditioning of Power System Matrices with Graphics Processing Units

Amirhassan Asgari Kamiabad and Joseph Euzebe Tate

Department of Electrical and Computer Engineering,
University of Toronto, Canada
`amirhassan.asgarikamiabad@utoronto.ca`
`zeb.tate@utoronto.ca`

Abstract. Programmable graphics processing units (GPUs) currently offer the best ratio of floating point computational throughput to price for commodity processors, outdistancing same-generation CPUs by an order of magnitude, which has in turn led to their widespread adoption in a variety of computationally demanding fields. Adapting power system simulations to these processors is complicated by the unique hardware architecture of GPUs, which precludes the usage of direct linear system solvers based on Gaussian elimination. Krylov subspace methods are better suited to the GPU architecture, yet the ill-conditioned nature of power system matrices requires substantial preconditioning to ensure robustness of these methods. To reduce the time spent on preconditioning, we have developed a GPU-based preconditioner designed specifically to handle the large, sparse matrices typically encountered in power system simulations. The preconditioning technique used, based on Chebyshev polynomials, is described in detail, as are the design decisions made in adapting this algorithm to the GPU. Evaluation of the performance of the GPU-based preconditioner on a variety of sparse matrices, ranging in size from 30 x 30 to 3948 x 3948, shows significant computational savings relative to a CPU-based implementation of the same preconditioner and a typical incomplete LU (ILU) preconditioner.

1 Introduction

The computational demands associated with power system simulations have been steadily increasing in recent decades due to a variety of factors, including: a consolidation of planning and control activities from individual control areas to large, regional organizations (such as MISO and PJM) [Andersson et al. 2005], an increased focus on wide-area system behavior, due in large part to the recommendations of the August 2003 blackout report [NERC Steering Group 2004]; an increase in the complexity of contingency analyses required by newer reliability standards such as NERC TPL-003 [NERC Board of Trustees 2005]; and more sophisticated commitment and real-time dispatch techniques arising from advanced market designs [Crow 2003, Shahidehpour and Wang 2003]. To keep up with these new computational challenges, it is imperative that power system simulation software is capable of taking advantage of the latest advances in computer hardware. Since 2003, the increase of clock frequency and the productive computation in each clock cycle within a single CPU have slowed down. The "power wall", which refers to a limit on clock frequencies due to an inability to handle corresponding increases in leakage heat, is the primary reason for the stall in processor clock speeds and represents a practical limit on serial computation [Kirk and Hwu 2010]. In order to increase computational throughput despite an inability to increase clock speeds, chip manufacturers have focused on bundling several processors together into one chip. In some cases, the number of cores put onto a single chip has been quite modest (e.g., the Intel Core2 Duo has only two processors on-chip), yet in other cases, most prominently in modern graphics processing units (GPUs), the trend has been towards an ever-increasing number of cores. By putting hundreds of processors within a single chip, GPU manufacturers have managed to move beyond the "power wall" and reach computational speeds that are well beyond those of same-generation CPUs. For example, the latest GPUs are capable of processing over 1 trillion floating point operations per second (teraflops or Tflops) on boards costing less than $500, whereas a similarly priced two- or four-core CPU has a peak computational throughput of 80 billion floating point operations per second (Gflops) [Kirk and Hwu 2010]. Research into the utilization of GPUs for simulation in a variety of fields, including power systems, has been primarily driven by the steadily increasing gap between the peak computational throughput of CPUs and GPUs.

In power system analysis tools ranging from dc power flow to transient stability, solution of the linear systems arising from the Newton-Raphson algorithm (and its derivatives) is often the most computationally demanding component [Koester et al. 1992 and Anupindi et al. 1993]. Two broad classes of linear system solvers have been used in power systems-direct solvers (of which the standard LU decomposition is the most prevalent) and iterative solvers. Within the class of iterative solvers, techniques include Gauss Jacobi and Gauss Seidel [Meijerink and van der Vorst 1977] and Krylov subspace methods such as the conjugate gradient (CG) and generalized minimum residual (GMRES) techniques. Krylov subspace methods have been used extensively in other disciplines where large, sparse linear systems are solved in parallel [Saad 2003] and can enable power system applications to fully utilize the current and next generation of parallel processors. While there has already been considerable research within the power systems community on Krylov subspace methods [Decker et al. 1996, Flueck and Chiang 1998], their widespread use in the power systems field has been limited by the highly ill-conditioned nature of the linear systems that arise in power system simulations [Dag and Semlyen 2003]. To combat this problem, recent research efforts have aimed at developing suitable preconditioners for power system matrices in order to improve the performance of Krylov subspace methods [Dag and Semlyen 2003, Mori et al. 1996, de Leon 2003].

This work presents the first preconditioner for use in power system applications that is designed for the massively parallel hardware architecture of modern GPUs. Several attempts have been made to utilize GPUs within the power systems field in the past few years, for both visualization [Tate and Overbye 2008] and simulation purposes [Gopal et al. 2007]. In [Gopal et al. 2007], a dc power flow solver based on the Gauss Jacobi and Gauss Seidel algorithms was implemented on an NVIDIA 7800 GTX chip using OpenGL (an API designed primarily for graphics rendering) to carry out the iterations on the GPU. The Gauss Seidel and Gauss Jacobi algorithms are rarely used in modern power system software due to their inability to converge in many situations (e.g., for heavily loaded systems, systems with negative reactances, and systems with poorly placed slack buses [Tinney and Hart 1967]), which limits this solver's utility. Implementations of more general linear solvers on GPUs have been developed [Bolz et al. 2005, Galoppo et al. 2005], yet these solvers either fail to account for ill-conditioned coefficient matrices or fail to take advantage of the matrix sparsity.

In the remainder of this chapter, an implementation of the Chebyshev polynomial preconditioner on the NVIDIA CUDA family of GPUs is described and performance data is provided for matrices ranging in size from 30 x 30 to 3948 x 3948. Section 2 presents an overview of general computing on GPUs, with specific details given regarding the target architecture of our preconditioner implementation; section 3 gives an overview of Krylov subspace methods and the Chebyshev preconditioner; section 4 details the implementation of the Chebyshev preconditioner on the GPU; section 5 presents numerical evaluation of the preconditioner and discussion of the results; and section 6 offers conclusions and avenues for future work related to the broader use of GPUs for power system simulation. The studies demonstrate that the Chebyshev preconditioner is an effective alternative to the traditional ILU preconditioner and can provide significant computational savings for preconditioning of sparse matrices.

2 General Purpose Computing on GPUs

Since 2003, GPUs from both NVIDIA and AMD/ATI have led the commodity processor market in floating-point performance due to a combination of hundreds of data-parallel cores and high memory bandwidth. On the NVIDIA GTX 280 processor used for our evaluations, there are 240 individual cores, known as stream processors (SPs). The SPs are organized into streaming multiprocessors (SMs) [NVIDIA 2008], which consists of 8 SPs, 64KB of register space, 16KB of low-latency memory (similar to the L1 cache of a CPU), a thread scheduler, and special function units for transcendental calculations. Within each SM, SPs are scheduled to run programs based using the single-program-multiple-data (SPMD) parallel processing paradigm [Kirk and Hwu 2010]. Programs executing on an SM can also access the much larger space of global memory (typically around 1GB), but this usually incurs a substantial latency penalty when compared to shared memory access. For the GTX 280, the latency in accessing global memory is approximately 500 clock cycles and less than 3 clock cycles for the per-SM shared memory.

Until 2006, programmers were forced to use graphics APIs such as OpenGL and DirectX, combined with custom fragment and vertex shaders written in proprietary languages such as Cg and GLSL, in order to carry out general processing tasks on GPUs. With the introduction of NVIDIA's CUDA [NVIDIA 2008], ATI's Stream [AMD 2009], and the cross-platform OpenCL [Munshi 2009] languages, programming GPUs has become increasingly similar to traditional C programming. The CUDA programming language used in our implementation includes a set of C language extensions and API function calls that enable the programming of NVIDIA'S CUDA-capable GPUs. The CUDA programming model consists of host (i.e., CPU) and device (i.e., GPU) executables (referred to as "kernels" in the CUDA programming model) that run on the CPU and GPU in parallel or sequentially depending on whether blocking or non-blocking calls are used to execute kernels on the GPU. In applications where the GPU does most of the computations, the host code is primarily used to initialize the GPU, copy data to and from GPU memory, and initiate execution of kernels on the GPU. The host code to perform these operations is straightforward and can be found in many references (e.g., [Kirk and Hwu 2010]), therefore the remainder of this chapter focuses on the development of the GPU kernels.

In the CUDA programming model, all threads associated with a particular kernel are allocated to a thread pool, known as a "grid" in CUDA nomenclature. The grid is subdivided into a set of "thread blocks", where each block has the same number of threads. The organization of threads into thread blocks is a key design choice, since all thread blocks must be able to run completely independently of one another without any restrictions on the order of execution. All threads within a block will execute on the same SM and have access to the same 16KB of shared, low-latency memory that is accessed through 16, 1KB banks. If the same bank is accessed by more than one thread simultaneously, a "bank conflict" occurs and the memory access becomes serial rather than parallel. Each SM can only process 32 threads (known as a "warp") at one time.

One important aspect of programming for the GPU is that full utilization of the GPU requires the number of thread blocks to be much larger than the number of processors, since this provides the GPU's scheduler with greater flexibility in hiding memory access latencies. Accordingly, developing programs that create a large number of compute-intensive threads is a key to ensuring that the individual processors on the GPU are fully utilized.

In certain applications, GPU implementations have yielded several orders of magnitude acceleration over conventional CPU implementations [Pharr and Fernando 2005]. While this acceleration potential is encouraging, attaining maximum computational efficiency for a particular algorithm depends on many factors. The most significant factors are: how much of the algorithm consists of steps which can be executed in parallel, how well the implementation maps to the GPU's unique combination of shared and global memory, and how well the organization of thread blocks supports the continuous utilization of the individual cores on the GPU. LU decomposition, the most commonly used direct solver for power systems, exhibits several troublesome issues with respect to GPU portability. First, LU decomposition needs the results of prior row eliminations to eliminate the current row, which is an inherently serial process. In addition, as the algorithm executes, significant inter-thread-block data transfer is likely to be needed (for example, if one thread block is assigned to each row or column) which is done through high-latency global memory. Another problem with adapting LU decomposition to the GPU is that the active region of decomposition shrinks as the algorithm goes on (i.e., the number of columns involved in a row elimination is reduced at each step of the factorization). The reduction in the active region makes it difficult to schedule a sufficient number of threads on the GPU. Pivoting, which is one of the main ways in which different LU decomposition algorithms are differentiated, also presents challenges for GPU implementation. On the CPU, the most computationally efficient way to swap rows is to just swap the pointers which determine the beginning of each row. This technique will result in poor performance on the GPU, since it requires independent access to each row's data and a large amount of global memory manipulation. The final step of solving a linear system with LU decomposition is forward and backward elimination, which are inherently sequential algorithms. Finally, if the factorization results in a large number of fill-ins, storing the LU factors in GPU memory can result in significant global memory overhead. Next, we present iterative algorithms for solving linear systems as an alternative to direct methods.

3 Krylov Subspace Methods and Preconditioning Techniques

Strong convergence characteristics have made the Newton-Raphson (NR) method and its derivatives (such as the fast decoupled power flow) the prevailing option for solving the nonlinear set of equations encountered in power systems analyses. The key to the standard NR method is its ability to transform the solution of a nonlinear set of equations into a sequence of linear systems. Krylov subspace methods are used to solve linear systems, such as those encountered during NR iterations, as shown in (1), where A is square, sparse (i.e., a significant computational advantage can be obtained by explicitly accounting for the number of nonzero entries [Chen 2005]), and has N columns.

$$Ax = b \tag{1}$$

In most power system software, the linear system in (1) is solved with direct methods that include optimal reordering and pivoting to reduce fill-ins and sensitivity to intermediate calculation errors. While LU factorization has been heavily optimized for use in power systems, the time to execute a single NR remains limited by the time to execute the matrix factorization step [Asgari and Tate 2009]. Therefore, a speed-up in the solution of (1) can immediately improve many of the tools that power engineers and researchers use on a daily basis. This work focuses on improving the efficiency of linear system solvers by implementing a GPU-accelerated preconditioner designed to improve iterative solvers in power system.

Iterative methods of solving (1), such as conjugate gradient, determine at each iteration k the update vector $\Delta x^{(k)}$ based on knowledge of the current and prior residual vectors (i.e., the set $\{b - Ax^{(j)}, j = 1, 2, \dots k\}$). Two iterative linear system solvers that are covered in many undergraduate textbooks (such as [Glover et al. 2007]) are the Gauss and Gauss Jacobi methods. These methods were heavily used before the reduction in processor and memory costs that enabled the widespread adoption of NR. The reason for this switch is that both the Gauss and Gauss Jacobi methods have trouble with convergence, particularly when dealing with power system matrices [Wood 1967]. Krylov subspace methods are another class of iterative solvers that have been studied extensively in a variety of fields, including power systems [Decker et al. 1996, Flueck and Chiang 1998, Dag and Semlyen 2003, Mori et al. 1996, de Leon 2003]. In Krylov subspace methods, each iteration involves an incremental addition to an orthogonal basis (the coefficient matrix raised to a power times the initial residual) and the solution is iteratively improved by minimizing the residual over the basis formed [Saad 2003].

One drawback of Krylov subspace methods is their strong dependence on the distribution of the eigenvalues of the coefficient matrix [Benzi 2002]. The condition number of the matrix is often used to quantify the eigenvalue spread of a matrix, and it is defined as the ratio of the maximum and minimum singular values of the matrix [Campos and Rollett 1995]. If the condition number is considerably more than unity, the eigenvalues of the coefficient matrix are widespread and the system is said to be ill-conditioned. On the other hand, if the condition number is close or equal to one, the matrix's eigenvalues are clustered tightly together and the system is well-conditioned. If the condition number is equal to one, the coefficient matrix is equal to a scalar multiple of the identity matrix and many iterative linear solvers can then converge in a single iteration [Benzi 2002].

Previous studies on the application of iterative linear solvers in power systems (and the results given below in section 5) have shown that the coefficient matrices obtained during power system analyses are often ill-conditioned [Asgari and Tate 2009]. To enable iterative solvers to still be used with these matrices, preconditioners, which are algorithms designed to improve clustering of the coefficient matrix eigenvalues, are employed. The ideal preconditioner is the inverse of the coefficient matrix, since this would result in a condition number of one; however, in practical preconditioners, the linear system is multiplied by an approximate inverse since calculation of the exact inverse would be too costly in both computation and memory utilization. After preconditioning, the subsequent iterative solver should be capable of solving the linear system in fewer iterations. Even more importantly, for linear systems where the coefficient matrix is very ill-conditioned, the iterative solver usually fails to converge without preconditioning; therefore, preconditioning is a necessary step to ensure that the iterative solution process is robust.

The most common preconditioning algorithm used with iterative solvers, incomplete LU (ILU) decomposition, has the same issues with GPU-based execution as LU decomposition, described above in section 2. Therefore, to enable GPU-accelerated preconditioning, alternative algorithms must be considered. The simplest method of preconditioning is diagonal scaling of the coefficient matrix. Since the scaling process needs only one multiplication by a diagonal matrix, it does not add a significant load to the whole process; however, it can greatly increase the clustering of the eigenvalues of the coefficient matrix [Campos and Rollett 1995]. This effect is studied in section 5.

The Chebyshev algorithm is an iterative method originally developed for approximating the inverse of a scalar number. Studies performed in [Dag and Semlyen 2003] have shown that the matrix-valued Chebyshev method can be used as an alternative preconditioning technique. In this method, Chebyshev polynomials are recursively calculated with the coefficient matrix of the linear system taken as the argument. As the number of iterations is increased, the linear combination of Chebyshev polynomials converges to the inverse of the coefficient matrix [Dag and Semlyen 2003]. The steps in the Chebyshev preconditioning process are described by the following equations [Axelsson 1996]:

$$D = \text{diag}(A), Z = \frac{2}{\beta - \alpha} A D^{-1} - I \qquad (2)$$

$$T_0 = I, T_1 = Z, M_0 = \frac{c_0}{2} I, T_k = 2ZT_{k-1} - T_{k-2} \qquad (3)$$

$$q = \frac{1 - \sqrt{\alpha/\beta}}{1 + \sqrt{\alpha/\beta}}, c_k = \frac{1}{\sqrt{\alpha\beta}} (-q)^k \qquad (4)$$

$$M_r = \frac{c_0}{2} I + \sum_1^r c_k T_k \approx \frac{A^{-1}}{2} \qquad (5)$$

where $\text{diag}(A)$ is a diagonal matrix created from the diagonal entries in matrix A.

The first step is the creation of Z (2), which shifts the eigenvalues of the diagonally-scaled A matrix into the range $[-1,1]$, provided that α and β are the minimum and maximum eigenvalues of the scaled A matrix. The value r is the upper bound on the summation of in equation (5), i.e., it is the number of Chebychev polynomials used to approximate the preconditioner matrix M_r. As described in [Dag and Semlyen 2003] and [Dag 2007], an estimate of β can be determined via the power method, and α is assigned based on typical performance of the Chebyshev preconditioner. Reference [Dag and Semlyen 2003] recommends setting $\alpha = \beta/5$ if $r < 3$ and $\alpha = \beta/\lfloor r/2 \rfloor \times 5$ if $r \geq 3$, where $\lfloor x \rfloor$ is a function that returns the largest integer less than x. Because the computational cost of the power method is much smaller than a single iteration of the Chebyshev algorithm [Dag 2007], the calculation of β was carried out before preconditioning using the host processor. If the coefficient matrix has known spectral characteristics, then the power method calculation can be omitted and an approximate β can be used instead [Dag and Semlyen 2003].

The preconditioner matrix M_r is defined in (5), and it depends on the recursively defined T_k matrices (3) and the coefficients c_k (4). The scalar r used as the upper bound of the summation in (5) denotes the number of Chebyshev iterations. The selection of r is described in more detail in section 5, including discussion of the sensitivity of the preconditioner effectiveness to r and the computational and storage costs associated with increasing r. In contrast to preconditioning methods based on Gaussian elimination, such as ILU, the Chebyshev method only uses matrix-matrix and matrix-vector multiplication and addition. Most parallel processing architectures, in particular GPUs, handle these operations very efficiently [Garland 2008], and the next section discusses how these operations were implemented on the GPU architecture.

4 Implementation of the Chebyshev Preconditioner on the NVIDIA GTX 280

Power system matrices are usually stored in sparse data structures, such as compressed row storage (CRS) or compressed column storage (CCS) [Saad 1994] format, in order to take advantage of sparsity. In our implementation, the **A**, **Z**, **T**, and **M** matrices are stored using the CRS format. Additional details regarding the data structures are provided below in the discussion of step 7. The Chebyshev preconditioner algorithm is divided into seven steps, as shown in Fig. 1. Steps 1, 2, and 7 are executed on the host: in step 1, the host reads the coefficient matrix **A** from memory and, if necessary, converts it to CRS format; in step 2, **A** is copied from the host to the GPU's global memory; and in step 7, the preconditioner matrix M_r is copied from the GPU to the host.

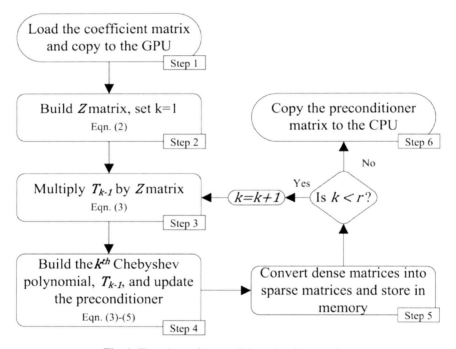

Fig. 1. Flowchart of preconditioner implementation

The remaining steps are performed entirely on the GPU. In step 3, the linear transformation described in (2) is performed on the GPU. The implementation of (2) requires scaling **Z** by a matrix and then subtracting values from its diagonal entries. These parallel matrix operations are implemented in a manner similar to the vector operations available in the CUBLAS library [NVIDIA Corporation 2008], a library of linear algebra functions provided by NVIDIA. In steps 4 to 6, GPU kernels perform the operations given in (3) to iteratively build the Chebyshev preconditioner.

The most computationally expensive kernel is the matrix-matrix multiplication performed in step 4 ($Z \times T_{k-1}$). First, the matrix-matrix multiplication is decomposed into a set of dot products:

$$XY = P \tag{6}$$

$$\begin{bmatrix} X_1Y_1 & \cdots & X_1Y_N \\ \vdots & \ddots & \vdots \\ X_NY_1 & \cdots & X_NY_N \end{bmatrix} = \begin{bmatrix} P_1 \\ \vdots \\ P_N \end{bmatrix} \tag{7}$$

$$P_{i,k} = X_iY_k = \sum_{v \in \{1,2,\ldots,N\}, X_{i,v} \neq 0, Y_{v,k} \neq 0} X_{i,v} Y_{v,k} \tag{8}$$

where $X_i \in \Re^{1 \times N}$ is the ith row vector of matrix X, $Y_i \in \Re^{N \times 1}$ is the ith column vector of matrix Y, and $P_i \in \Re^{1 \times N}$ is the ith row vector of matrix P, and $\cdot_{i,k}$ refers to the element at row i and column k of the matrix $\cdot$. Highly optimized matrix-vector kernels already exist for sparse matrices multiplying dense vectors, but these kernels result in wasteful multiplications, additions, and memory accesses due to the unnecessary multiplication and addition calculations performed on the elements which are zero (e.g., in [Bell and Garland 2008], the summation in (8) is carried out without considering which entries in Y_k are non-zero). As indicated in the summation index given in (8), only calculations involving non-zero elements of X and Y are carried in determining each product matrix entry $P_{i,k}$. This explicit consideration of the matrix sparsity results in fewer multiply and add operations and fewer memory writes. In the GPU implementation of (7), 256 threads are assigned to each thread block, and inside each thread block, a group of 16 threads (referred to as a "half warp") is in charge of multiplying a single row of the sparse matrix X with the corresponding sparse columns of Y (i.e., each half warp is responsible for computing one row of the product matrix, P_i). In addition, consecutive threads within a half warp can coalesce global memory accesses to reduce latency if the global memory data is sequential. To carry out the summation in (8), each thread reads in the appropriate columns of Y from global memory. This kernel is executed for all N rows in the multiplier matrix X, so there are as many as N thread blocks pending execution at any given time.

When step 4 is completed, each row of the product matrix is stored in shared memory as a dense N-element vector, as shown at the top of Fig. 2. Dense storage is used within each block so that each thread within the block can write to the product row without having to synchronize with any other threads. The standard data structures used in power system software, single- and double-linked lists, would require synchronization on each read and write operation in order to ensure that the linked list integrity is maintained; this in turn would lead to an increase in the amount of per-thread idle time. Although each row of the product matrix is stored in shared memory as a dense vector, the row data is reduced to a sparse format before it is written to global memory and the next iteration is performed. More details on the dense to sparse vector conversion are provided below in the description of step 6.

The input vector I arrives at step 6 in a dense format:

| $I =$ | 0 | 0 | 0 | A | 0 | 0 | B | C | 0 | 0 | D |

An intermediate "pointer vector" Ptr is then constructed according to the following rule:

$$Ptr_j = \begin{cases} 1 & \text{if } I_j \neq 0 \\ 0 & \text{otherwise} \end{cases}$$

| $Ptr =$ | 0 | 0 | 0 | 1 | 0 | 0 | 1 | 1 | 0 | 0 | 1 |

A prefix sum is then applied to the pointer vector, with entries given according to the following relation:

$$PS_j = \sum_{i=0}^{j} Ptr_i$$

| $PS =$ | 0 | 0 | 0 | 1 | 1 | 1 | 2 | 3 | 3 | 3 | 4 |

The sparse data structure, consisting of a vector of values V and a vector of indices into the dense vector, Ix, is then constructed based on the following operations:

For $k = 1, 2, \ldots, N$
If $PS_k - PS_{k-1} = 1$
$V_{PS_{k-1}} = I_k$
$Ix_{PS_{k-1}} = k$

Resulting in the final sparse representation of I as:

$V =$	A	B	C	D
$Ix =$	3	6	7	10

Fig. 2. Conversion of the matrix row vectors from dense to sparse formats in step 6 of the algorithm

Upon completion of step 4, step 5 is carried out by determining the constant c_k according to (4) and performing the matrix additions according to (3) and (5). The kernel in step 6 then converts each dense row vector of the Chebyshev polynomial and preconditioner matrices into a sparse, CRS-formatted vector. The benefits of this reduction are threefold: first, it reduces the number of high-cost global memory writes that must be performed at the end of each iteration; secondly, it limits the amount of global memory that must be allocated for storage of matrices between iterations; and thirdly, it reduces the number of costly global memory reads that must be performed in step 4 in the subsequent iteration. The basic algorithm for this kernel is illustrated in Fig. 2. An example of a dense vector that is obtained as the output of step 5 is shown at the top of Fig. 2. This vector has 4 non-zero elements, shown by the letters A to D, at columns 3, 6, 7, and 10, respectively, with the convention that the first element of the vector is indexed as 0. The first step in the conversion to a sparse vector is to create an intermediate "pointer vector", Ptr, based on which entries are non-zero in the input vector. Each element of Ptr is set to 1 for any non-zero entries in the input vector I and zero otherwise. The next step in the conversion is to perform a prefix sum [Garland 2008] on the pointer vector. To perform the prefix sum, an optimized kernel included in the CUDA libraries [NVIDIA Corporation 2008] is used. The output of the prefix sum, **PS**, is described and illustrated in Fig. 2. Within the **PS** vector, every incremental step between columns corresponds to the existence of a non-zero element in the input vector (e.g., an increment occurs between PS_2 and PS_3 in the example of Fig. 2., indicating there is a non-zero entry in column 3 of the input vector). The final step of the conversion uses the entries in **PS** to allocate the sparse data structure (note that the last entry of **PS** contains the number of nonzero entries to be stored in the sparse data structure) and to populate it with the appropriate values. Pseudo code for this final operation, along with the resulting sparse vector data structure, is shown at the bottom of Fig. 2.

Steps 4 to 6 are repeated r times and generate r Chebyshev polynomials in order to carry out the summation in (5). After each iteration, including the final iteration, the matrices are stored in CRS format. If the iterative solver is implemented on the GPU, this matrix can be read from global memory within the solver kernel(s). Otherwise, the preconditioner matrix can be sent back to the host for usage in a CPU-based iterative solver (as indicated in step 7 of Fig. 1.).

5 Evaluation

We tested our preconditioner with several sparse matrices to ensure it is a viable alternative to CPU-based preconditioning. The first set of matrices studied were the dc power flow matrices associated with the IEEE 30-, 57-, 118-, and 300-bus test cases [University of Washington (1993)]. To examine the performance on a practically-sized system, we also evaluated the performance of the algorithm on the dc power flow matrix of a 1243-bus European case [Qing and Bialek 2005]. Additional test matrices were obtained from the NIST Matrix Market [Campos and Rollett 1995], including two matrices from the PSADMIT set (with N values of 494 and 685) and two matrices from the Harwell-Boeing Sparse Matrix Collection (with N values of 1801 and 3948) since these two matrices are frequently used to evaluate preconditioning algorithms [Campos and Rollett 1995]. The experiments were performed on an NVIDIA GTX 280-based graphics card. This graphics card has 240 SPs running at a 602MHz core clock speed and 1 GB of global memory accessible at a peak bandwidth of 155.52 GB/s. The host side is a quad core Intel processor running with a core clock speed of 2.66GHz and 12MB of L2 cache.

One important parameter of the Chebyshev preconditioner is r, the number of Chebyshev polynomials used to calculate the preconditioner matrix in (5) and the number of times the inner loop of Fig. 1 is traversed. At one extreme, if the number of Chebyshev iterations is taken to be a very large number (e.g., close to the dimensions of the coefficient matrix), the preconditioner calculations would result in a very close approximation to the coefficient matrix inverse. As a result, the expected effect of a high-r preconditioners is to bring the condition number of the preconditioned system close to unity and, as a result, drastically reduce the number of solver iterations. The reduction in the solver iterations does come at a price, however, since the number of calculations that must be performed in the preconditioning step is proportional to r^2 and storage of the full, non-sparse inverse on the GPU could exhaust its memory capacity. On the other hand, choosing a value of r that is too small can result in an ineffectual preconditioner that provides no real benefit to the iterative solver.

To determine an appropriate value of r for use in our implementation, sensitivity analyses were carried out in which changes in r were compared with the resulting changes in the condition number of the coefficient matrix. For example, Fig. 3 shows the trend in the condition number as r is increased for the dc power flow matrices from the 57- and 118-bus IEEE test cases. The first step shows the reduction in condition number by scaling the coefficient matrix with its main diagonal; as shown in the figure, the diagonal scaling results in a half-decade reduction in the condition number, and the Chebyshev iterations result in another decade of reduction for values of r up to 3. Afterward, the gains of increasing r are much less pronounced, indicating that the greatest gain of using the preconditioner comes from the first few iterations. Because this same phenomenon was observed for the other matrices under study, we concluded that the optimum number of Chebyshev iterations should be between two and four.

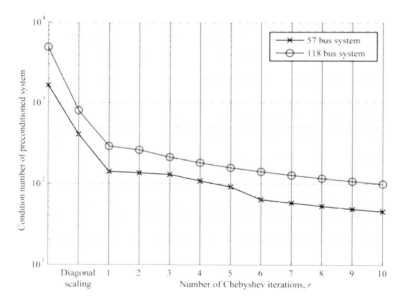

Fig. 3. Effect of r values on the preconditioned system's condition number for the 57- and 118-bus systems

To understand the impact of increasing r on the sizes of the preconditioner matrices, (3) is rewritten in terms of $\mathbf{Z}$:

$$\begin{aligned} T_0 &= I, T_1 = \mathbf{Z} \\ T_2 &= 2\mathbf{Z}T_1 - T_0 = 2\mathbf{Z}^2 - I \\ T_3 &= 2\mathbf{Z}T_2 - T_1 = 4\mathbf{Z}^3 - 3\mathbf{Z} \\ &\vdots \\ T_k &\propto \mathbf{Z}^k \end{aligned} \quad (9)$$

Therefore, in each Chebyshev iteration, the sparsity structure of T_k and the associated preconditioner matrix are determined by the structure of $\mathbf{Z}^k$. The coefficient matrices associated with power system analysis (e.g., from dc power flow analyses) have a sparsity structure that is closely tied to the network topology of the study system. According to 2, $\mathbf{Z}$ has the same sparsity structure as the coefficient matrix. One way to analyze the behavior of $\mathbf{Z}^k$, then, is to consider the effect of raising the adjacency matrix, defined element-wise as:

$$Adj_{m,n} = \begin{cases} 1 & \text{if } Y_{m,n} \neq 0 \\ 0 & \text{otherwise} \end{cases} \quad (10)$$

to successive powers of k. From graph theory, it is known that taking the kth power of the adjacency matrix introduces a non-zero entry at row m and column n if buses m and n are reachable by traversing k or fewer edges (i.e., $l_{m,n} \leq k$, where $l_{m,n}$ is the minimum path length between buses m and n). The diameter of the adjacency matrix, $Diam(\mathbf{Adj})$, is defined as the maximum value of $l_{m,n}$ over all possible row and column combinations:

$$Diam(\mathbf{Adj}) = \max_{m \in \{1,2,\ldots,N\}, n \in \{1,2,\ldots,N\}} l_{m,n} \quad (11)$$

Returning to (9), this suggests that any value of r equal to or greater than the diameter of a given power network will result in a matrix that is fully dense. For many real-world power systems, such as the Western North America, Northern China, and Central China power grids, the diameter is significantly lower than the number of system buses [Sun 2005]; therefore, a small value of r is needed to ensure the resulting preconditioner matrix is kept sparse.

The effect of the Chebyshev preconditioner on each of the test matrices is provided in Table I. The data reported in the table are the number of row for each matrix ("Size"), the condition number ("Cond#") and the number of iterations it took to solve a system with the given coefficient matrix ("CG#"). The columns containing the condition number and iterative solver's performance data for the original, non-preconditioned matrix are labeled "$r = 0$". The columns labeled "$r = 3$" contain the performance data obtained after applying the Chebyshev preconditioner with three iterations. The most significant result shown in the table is that all of the cases with more than 300 buses resulted in non-convergence of the iterative solver (indicated by a superscript "nc" in the corresponding table entries). Application of the Chebyshev preconditioner with $r = 3$ resulted in a significant reduction in the condition number of all nine matrices and, most importantly, was able to adjust the spectral properties sufficiently to get convergence in the iterative solver. The conjugate gradient squared (CGS) method was used to test the iterative solver performance, which was executed on the host using the cgs function in MATLAB.

Table 1. The impact of the third-order Chebyshev preconditioner on the condition number and CGS iteration count for several test matrices

Matrix Name	Size	Cond# ($r = 0$)	Cond# ($r = 3$)	CG # ($r = 0$)	CG# ($r = 3$)
30-Bus	30	913.23	71.80	37	14
57-Bus	57	1.6e+03	147.14	57^{nc}	22
118-Bus	118	3.9e+03	288.47	112	30
300-Bus	300	1.1e+05	5732.5	300^{nc}	72
494-Bus	494	3.9e+06	32855.1	494^{nc}	216
685-Bus	685	5.3e+05	3542.51	685^{nc}	112
EU case	1243	3.2e+05	14244.3	1243^{nc}	117
bcsstk14	1806	1.3e+10	1.2e+05	1806^{nc}	52
bcsstk15	3948	7.9e+09	8.9e+05	3948^{nc}	247

We also implemented a sequential version of the Chebyshev preconditioner in MATLAB for comparison of the parallel, GPU and serial, CPU implementations. The sequential code accepts sparse matrices in compressed column storage or sparse coordinate storage formats. We optimized our serial implementation as much as possible to have a meaningful comparison of the two implementations. Table II presents the process time of the GPU implementation versus the CPU implementation. The results indicate that the speed-up increases as the matrix side increases, up to a maximum speed-up of 8.93x for the largest matrix tested. The relationship between matrix size and speed-up is especially important if the GPU is used as a coprocessor for large power system matrices, since it indicates that the GPU implementation scales appropriately for the large linear systems that are likely to appear in real power systems applications.

We are also comparing our parallel iterative preconditioner with direct preconditioners commonly used in the literature. In order to compare our iterative preconditioner on the GPU to direct preconditioners on the CPU, the MATLAB implementation of the sparse ILU factorization was used. This method performs a LU factorization on the system; however, if an element is encountered during the factorization that is less than a specified threshold, that value is discarded [Campos and Rollett 1995]. By increasing this threshold, the ILU becomes less incomplete and more like a full LU factorization. Other than the threshold method, various schemes for ILU preconditioning are available in the literature [Benzi 2002]. We preconditioned each system with both ILU and the three-iteration Chebyshev preconditioner then solved each system with the CGS solver. For each test case, the ILU threshold was set such that the number of CGS iterations used to solve the ILU preconditioned system was the same as the Chebyshev preconditioned

Table 2. A comparison of the GPU Chebyshev preconditioner implementation with CPU-based Chebyshev and ILU preconditioners

Test Matrix		ILU			Chebyshev			Speed-up	
Name	nnz	drop tol.	nnz	Time CPU(ms)	nnz	Time CPU(ms)	Time GPU(ms)	vs. ILU on CPU	vs. Cheb. on CPU
30-Bus	108	6e-02	170	0.299	258	0.27	0.47	0.63	0.57
57-Bus	205	4e-02	322	0.365	461	0.68	1.25	0.29	0.54
118-Bus	464	4e-02	812	0.488	1198	1.55	2.30	0.21	0.67
300-Bus	1121	5e-04	7088	2.110	2890	7.69	6.47	0.32	1.18
494-Bus	1080	4e-04	6534	8.096	4062	19.7	8.06	0.99	1.83
685-Bus	1967	4e-03	18386	23.33	9337	44.1	13.72	1.70	3.21
EU case	4872	1e-03	23976	43.42	13873	124.2	21.03	2.06	5.91
bcsstk14	32630	6e-06	272878	90.12	195654	262.2	33.46	2.69	7.85
bcsstk15	60882	4e-06	1029908	547.1	527666	591.1	66.18	8.26	8.93

system. A comparison of the ILU preconditioner (executed on the CPU) and Chebyshev preconditioner (executed on the GPU) computation times are given in Table 2. For small system sizes, the GPU-based Chebyshev preconditioner uses more calculation time than ILU. In these cases, the amount of computation is not large enough to outweigh the overhead of the GPU execution, such as the time spent to allocate the executable kernels on the GPU. In addition, for smaller matrices, the computational resources of the GPU are not saturated by the preconditioner calculations. As the number of calculations increases, the overhead is amortized over more calculations and there are more opportunities for latency masking; therefore, there should be a relative increase in performance for larger matrices. Table 2 confirms that as the matrix size increases, the GPU-based Chebyshev preconditioner requires less computation time relative to the ILU preconditioner.

Besides the time spent on calculation in the preconditioning process, the number of non-zeros of the preconditioned matrix is an important factor in evaluation of the preconditioning technique. The sparsity of the preconditioned matrix has a great effect on the computation time and memory requirements for the rest of the linear solver process. As shown in Table 2, the iterative preconditioner results in a lower number of nonzero entries for larger matrices.

6 Concluding Remarks

We have implemented an optimized, parallel preconditioner for power system matrices that can take advantage of the current generation of massively parallel GPUs and is easily adaptable to future processors that are likely to have even more cores. Our method uses an efficient polynomial preconditioning technique, which favors the SPMD parallel structure of GPUs. The preconditioner is capable of processing large sparse matrices and performs well with ill-conditioned linear systems encountered in power systems analyses. Our implementation is shown to perform well with Krylov subspace methods, in particular the CGS iterative solver. The process time of the preconditioner is comparable with production-grade, direct serial algorithms, and the results from Table 2 show that the method exhibits good scalability in both computation time and memory requirements.

This work focused exclusively on the preconditioner implementation and performance, and we have begun developing a robust GPU-based iterative solver that relies on the Chebyshev preconditioner described herein. Many of the Krylov subspace solvers, such as BiCG and GMRES, have similar computational needs at the kernel level (e.g., both BiCG and GMRES require optimized sparse matrix-vector products to perform well), and we are hopeful that the kernels described in section 4 can be repurposed for the iterative solvers. For readers interested in more details regarding the CUDA implementation, the source code for the kernels implemented in this work are available in [Asgari 2011].

References

AMD/ATI, ATI Stream Computing: Technical Overview (2009)

Andersson, G., Donalek, P., Farmer, R., Hatziargyriou, N., Kamwa, I., Kundur, P., Martins, N., Paserba, J., Pourbeik, P., Sanchez-Gasca, J., Schulz, R., Stankovic, A., Taylor, C., Vittal, V.: Causes of the 2003 major grid blackouts in North America and Europe, and recommended means to improve system dynamic performance. IEEE Trans. Power Syst. 20(4), 1922–1928 (2005)

Anupindi, K., Skjellum, A., Coddington, P., Fox, G.: Parallel deferential-algebraic equation solvers for power system transient stability analysis. In: Proc. Scalable Parallel Libraries Conference, pp. 240–244 (1993)

Arrillaga, J., Smith, B.: AC-DC Power System Analysis (IEEE Power Engineering Se-ries). Institute of Engineering and Technology, London, United Kingdom (1998)

Asgari, A.: Implementing a Preconditioned Iterative Linear Solver Using Massively Parallel Graphics Processing Units. Master's thesis, University of Toronto (2011)

Asgari, A., Tate, J.E.: Implementing the Chebyshev polynomial preconditioner for the iterative solution of linear systems on massively parallel graphics processors. In: CIGRE Canada Conference on Power Systems, Toronto (2009)

Axelsson, O.: Iterative Solution Methods, p. 357. Cambridge University Press, New York (1996)

Bell, N., Garland, M.: Efficient sparse matrix-vector multiplication on CUDA. Technical Report NVR-2008-004, NVIDIA (2008)

Benzi, M.: Preconditioning techniques for large linear systems: a survey. Journal of Computational Physics 182, 418–477 (2002)

Bolz, J., Farmer, I., Grinspun, E., Schroder, P.: Sparse matrix solvers on the GPU: conjugate gradients and multigrid. In: ACM SIGGRAPH 2005 Courses, Los Angeles, California, vol. 171 (2005)

Campos, F.F., Rollett, J.S.: Analysis of preconditioners for conjugate gradients through distribution of eigenvalues. International Journal of Computer Mathematics 58(3), 135–158 (1995)

Chen, K.: Matrix Preconditioning Techniques and Applications. Cambridge Monographs on Applied and Computational Mathematics. Cambridge University Press, New York (2005)

Crow, M.L.: Computational Methods for Electric Power Systems. CRC Press, Boca Raton (2003)

Dag, H.: An approximate inverse preconditioner and its implementation for conjugate gradient method. Parallel Computing 33(2), 83–91 (2007)

Dag, H., Semlyen, A.: A new preconditioned conjugate gradient power flow. IEEE Trans. Power Syst. 18(4), 1248–1255 (2003)

De Leon, F.: Discussion of A new preconditioned conjugate gradient power flow. IEEE Trans. Power Syst. 18(4), 1601 (2003)

Decker, I.C., Falcao, D.M., Kaszkurewicz, E.: Conjugate gradient methods for power system dynamic simulation on parallel computers. IEEE Trans. Power Syst. 11(3), 1218–1227 (1996)

Flueck, A.J., Chiang, H.-D.: Solving the nonlinear power flow equations with an inexact Newton method using GMRES. IEEE Trans. Power Syst. 13(2), 267–273 (1998)

Galoppo, N., Govindaraju, N.K., Henson, M., Manocha, D.: LU-GPU: efficient algo-rithms for solving dense linear systems on graphics hardware. In: Proceedings of the 2005 ACM/IEEE Conference on Supercomputing, Seattle, WA (2005)

Garland, M.: Sparse matrix computations on many core GPU's. In: Proc. 45th ACM IEEE Design Automation Conference, pp. 2–6 (2008)

Glover, J.D., Sarma, M.S., Overbye, T.J.: Power Systems Analysis and Design, 4th edn., CL Engineering, Stamford, Connecticut, USA (2007)

Gopal, A., Niebur, D., Venkatasubramanian, S.: Dc power flow based contingency analysis using graphics processing units. In: Proc. 2007 IEEE Power Tech., pp. 731–736 (2007)

Kirk, D., Hwu, W.: Programming Massively Parallel Processors: A Hands-on Approach. Morgan Kaufmann Publishers, Burlington (2010)

Koester, D.P., Ranka, S., Fox, G.C.: Power Systems Transient Stability-A Grand Computing Challenge. Technical Report SCCS 549, Northeast Parallel Architectures Center, Syracuse, NY (1992)

Meijerink, J.A., van der Vorst, H.A.: An iterative solution method for linear systems of which the coefficient matrix is a symmetric M-Matrix. Mathematics of Computation 31(137), 148–162 (1977)

Mori, H., Tanaka, H., Kanno, J.: A preconditioned fast decoupled power flow method for contingency screening. IEEE Trans. Power Syst. 11(1), 357–363 (1996)

Munshi, A.: The OpenCL Specification: Version 1.0. Khronos OpenCL Working Group (2009)

NERC Board of Trustees, Standard TPL-003-0—System Performance Following Loss of Two or More BES Elements. Technical standard, North American Electric Reliability Corporation (2005)

NERC Steering Group, Technical Analysis of the August 14, Blackout: What Happened, Why, and What Did We Learn? Technical report, North American Electric Reliability Corporation (2004)

NVIDIA Corporation, CUDA CUBLAS library (2008)

NVIDIA Corporation, NVIDIA CUDA Programming Guide: Version 2.0 (2008)

Pharr, M., Fernando, R.: GPU Gems 2: Programming Techniques for High-Performance Graphics and General-Purpose Computation. Addison-Wesley, Reading (2005)

Qing, Z., Bialek, J.W.: Approximate model of European interconnected system as a benchmark system to study effects of cross-border trades. IEEE Trans. Power Syst. 20(2), 782–788 (2005)

Saad, Y.: SPARSKIT: A basic toolkit for sparse matrix computations (1994)

Saad, Y.: Iterative Methods for Sparse Linear Systems, 2nd edn. SIAM, Philadelphia (2003)

Shahidehpour, M., Wang, Y.: Communication and Control in Electric Power Systems: Applications of Parallel and Distributed Processing. Wiley Interscience, Piscataway (2003)

Sun, K.: Complex networks theory: A new method of research in power grid. In: Transmission and Distribution Conference and Exhibition: Asia and Pacific, 2005 IEEE/PES (2005)

Tate, J.E., Overbye, T.J.: Contouring for power systems using graphical processing units. In: Proc. 41st Annual Hawaii International Conference on System Sciences, Hawaii, USA (2008)

University of Washington, Power Systems Test Case Archive, Seattle, Washington, USA (1993)

Wood, H.: Discussion of power flow solution by Newton's method. IEEE Trans. Power App. Syst. PAS-86, 1458–1459 (1967)

Reference Network Models: A Computational Tool for Planning and Designing Large-Scale Smart Electricity Distribution Grids

Tomás Gómez, Carlos Mateo, Álvaro Sánchez,
Pablo Frías, and Rafael Cossent

Institute for Research in Technology (IIT), ICAI School of Engineering,
Comillas Pontifical University, Spain
tomas.gomez@iit.upcomillas.es
carlos.mateo@iit.upcomillas.es
alvaro.sanchez@iit.upcomillas.es
Pablo.frias@iit.upcomillas.es
Rafael.Cossent@iit.upcomillas.es

Abstract. Reference Network Models (RNMs) are large-scale distribution network planning tools. RNMs can be used by policy makers and regulators to estimate efficient distribution costs. This is a very challenging task, particularly being network planning a combinatorial problem, which is especially difficult to solve due to the vast size of the distribution areas, and the use of several voltage levels. This chapter presents the main features of RNMs developed by the authors, including high performance requirements related to the type and size of the problem. The model can be used to plan distribution networks either from scratch or incrementally from existing grids. Different case studies illustrate the applicability of these models to the assessment of the impact of massive deployment of renewable distributed generation, demand response actions, and plug-in electric vehicle penetration on distribution costs. The results obtained provide valuable information to guide strategic policy-making decisions regarding the implementation of renewable energy programs and smart grid initiatives.

1 Introduction

Reference Network Models (RNMs) were originally created to respond to a need experienced by energy regulators when assessing efficient network costs in order to fix regulated revenues of electricity distribution utilities. After the electric power sector restructuring that took place in many countries during the 1990s, electricity transmission and distribution remained as regulated activities. In this context, a new regulatory approach, known as "incentive regulation", was introduced. The main purpose of incentive regulation is to promote cost reduction and higher efficiency in those activities which are still considered natural monopolies [Rothwell and Gomez 2003]. Nonetheless, incentive regulation required energy regulators to apply new regulatory tools, such as regulatory accounting, benchmarking techniques [Jamasb and Pollit 2003], and RNMs to build a reference utility [Rudnick and Donoso 2000]. RNMs are also being used to assess costs and benefits of smart grids in future scenarios with increasing penetration of new technologies.

A RNM is a large-scale network distribution planning tool. This tool is highly computational demanding since it requires extensive data and uses many algorithms to design the network. Network planning is a combinatorial optimization problem which is non-deterministic polynomial-time hard (NP-hard), and therefore inherently difficult to solve [Mori and Iimura 2003], especially as the size of the problem grows. The RNM developed deals with high, medium and low voltage levels, in vast distribution areas, which further increase the complexity of the problem. The RNM addresses the problem through the use of heuristics, and considering several distribution areas, so that each of them can be planned separately, in order to diminish computing times by paralleling processes.

Two types of RNMs have been developed: i) A "green-field" model which designs the whole distribution grid, including substations and power lines, from scratch, and ii) an "expansion" model which designs the reinforcement and new additions to the existing network needed to face future situations, for instance growth demand or new distributed generation. In both models, the design problem consists in minimizing the total investment in new installations plus associated operational costs, mainly energy losses, in order to supply the expected demand while meeting reliability and quality of supply criteria.

The main characteristics of the RNMs that make them significantly different from the current distribution planning tools are:

- The size of the problem: The RNM designs the distribution grid in a considered area of service from the transmission substations to the final consumers and distributed generators considering the three hierarchical levels: high voltage or subtransmission, medium voltage or primary distribution feeders, and low voltage or secondary distribution.
- The geographical differences: Both rural and urban zones included in the considered area of service are simultaneously optimized by designing the grid to supply all the customers (thousands to millions) and connect distributed generators. Geographical constraints regarding the street map in urban areas and environmental factors, such as coastline, nature reserves, and mountains are taken into account. The continuous interaction between the electrical network design algorithms and the Geographical Information System (GIS) ensures the feasibility and optimality of the layout and location of the new installations.
- The ability to propose new network alternatives: In traditional planning tools it is the user, the planner, who provides the network expansion candidates. However, the RNM is capable of generating network reinforcements automatically and locating and sizing new installations, such as substations and transformers.

In Figure 1, a schematic representation of a distribution grid is shown. The investment decisions made by the RNM refer to the location and size of new high voltage/medium voltage (HV/MV) substations, new medium voltage/low voltage (MV/LV) transformers, and new HV lines, and MV and LV feeders. The dimension of the service area can cover a territory of several hundreds of square kilometers with millions of residential customers and thousands of medium commercial or industrial customers.

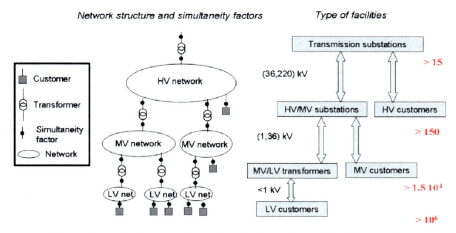

Fig. 1. Hierarchical structure by voltage levels and size of distribution grids

In this chapter, a review of distribution planning models which have been proposed in the literature is first made, followed by the description of the main features and algorithms that compose the developed RNMs, including high performance requirements. Finally, two sections regarding the application of RNM for the assessment of the impact of massive deployment of renewable distributed generation, demand response actions, and plug-in electric vehicle penetration on distribution costs are detailed.

2 Distribution Planning Methods: A Review

Network planning is a traditional decision problem where new installations, i.e. power lines and substations, are programmed along the study horizon in order to supply the expected future demand complying with reliability of supply criteria and minimizing total investment and operational costs. In the short-term, network planning mainly deals with network reinforcements of the existing grid to meet stricter reliability criteria or to replace aged installations, i.e. expansion planning. In the long-term, new substations, required to supply growing demands, and connect new power generators are the main drivers which determine the future network topology, i.e. expansion planning and green-field planning in the new growing areas.

The planning problem can be mathematically formulated as an optimization with the objective of minimizing the total cost subject to technical constraints and operational limits (lines and transformers capacity limits, suitable range of voltages, etc). Network planning is a complex, combinatorial, non-linear problem, with a high number of binary variables (investment decisions). Moreover, the risk of falling in non-optimum solutions is high since it is also a non-convex problem.

An important number of methods to solve this problem can be found in the literature, both for transmission and distribution networks [Khator and Leung 1997] and [Latorre et al 2003]. These methods can be broadly categorized into those based on mathematical optimization algorithms, and those based on heuristics or metaheuristic methods. In many cases, different simplified network models, such as the DC load flow or the Garver model [Garver 1970], are used in order to shorten computation times.

Resolution with mathematical optimization programs is inefficient as it requires large computational times. In order to overcome these difficulties, decomposition methods like Benders [Binato et al 2001] or hierarchical decomposition [Romero and Monticelli 1994] have been applied to transmission networks in order to facilitate the resolution. Nevertheless, these methods are limited to small or medium networks. They cannot be used in real-size networks because they require a high computational burden. In these cases, metaheuristic optimization methods have proven to be more efficient in combinatorial problems resolution. The metaheuristic methods applied to network planning comprise genetic algorithms [Ramirez-Rosado and Bernal-Agustin 1998], [Carrano et al 2008], simulated annealing [Gallego et al 1997], tabu search [Ramirez-Rosado 2006] or particle swarm [Al-muhaini and Arabia 2007], [Ping et al 2009]. Metaheuristics do not guarantee obtaining the global optimum; although in practice, they reach near optimal solutions.

The RNM described in this chapter is a single-period horizon model[1], as opposed to multi-period models, that uses heuristic algorithms and includes contingency analysis to design the distribution grid. In addition, it deals with the integration of distributed generation (DG) in distribution planning. This issue has been analyzed in [Chowdhury et al 2003] considering reliability. A heuristic approach for DG planning from the perspective of a distribution company is also presented in [El-Khattam et al 2004]. Recently [Navarro and Rudnick 2009a, 2009b] have described a large-scale distribution planning model which has some similarities with the developed RNM. However, their model is only focused on low-voltage network planning and uses a Tabu search for optimization. In [Díaz-Dorado et al 2003] a method based on evolutionary strategies for large low voltage rural networks design is also presented. The RNM described in this chapter deals with high, medium, and low-voltage networks instead and uses a branch-exchange technique as optimization [Míguez et al 2002]. More examples of application of branch exchange techniques to distribution planning can be found in [Nara et al 1992], [Goswami 1997], and [Peponis and Papadopoulos 1997].

---

[1] Single-period refers to the fact that RNMs perform a static distribution network planning. Notwithstanding, within a single-period planning approach it is possible to consider several situations or snapshots that may influence the grid design; for example, peak demand and peak generation.

3 RNM Features and Algorithms

This section provides an overview of the functionalities of RNMs and a more detailed description of the architecture and algorithms used by the models as well as the necessary inputs and the outputs obtained from the models.

3.1 Functionalities of RNMs

RNMs are optimization models able to design an electrical reference or adapted network for very large distribution areas comprising up to a few million consumers. A reference network is a theoretical network that complies with the same geographical, reliability and technical constraints as the actual grid at a minimum cost. Reference networks can be used by regulators as a benchmark for actual distribution networks in order to set the allowed revenues of distribution system operators (DSOs). Spain is one of the pioneers in the application of RNMs to regulate DSOs. The general principles concerning the economic regulation of electricity distribution in Spain using RNMs are set in RD 222/2008 [Ministry of Industry, Tourism and Trade, 2008].

There are two different approaches to obtain a reference network: i) green-field models and ii) expansion planning models. The former approach builds an optimal network from scratch, hence disregarding historic evolution of the networks. The latter approach takes the current network as the starting point and then builds the necessary reinforcements to cope with both horizontal (new network users) and vertical (changes in load demands and/or generation capacities of existing users) growth in demand and DG production. Furthermore, the initial network considered by the expansion-planning RNM can be obtained with the green-field RNM, thus allowing to feed one of the models with the results of the other. This capability has been essential in analyses aiming at assessing the impact that DG can have on distribution network costs and to evaluate how the implementation of more active grid operational strategies can mitigate this impact. This analysis will be described later in this chapter.

3.2 Distribution Network Planning with RNMs

Distribution networks comprise several layers or voltage levels. Distribution grids start at the transmission substations, from where the HV subtransmission networks feed electricity to the distribution substations and HV consumers, typically large industries. Then, the MV grid supplies medium and small industries and commercial consumers as well as the MV/LV transformers. Finally, the LV grid feeds the smallest consumers, that include small commercial and residential consumers (see Figure 1). Subtransmission networks are highly meshed. MV grids are normally built with a certain degree of meshing, particularly in urban areas, although they are operated radially. Finally, LV networks are usually totally radial. As one approaches lower voltage levels, the number of elements grows exponentially. This is one of the main differences with transmission networks.

Figure 1 shows the location of simultaneity factors on different points of the distribution network. Simultaneity factors are needed for planning purposes, in order to account for the fact that the maximum power flow in the different network components does not occur at the same moment in time. As the grid voltage level rises, more downstream customers and installations are aggregated. However, the peak of an upstream network element is lower than the sum of the peaks of its downstream fed network components, because they do not all occur at the same time. Therefore a simultaneity factor has to be considered when peak power flows are aggregated. Without simultaneity factors, network components may be assigned a much bigger size than necessary. For example, if DSOs assumed that all LV consumers consume their maximum power at the same time, LV grids and MV/LV transformers would be much bigger in terms of capacity than what it would be actually required. Similarly, MV/LV transformers and distribution substations have two different simultaneity factors, one upstream of the transformer and another downstream. The upstream simultaneity factor models the fact that not all transformers are at their peak at the same time, whereas the downstream one accounts for the fact that not all the lines connected to them will be loaded at their maximum simultaneously.

Table 1 provides typical values for simultaneity factors in the Spanish distribution networks. It can be observed that the values of the simultaneity factors increase with the voltage level. This is due to the fact that the higher the voltage level, the lower the number of network users and installations that are aggregated to compute peak flows. It will be explained below how the algorithms take into account the effects of simultaneity factors. Further information can be found in [Mateo et al 2010].

Table 1. Typical values for simultaneity factors in Spain

Simultaneity factors	Typical values
LV customers	0.2 – 0.4
LV feeders	0.8
MV/LV transformers & MV customers	0.8
MV feeders	0.85
HV/MV substations & HV customers	1

It is noteworthy that the distribution network is built taking into account the actual location of network users and network components (in the case of the expansion-planning RNM), as well as other geographical constraints such as environmental factors or street maps within urban areas. This is possible owing to the interaction between the planning algorithms and a geographical information system (GIS). Figure 2 shows how the planning algorithms interact with the GIS in order to include the cost increase caused by geography and topography, and to optimize the network layout.

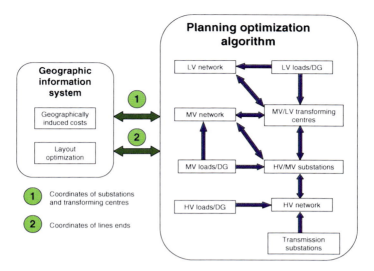

Fig. 2. Interaction between the planning algorithms and the GIS

In urban areas, actual distribution networks must be built following the streets since they cannot cross buildings or parks. If necessary, electrical lines may cross the streets, mainly in large avenues, perpendicularly to the road. RNMs mimic this behavior by building an approximate street map based on the location of electricity consumers. Both urban regions and street maps are endogenously detected and generated based on the number and density of consumers. Lines are forced to follow these street maps as shown in Figure 3, where blue triangles represent the HV/MV substations and yellow squares the MV/LV transformers. Thick black lines represent the MV feeders and thin green lines correspond to the approximate street map automatically generated.

[Mateo et al 2010] presents an assessment of the impact of street maps on the length of the distribution networks. In order to perform this analysis, an urban area serving above one million consumers was planned by the green-field RNM. Results showed that, for the same distribution area, the LV grid calculated by the RNMs was 16.8% longer when street maps were considered. On the other hand, the length of the MV network obtained increased by 37.5% as compared to the situation where street maps are not considered. Naturally, this will have a significant impact on the distribution network costs.

Outside urban areas, distribution networks must observe some geographical constraints as well. For example, electrical lines cannot cross certain regions such as protected natural areas, rivers or seas. These regions are considered by the RNM by introducing the geographical coordinates of the vertexes of polygonal outlines containing the forbidden regions. Furthermore, orography may also influence the design of distribution networks. Therefore, RNMs can interpret raster maps so as to avoid and skirt mountains or steep regions. Figure 4 displays two reference networks obtained for the same distribution area with and without considering geographical constraints. It can be seen that some lines have been modified and consumers transferred to a different feeder so as to avoid mountains (line C) or steep areas (lines A and B). Contrary to street maps, environmental factors must be introduced as an exogenous input to the models.

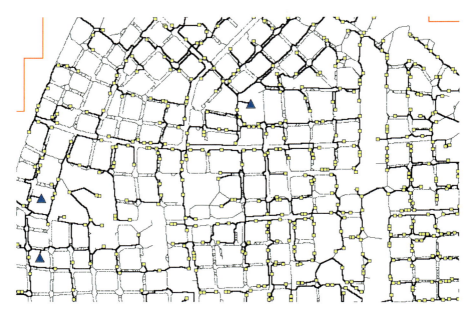

Fig. 3. Distribution network design following street maps

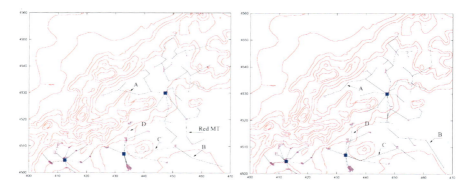

Fig. 4. Distribution network design considering orography

3.3 RNM Inputs

Given its complexity, the RNMs require extensive input data. The quality of these inputs greatly determines the quality of the final results. Thus, it is of utmost importance to correctly fine-tune this information so as to attain the desired results. The following information constitutes the most relevant inputs that must be provided to the RNMs:

i) Network users: loads, DG, EVs and storage
It is required to specify the exact location of every single user through its X, Y, and Z coordinates, voltage level at the point of connection, contracted power or installed capacity and power factor. For large distribution areas, the number of consumers, particularly small LV consumers, can be very large. The networks studied can range from several thousand users up to a few million connections. Gathering all this information is one of the main difficulties of using the RNMs.

For those network users that can behave either as generators or as loads depending on their operation mode, i.e. EVs and/or storage, it must be specified in the input files whether the capacity assigned to them is being consumed or injected into the grid.

ii) Transmission substations
The RNMs do not optimize the location of transmission substations as this is generally out of the control of DSOs. Therefore, the location and capacity of these substations must be provided as an input to the models in any case. Notwithstanding, the expansion-planning model may indicate that it is necessary to reinforce a transmission substation in order to avoid an unfeasible solution.

iii) Library of standardized network components
The RNM take into account the lumpiness of network investments as their decisions are made on the basis of a library of standard components. The library must comprise: HV power lines, MV and LV feeders, HV/MV substations, MV/LV transformers, protection equipment (breakers, fault detectors and switches), maintenance crews, capacitor banks and voltage regulators. Whenever necessary, these elements must be differentiated per voltage level, type of area and whether they are built overhead or underground.

Several data ought to be provided for each one of the possible network components. These basically comprise the following: investment and maintenance costs, rated capacity, electrical properties such as impedances, and useful life. Moreover, in order to compute the expected reliability indices it is necessary to provide the models with failure rates and a standard annual duration of preventive maintenance actions that are carried out on each type of component (HV line, MV/LV transformer, etc.), for overhead and underground elements and in each kind of area (urban or non-urban). Finally, fuzzy repair times in case of network component failures are assumed. Thus, the minimum, maximum, and mean repair times must be provided with the same structure to that of preventive maintenance times.

The library of standardized elements determines, to a great extent, the quality of the results. A library which does not sufficiently reflect the real options that DSOs have when designing their networks will yield rather insignificant results. Hence, a lot of effort should be placed in the development of the library.

iv) Other modeling parameters:
In addition to the previous information, RNMs need various parameters in order to perform all the computations involved. The most relevant ones are included in the following list:

- Simultaneity factors. As explained above, the RNMs require simultaneity factors to adequately design distribution networks.
- Economic parameters needed to calculate the present value of network costs and evaluate investment options. These comprise the cost of energy losses, the weighted average cost of capital (WACC) assumed, and costs of ditches and posts to install conductors in different types of areas.
- Load modeling and GIS related parameters. Density and minimum number of consumers to classify them into different areas and identify settlements, degree of undergrounding required within settlements per voltage level, impact of ground slope and height on costs, and street maps parameters.
- Technical and quality constraints. In addition to capacity constraints, the RNMs must observe the maximum and minimum bus voltages and the limits imposed on reliability of supply indices. The RNMs described in this report use the ASIDI and ASIFI indices as defined in [IEEE 2001]. Bus voltage limits are set per voltage level. MV network must comply with zonal and individual reliability indices, which are separately fixed for urban, semi-urban, concentrated rural, scattered rural and industrial areas.

Moreover, all the data corresponding to the initial network ought to be provided in the case of an expansion-planning RNM. This must include the topological as well as the electrical data for the existing network. Furthermore, the initial network users and the "new" incremental ones, either horizontal or vertical growths, must be differentiated.

3.4 RNM Outputs

The results obtained by the RNMs are twofold. On the one hand, a summary of the most relevant information of the network designed and the corresponding costs adequately broken down per type of network component. Furthermore, the reliability or continuity of supply indices attained with the given grid are provided.

On the other hand, detailed graphical output files are created by the RNMs. Each of these files corresponds to a type of network component which includes not only geographical information of the GIS for the elements of that type, but also electrical information such as impedances, thermal capacity or peak power flow. The expansion-planning RNM provides all the former information differentiating between the initial network and the increments needed to accommodate the horizontal and vertical increases in network users.

Figure 5 shows two examples of reference networks, particularly HV and MV levels, obtained from scratch with the green-field RNM. More information concerning these distribution areas can be found in [Cossent et al 2009]. Additionally, it can be seen in the image on the right in Figure 5 the population settlements that were identified by the model. Within these settlements, street maps were built in order to force the distribution network to follow it as explained above.

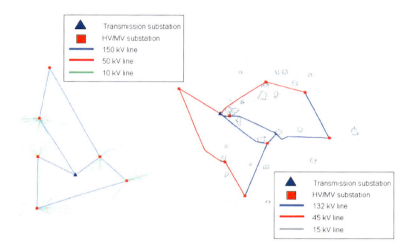

Fig. 5. Examples of geographical representation of the outputs of RNMs

3.5 RNM Architecture and Algorithms: Overview

The functional architecture of the RNMs is displayed in Figure 6. Furthermore, the main inputs needed for each one of the stages are shown in the figure. In the end, the objective function minimizes the one-off investment costs plus the present value of energy losses and maintenance costs for a specified number of years. The present value of annual costs is computed through a given WACC, taken as discount factor, considered the same for all costs.

Firstly, a DG/Loads modeling stage is performed in order to identify cities/towns and classify consumers into five categories: urban, sub-urban, concentrated rural, scattered rural and industrial areas. This classification is carried out according to the load density and number of customers of each kind. This affects different aspects such as reliability or continuity of supply requirements for the consumers located in each type of area or whether overhead or underground lines are built. Additionally, the street maps within densely populated areas are automatically generated.

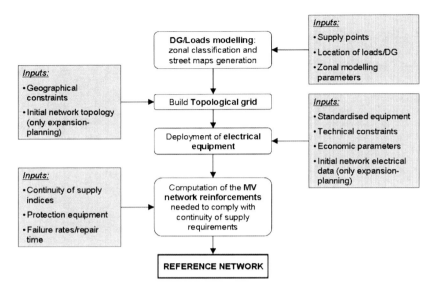

Fig. 6. Logical architecture of RNMs: steps involved and relevant input data

Secondly, an optimal network layout is computed. This topological network takes into account geographical constraints such as forbidden ways through, orography, street maps and, in the case of the expansion-planning RNM, the topology of the initial network. All geographical constraints except for street maps, are external inputs to the models. Note that, at this stage, the resulting network possesses electrical characteristics, albeit these are not optimized yet.

At this stage, possible infeasibilities in future steps are avoided by means of a simplified preliminary electrical test. This is done by assigning to each branch and node of the topological network the network element with the largest capacity that can be found in the library of network elements, taking into account the type of area where each element of the grid is located, its voltage level and whether overhead and/or underground elements are possible. For instance, a long branch of the topological network may be removed if the voltage drop would surpass the maximum allowable value provided that that branch was built with the largest possible conductor found in the library loaded at its thermal rating.

The topological network is built through a bottom up approach, starting from the lower voltage levels to the higher ones. The topology of initial MV and LV grids is radial. Note that the topology of the final MV grid is only determined after reliability or continuity of supply is taken into account in a subsequent stage. On the other hand, the initial HV network is designed according to an N-1 reliability criterion, i.e. every load and substations must be supplied through at least two paths. More details about how the bottom-up whole network optimization is carried out and the interactions with the electrical equipment sizing are presented in Section 3.6.

The electrical equipment deployment and sizing stage involves assigning to each segment or node of the topological network an optimally sized network element (line, transformer, etc.) by running a power flow for the network users given as input. At this step, technical constraints such as voltage and capacity limits are considered. Different power flow algorithms are used for HV meshed networks and MV/LV radial networks. The use of simultaneity factors at both ends of HV/MV substations and MV/LV transformers involves that the power entering one voltage level is not equal to the power supplied by the remaining voltage level. This requires modifying the modeling of these elements for power flow calculations as shown in Figure 7 and equation (1) where currents, I, voltages, V, simultaneity factors, F, and the elements of the admittance matrix, Y, are interrelated [Mateo et al 2010].

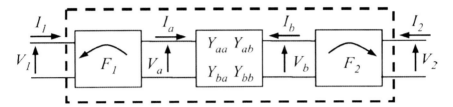

Fig. 7. Two-port power flow model including simultaneity factors F1 and F2. Source: [Mateo et al 2010]

$$\begin{pmatrix} I_1 \\ I_2 \end{pmatrix} = \begin{pmatrix} Y_{aa} \times F_1 & Y_{ab} \times F_1 \\ Y_{ba} \times F_2 & Y_{bb} \times F_2 \end{pmatrix} \begin{pmatrix} V_1 \\ V_2 \end{pmatrix} \quad (1)$$

The deployment of network components is done minimizing distribution network costs, including an estimate of energy losses computed as the product of power losses at peak demand by a representative loss factor. It is important to remark that, despite the fact that energy losses can only be roughly estimated, they must be taken into account in order to adequately dimension grid components. For example, a specific conductor may suffice to support a certain power flow given its thermal capacity; however, the cost of losses may justify a thicker conductor as a more economic solution over the lifetime of the asset [Peco 2001].

The computation of the energy losses must take into account the use of simultaneity factors. Otherwise, the difference between the power entering and exiting the substation due to simultaneity factors may be wrongly attributed to internal energy losses.

Real networks are not dimensioned merely according to current demand, but considering a future expected load growth. Hence, the RNMs assume a certain annual load growth for a number of years. Network components are thus sized in such a way that they may cope with the demand at the end of the growth period. Note that, as mentioned in Section 2, the RNMs are single-period planning models. Notwithstanding, this does not prevent the models to consider the expected future growth in demand.

The final stage is focused on designing the grid to meet continuity of supply constraints. Those requirements are incorporated to the initially designed radial MV grid. The final MV network must comply with the minimum continuity of supply indices set regarding average frequency and duration times for supply interruptions. These RNMs use ASIDI and ASIFI indices as defined in [IEEE 2001]. Both zonal and individual indices are taken into account. The failure rates of network elements are aggregated to compute the frequency of interruption of every load. Fault location and repair times are simulated taking into account the location (urban or rural) and type of network (overhead or underground). Additional equipment such as normally open meshing feeders, circuit breakers, maintenance crews, or fault detectors may be placed to comply with the continuity of supply requirements if needed.

3.6 RNM Architecture: Implementation

Both RNMs that are presented in this chapter, green-field and expansion-planning, share a common kernel structure. This structure is composed of several abstraction levels: i) containers, ii) topological, iii) electrical, and iv) quality of service. Each abstraction level is represented by a row in the matrix shown in Figure 8. Moreover, the following three information layers are represented by columns in the aforementioned Figure 8:

- Input/Output interfaces enable data exchange between RNM modules and users or applications. At present, they support two different formats: CIM (Common Information Model, EPRI) and the format developed by the Spanish Energy National Commission to demand data to the electric utilities.
- Data Models consist of structures and data required to represent electrical networks.
- Algorithms are functions that take advantage of data models to obtain specific results.

The first level, "containers", represents the lowest level in the hierarchy of abstraction. They depict networks as a collection of nodes and edges which interconnect them (see representation in Figure 9a). It could be regarded as a graph without any topological information, i.e. each node location is relative to other nodes. Data models implemented in this level comprise graphs, trees, vectors, matrixes, sparse matrixes, raster maps and graph slices. Special mention should be made of graph slices, which allow selecting a portion of a graph to apply localized algorithms. There are many algorithms which have been implemented at this level. To cite some of them, Depth-first search (DFS) and Breadth-first search (BFS) algorithms, whose importance lie on their ability to detect radial and meshed networks, or Dijkstra's algorithm, which determines the lowest cost path in a graph.

The second level of abstraction corresponds to the topological level. It incorporates topological information for each node and edge (see representation in Figure 9b). Under this type of representation, every node is localized in UTM coordinates

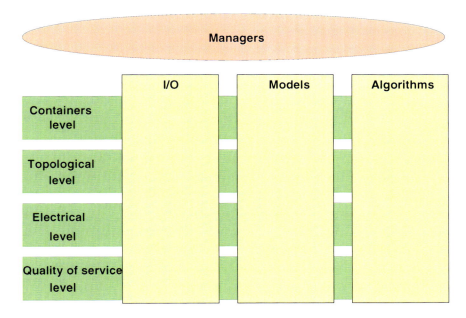

Fig. 8. Kernel structure of RNMs

and, thus, geo-referenced. This approach permits to take into account environmental factors, such as natural reserves or oceans, and topographic data (e.g. mountains, rivers) reason for which this level requires a Geographic Information System (GIS) that models street maps with graphs and topographic information with raster maps. For further information on GIS, refer to [Mateo et al 2010]. Some topological data that could be included at this stage are: forbidden regions, industrial areas, street maps, topographical maps, or urban cities and towns. Furthermore, as far as algorithms are concerned, the following have been implemented: the Delaunay triangulation algorithm (to join near nodes achieving a mesh of triangles), the PRIM algorithm to obtain minimum spanning trees (it is based on the mesh obtained with Delaunay triangulation), and the automatic street map generation algorithm described in [Mateo et al 2010].

The electrical level adds the corresponding electrical characteristics and parameters associated to each node and edge (see representation in Figure 9c) in order to build the electrical network. On the one hand, nodes could be interconnection nodes, generators, loads corresponding to final customers, or capacitors. On the other hand, edges could be power line overhead or underground sections, power transformers, and switching and protection equipments. According to this, when solving the corresponding power flow of the particular network configuration the electrical state is given by voltage magnitudes at the nodes and power flows in the branches or circuits. Some data models implemented in this level are: generators, substations, final customer loads, and capacitors, among others. The main algorithm at this level is the power-flow, which requires a two-port power model for each electrical device (see Figure 7). Three power-flow algorithms have been implemented: radial power flow, Newton-Raphson and Gauss-Seidel. Another important group of algorithms are the ones used for distribution network planning, which have been already introduced in section 3.5 and will be further detailed in section 3.7.

Finally, the quality of service level incorporates to the nodes and edges, represented as electrical equipment, their quality of service attributes. These attributes are mainly the already commented reliability of supply parameters, i.e. failure rates and repair times. Specific equipment regarding switching and protection systems together with maintenance crews are modeled at this level. Regarding developed algorithms, there is a module for reliability of supply assessment including fault location, isolation and restoration, and evaluation of continuity of supply indexes, ASIDI and ASIFI, computed in the different distribution regions. In addition, algorithms for network planning, including network reinforcement, and location of new protection and switching equipment in order to achieve the required continuity of supply levels, have also been developed and implemented.

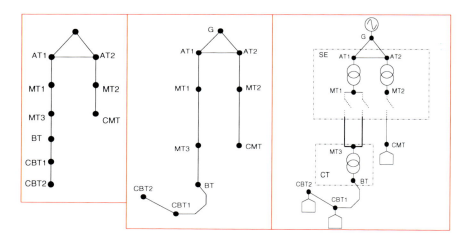

Fig. 9. Grids representations: a) containers level, b) topological level, and c) electrical level

The kernel structure includes other additional modules like memory and error managers which permit to execute, in a reasonable time, large scale networks with millions of final customers, and to monitor and debug the execution of any application that runs over this structure.

3.7 RNM Planning Algorithms

In the following, for the sake of illustration, the algorithms for network planning implemented in the green-field RNM are described following the bottom-up sequence represented in Figure 10.

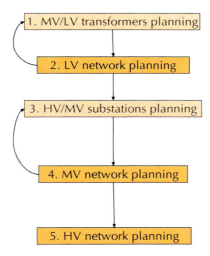

Fig. 10. Sequence of network planning algorithms for green-field planning

The first step is to decide the number, size and location of MV/LV power transformers taking as input the LV final customer and distributed generation data. An initial estimation of the number of MV/LV transformers is made on the basis of the power density of each identified city or town. A modified K-means algorithm is then used to locate the proposed MV/LV transformers.

After locating MV/LV transformers, the LV network can be planned connecting them with LV final customers and distributed generation (see 2 in figure 10). The process is carried out as follows. First, the Delaunay algorithm is run with all input data. Second, each final customer and distributed generator is associated to a MV/LV transformer, applying an electric momentum criterion, to obtain several clusters with a unique MV/LV transformer. Third, a minimum spanning tree is run in each cluster, where MV/LV transformers are the root nodes of the tree. Fourth, a branch-exchange optimization algorithm is executed to estimate a quasi-optimum LV network, subject to voltage and current constraints, minimizing investment, and operation and maintenance costs. This optimization sometimes implies relocating the LV/MV transformers and thus, returning to the first step. Finally, the conductor size optimization is performed to select the optimum for each LV overhead or underground line section. To this end, an additional term that takes losses into account is incorporated to the objective function.

The third stage consists of deciding the number, size and location of HV/MV substations (see 3 in figure 10). The logic for algorithms at this stage is similar to the one described for LV/MV transformers planning. However in this case a supplementary feature has been introduced to determine the voltage level of each substation, since the HV network can have several voltage levels. High voltage levels are estimated by means of a gross planning of a radial HV network, similar to the one designed for the LV network, which imposes voltage requirements to each conductor in each cluster.

The fourth step of the algorithm is to map out the MV network, which will link MV/LV transformers to HV/MV substations (see 4 in figure 10). The process resembles LV network planning, despite including new features that allow taking the quality of service level into account. Once conductor optimization is accomplished, some network design actions are adopted to achieve the required quality of service levels. As far as ASIFI is concerned, if the target is close to the actual one presented by the already designed grid, switching equipment is included in optimal locations of the MV network in order to meet the target. Otherwise, if the target is far enough, more HV/MV substations would be needed and, hence, HV/MV substation planning must be run again. Moreover, from the ASIDI point of view, the MV network is meshed with normally open meshing feeders, in addition to the inclusion of maintenance crews and additional switching equipment. Algorithms to determine the number and grid location of those elements have been implemented.

The fifth step deals with HV network planning, which links transmission substations to MV/LV substations (see 5 in figure 10). After running the Delaunay algorithm taking as input the location of HV/MV substations, HV customers, and transmission substations; each HV customer and HV/MV substation is associated to a single transmission substation. Considering rough estimations of the electrical momentum and voltages at those load points, several load point clusters are associated with each one of the transmission substations. Subsequently, a minimum spanning tree is computed in each cluster considering transmission substations as root nodes. An optimization algorithm is executed to estimate a quasi-optimum HV network, meeting voltage and current constraints, and the N-1 reliability criterion. This optimization is performed following a guided search tree where the different network options are classified according to the minimization of the weighted sum of the energy not supplied (ENS), investment, and operation and maintenance costs. The resulting output is a HV meshed network. In some occasions, the optimization process requires re-planning the HV/MV substations location; i.e. going back to the third step. Finally, the conductor size of HV overhead or underground power lines are optimized, taking into account energy losses costs.

On the other hand, the expansion-planning RNM applies additional algorithms to decide whether new network users are connected downstream of existing transformation capacity or, alternatively, new substation or transformers are required. Thus, new network users are classified into so-called feasible and unfeasible. Unfeasible network users are those that cannot be supplied by the initial network even when connected directly to a nearby substation because of capacity or voltage constraints. Furthermore, there are some users which could be supplied trough nearby transformation capacity without violating technical constraints, but are considered unfeasible since it would be less expensive to supply them from higher voltage levels. In order to determine this, the cost of connecting these users to their voltage level in the initial network is compared to an estimation of the cost of building new transformation capacity and the necessary lines to connect these users to a higher voltage level. Once new network users have been classified into these two categories:

- Within each voltage level, HV/MV substations or MV/LV transformers are installed for unfeasible consumers to make them feasible. Once they have been installed, all consumers turn feasible.
- Then, feasible consumers are initially connected directly to the nearest substation or transformer and a heuristic branch-exchange algorithm is applied to optimize their connection to the initial grid. This algorithm seeks to minimize the present value of investment and maintenance costs as well as the cost of losses associated with the new connections.

3.8 High Performance Requirements of the RNM

The RNM requires planning electrical networks in distribution areas. As distribution network planning is a NP-hard combinatorial optimization problem, it is inherently difficult to solve [Mori and Iimura 2003]. Computational complexity accounts for the fact that the computational burden and the resources required for solving a problem rapidly increase as the size of the problem grows. Furthermore, RNMs are very large scale planning tools which handle data concerning customers, substations, conductors, etc. For example, in Spain, the input data file corresponding to customers is a plain text file of approximately 2GB, which amounts to over 27 million customers. Besides, these models do not deal with a single voltage level. Instead, they have to plan low, medium and high voltage networks.

The RNMs are programmed in C++. Mathematical programming is not appropriate for this type of problems due to the huge computational burden derived from a real-world situation [Mori and Iimura 2003]. Moreover, some information sources such as the orography, the street maps or the standard equipment library may not be adequate for this type of techniques. Bearing these considerations in mind the RNM addresses the problem applying heuristics, such as branch-exchange [Míguez et al 2002]. Although this approach does not guarantee optimality the model is able to achieve reasonably efficient solutions, within a finite computation time.

Nevertheless, even with the use of heuristics, the total computation time required to plan the Spanish electrical distribution networks is about 324 hours. In order to decompose the problem and obtain the results with a reasonable amount of resources, distribution networks are planned separately for each company, covering areas of the size of a province. In the case of Spain, about 80 areas are considered. As the different areas are planned separately, they can be computed in parallel. For this purpose a launcher program is used. Data for each distribution area is extracted from the input global files and then different threads are created for each area. Each one of them executes the RNM separately. Figure 11 shows the launcher on the left, and the threads on the right. This approach permits to parallelize the processes. A server of 128GB RAM, 4 cores, each of them with 4 threads allows parallelizing up to sixteen distribution areas. With this approach the time restriction is the slowest case, which is about 33hours.

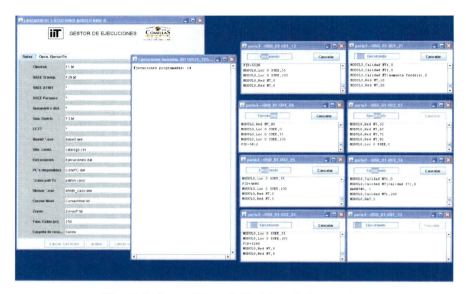

Fig. 11. Reference Network Model parallel computing

4 Impact of Distributed Generation on Grid Investment: Mitigation Options

The introductory section of this chapter highlighted the need to assess the cost and benefits of integrating renewable and distributed generation (DG) within the current energy context. The capabilities of RNMs make them a suitable tool to quantify the impact that new distributed energy resources can have on distribution network costs. The flexibility of RNMs also enhances the analysis of different operational approaches, from a passive grid operation to a more active network management (ANM) or smart grid. This information can be used by regulators to set the allowed revenues of distribution companies including the effects of DG or to design regulatory incentives to connect DG in an efficient manner.

This section will describe how RNMs can be used to assess distribution network costs under growing DG penetration levels including the effects of ANM strategies. This methodology will be applied to an actual distribution area located in The Netherlands. The results obtained will be presented and discussed. This section strongly feeds from [Cossent et al 2010a] and [Cossent et al 2010b].

4.1 Computing the Impact of DG on Distribution Costs through RNMs

Traditionally, distribution networks were designed to support the peak demand. Therefore, conventional planning models tend to consider only this situation when sizing network components. However, under significant DG penetration levels, it

Reference Network Models 267

may be necessary to consider a wider range of possible situations. Hereinafter, the concepts of scenario and snapshot will be used for the sake of clarity:

- Scenario: Characterized by the location (geographical coordinates and voltage level) of all network users (loads and DG), rated capacity of DG units and contracted power of consumers in a certain area.
- Snapshot: Any of the possible simultaneous energy consumption and DG production situations within a given scenario. According to this definition, a scenario can be composed of several snapshots. Herein, two snapshots per scenario will be considered: maximum net demand and maximum net generation. These are assumed to be the most relevant situations.

Since the RNMs available at the time this study was carried out were only prepared to consider one snapshot per scenario, it was necessary to analyze both snapshots separately. Nonetheless, future work will allow the RNMs to simultaneously consider several snapshots, thus overcoming this particular difficulty and simplifying the methodology.

Either the existing distribution network or a green-field network can be taken as a starting point for the analyses. The former alternative can introduce distortions in the results owing to the historic evolution of actual distribution networks (either over-investment or sub-investment is possible). On the other hand, disregarding the real network can yield totally insignificant results. The appropriate choice depends on the purpose of the study. In this case, the latter option was chosen because the study presented was not intended to provide useful information to a particular distribution company, but to draw general conclusions about the impact of DG on distribution costs. Notwithstanding, results are of interest since the real location of all consumers, DG units and supply points, the rated capacity of DG units and the contracted power of consumers can be used in any case.

Once some preliminary issues have been discussed, the methodology schematically displayed in Figure 12 will be described. For each scenario, one of the snapshots is used to obtain an initial network with the green-field RNM. The necessary reinforcements to this initial network that are required to support the power flows existing in the second snapshot are computed with the expansion-planning RNM[2]. Thus, the ability of the resulting network to handle any situation that may take place under any particular set of conditions of load and DG is ensured.

It is noteworthy that the order in which the snapshots are considered is relevant. The initial network determines to a great extent the final results. Hence, this choice cannot be made fully arbitrarily. It is deemed sensible to input first the most capacity demanding snapshot, i.e. the maximum generation one if peak net

---

[2] By doing this, the expansion-planning RNM is not applied to compute the reinforcements required to supply new load points or connect new DG units, as the expansion planning problem is generally understood, but to compute the reinforcements that are required to support the variation in simultaneous consumption and generation that may take place from one snapshot to another.

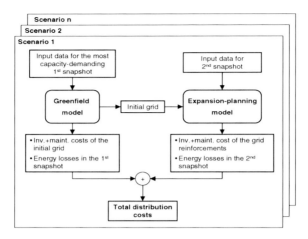

Fig. 12. Methodology to assess the impact of DG on distribution network costs

generation was higher than peak net demand and vice versa. This choice can be very clear in some areas, whereas in others peak net demand can be comparable to peak net generation. Being this the case, network costs can be computed under the two possible orders and the alternative with the lowest costs can be selected.

Finally, the outputs from both models are combined in order to calculate total distribution costs. Investment costs are the sum of those of the initial network and the reinforcements. Maintenance costs are directly obtained as those associated with the total investments. Finally, energy losses are calculated by taking into account the power losses occurring in the two snapshots considered. Note that energy losses can only be roughly estimated due to the uncertainties that exist about the evolution of demand and DG over a number of years. However, it is a relevant factor to size network components that cannot be neglected. The higher the number of snapshots considered, the more accurate the estimation of losses will be.

Following this process for several scenarios of DG and demand and comparing the results, it is possible to evaluate the impact that DG can have on distribution costs. Moreover, the analyses can be extended to several types of distribution areas (e.g. urban, rural, etc.) and different production technologies (e.g. photovoltaic, wind power, or combined heat and power (CHP)).

The transition towards smart grids will allow DSOs to implement advanced response options such as enhanced DG controllability or demand side management to achieve an ANM. In real life, this could be achieved through bilateral contracts between producers and DSOs, prices with some location/temporal differentiation, interruptibility contracts or real-time generation curtailment. RNMs can take into account these response options by modifying the simultaneity factors for loads and DG units[3]. A differentiation between DG technologies and types of loads can be made to account for the particular features of each of them. Thus, the impact of

---

[3] Note that simultaneity factors for DG units during peak hours can be viewed as some kind of capacity value of DG, i.e. its contribution to meet to peak demand.

DG on distribution network costs under a conventional passive approach and an ANM planning paradigm can be compared.

The main drawback of this approach is that the relevant response options and their potential have to be identified and quantified exogenously to the model for each case study. Note that the precise ANM strategies actually implemented is not relevant to the RNMs, only their effect on the simultaneity factors in the snapshots considered. Notwithstanding, this is not essentially different from the load forecasts that have been traditionally required to plan distribution networks.

4.2 Case Study: The Kop van Noord Holland Region

Kop van Noord Holland is a rural/sub-urban distribution area serving approximately 80,000 consumers spread over approximately 675 km^2. Most of the population lives in the south, whereas horticultural exploitations are present all over the region. This region is very favourable for the installation of wind farms and CHP units are widely used to provide heat to the numerous horticultural greenhouses. The distribution network is composed of a 150-50kV HV grid and a 10kV MV grid. The LV network will not be considered; LV consumers will be aggregated at the corresponding MV/LV transformers instead. This distribution area is displayed in Figure 13.

Fig. 13. Kop van Noord Holland. Source: [Cossent et al 2009]

This region is very interesting for studying the effects of DG because today DG capacity (CHP and wind power) is already comparable to peak demand. What is more, a considerable growth in DG capacity, above that of load, is expected in the near future. Consequently, generation will exceed local demand at certain times. In order to analyze this, eight scenarios have been constructed by combining two levels of demand (2008 and 2020) and four DG levels (no DG, 2008 DG, 2020 DG medium and 2020 DG high). The existing consumers and DG units at the beginning of 2008 were taken, whereas data for 2020 are estimates provided by the local distribution company. More details are provided in Table 2.

Table 2. Number and capacity of consumers and DG

Demand			Distributed Generation		
2008	MV load points	1307	2008	MV DG units	147
	MV load [MW]	317.27		MV DG capacity [MW]	202.45
2010	MV load points	1628		HV DG units	1
	MV load [MW]	856.15		HV DG capacity [MW]	23.65
			2020 medium penetration	MV DG units	321
				MV DG capacity [MW]	756.5
				HV DG units	1
				HV DG capacity [MW]	31
			2020 high penetration	MV DG	386
				MV DG capacity [MW]	1083.48
				HV DG	475
				HV DG capacity [MW]	1329.65

The advanced response options studied comprise shifting lighting demand of greenhouses (required to control the growth of flowers) to those periods with higher DG production, curtailing wind output at specific times and controlling CHP production through the use of thermal storage or back-up gas boilers. Table 3 and Table 4 show the power that is controlled through each one these response options and the corresponding change in the simultaneity factors respectively.

Table 3. Capacity managed through the different response options

Relevant snapshot	Wind curtailment [MW] Maximum generation	Rise in CHP generation [MW] Minimum generation	Reduction in CHP generation [MW] Maximum generation
2008 DG	n.a	30	40
2020 DG medium	80	120	180
2020 DG high	200	150	265
		2008 demand	2020 demand
Shift in demand of horticulturists [MW]		15	100

Table 4. Simultaneity factors

	Maximum net demand snapshot			Maximum net generation snapshot		
	MV loads	Wind	CHP	MV loads	Wind	CHP
BAU	0.7	0	0	0.24 (2008) 0.12 (2020)	1	1
ANM	GHs 0.5 Other 0.7	0	0.21	GHs 0.44(2008) 0.32(2020) Other 0.24(2008) 0.12(2020)	1 (2008) 0.6 (2020)	0.7

4.2.1 Results Obtained from the RNMs

In Figure 14, it can be seen that owing to the extraordinary high DG penetration levels that are expected in this area, distribution network costs increase in spite of the implementation of advanced response options. Notwithstanding, ANM allows attaining savings of over 30% of total costs as compared to a conventional or business-as-usual (BAU) planning approach.

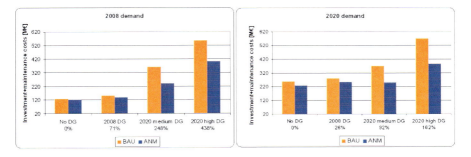

Fig. 14. Investment and maintenance costs: BAU versus ANM

This huge savings can be explained by the wide variety of response options and the extreme planning assumptions considered for the BAU situation. Note that under a BAU paradigm it has been assumed that DG did not produce at all at peak load hours whereas all units were producing at rated capacity during valley hours.

It should be remarked that the former conclusions are case specific. The potential benefits of ANM greatly depend on the characteristics of each distribution area (DG penetration, relative location of loads and DG, DG technologies), planning assumptions and the response options implemented. For example, cost savings can be expected to be greater in areas with high load and DG concentration or where a wider range of response options is considered. Therefore, RNMs may also allow distribution companies or regulators to identify the most beneficial ANM strategies for each distribution area, with regard to distribution network costs. This could constitute a valuable input for a broader cost-benefit analysis.

5 Plug-in Electric Vehicles and the Need of Smart Charging Strategies

The increasing environmental concern and the growing oil prices are driving the research efforts towards finding new transport fuels, such as biofuel, hydrogen or electricity. Among them, EVs seem to be the most promising option. Regarding EVs, two alternatives can be defined: pure battery electric vehicles (BEV) and plug-in hybrid electric vehicles (PHEV) which essentially work with a combination of two power sources, i.e. batteries and gasoline. The latter are an extended version of current hybrid electric vehicles including a battery with larger autonomy and able to connect to the grid to be charged. PHEVs might be a fast and temporary solution until more advanced EVs technologies become mature. The first generations of PEVs are expected to be connected to the grid just for charging the battery. However, as the technology matures, PEVs will also be able to inject energy into the grid when connected [Valentine-Urbschat 2009].

The grid access of PEV charging points is a new responsibility of the DSOs, as it occurs with other consumers or distributed generation. Similarly to DG, DSOs will have no control over the location of future PEV charging points, or the periods and frequency of PEV charging. Therefore, the network expansion planning problem will become more complicated due to a higher degree of uncertainty. The purpose of this section is to present a realistic analysis of the impact of PEV penetration on DSO investments and how different charging strategies can help to reduce the need of network reinforcements.

5.1 Description of the Approach

The objective of the analysis is to assess the impact of the EVs' penetration in the investments in a real distribution network. Initially, a base distribution network for the area is built considering that there are no PEVs connected. For this purpose, the green-field RNM is used to design the optimal network, using the size and location of consumers and DG. Then, different scenarios for PEV penetration, including three penetration and two demand levels, are defined. Finally, the additional investments are computed, for each of the PEV scenarios, and the different PEV charging strategies assessed.

5.1.1 Base Distribution Network

The area of study covers a region of 20 km^2, including three residential towns. The resulting base case distribution network is represented in Figure 15 [Cossent et al 2009]. In the area there is only one high voltage (HV) substation, whose capacity is assumed to be large enough to supply the required load in all scenarios. The medium voltage (MV) network is composed of 20 kV underground feeders that connect the substation with medium/low voltage (MV/LV) transformation centers. Low voltage (LV) cables connect LV electricity consumers to MV/LV transformers. There are 6,121 LV electricity consumers with a total contracted demand of 34.21 MW, and 15 MV customers with a total contracted demand of 37.54 MW. There are also some DG units: solar photovoltaic cells and micro CHP connected to the network.

The total population of vehicles owned by residents in the area has been estimated in 3,676 with a ratio of 0.6 vehicles per home or LV supply point.

5.1.2 Scenarios for EVs

The batteries used by PEVs are characterized by its storage capacity and by the power required to charge them in a predefined period of time, as presented in Table 5 [Kromer 2007], [Axsen et al 2008]. Charging rates (0.2C, 1C, or 2C) indicate the time to reach the full storage capacity from empty battery (e.g. at 0.2C, 1/0.2=5 hours are required). The possibility of V2G has been also considered in the study, where a percentage of the total PEVs are modeled as injecting energy into the grid at peak hours.

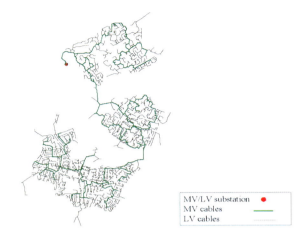

Fig. 15. Base case distribution network

PEVs will be connected to the distribution network at individual charging points (located in residential or parking areas) or in charging stations (comprise several connection points for fast charging). Generally, individual charging points will be in the LV network, while charging stations would be at the MV network.

Table 5. Characterization of different PHEV and BEVs batteries

	PHEV30	PHEV40	PHEV60	BEV
Real power [kW]	44	46	99	80
Energy capacity [kWh]	8	17	18	48
Charge at 0.2C [kW]	1.6	3.4	3.6	9.6
Charge at 1C [kW]	8	17	18	48
Charge at 2C [kW]	16	34	36	96

The level of uncertainty regarding the technological development and how these technologies will penetrate the market is high. Therefore, different PEVs penetration scenarios are defined, in line with other studies [EPRI&NRDC 2007]:

- Scenario 1: 35% of total vehicles are assumed to be PEV. PHEV 30, 40, and 60 models with slow and fast charging (1C) possibilities are considered. In addition, 10% of the PEV connected at peak hours can operate in V2G mode.
- Scenario 2: 51% of total vehicles are assumed to be PEV. In addition to PHEVs, BEVs are also included.
- Scenarios 3: 62% of total vehicles are assumed to be PEV. Fast charging (2C) is only considered in this scenario.

Finally, the impact of EVs on distribution networks will be analyzed in two demand scenarios: peak (from 4 pm to 9 pm) and off-peak hours. It is assumed that 40% of the PEVs are charging at peak hours, when 25% of the EVs perform a fast charge mode, and 10% are injecting power into the grid. At off-peak hours, it is assumed that 85% of the PEVs are connected, and only 5% are in fast charging mode. It is assumed that time-of-use tariffs or even real-time pricing with smart meters will be adopted for PEV charging.

5.2 Impact of PEV in DSO Costs

In the following, the needs of investments in the distribution network due to the penetration of PEVs is presented, together with the benefits for using two different smart charging strategies.

5.2.1 Investment Needs at Peak Hours

Peak load is the most relevant parameter for network planning. Consumers' peak demand is obtained by applying different simultaneity factors to the contracted demand, as presented in Table 1. For this analysis, a simultaneity factor 1 for PEVs is initially defined, which indicates that all PEVs charge at the same time. This constitutes a worst-case scenario approach, which may be used as a proxy for a scenario in the absence of smart charging strategies.

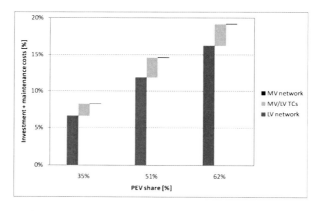

Fig. 16. Incremental investment and maintenance costs

Figure 16 presents the incremental investments, as a percentage of the investment cost for a distribution network without PEVs, needed in peak hours to integrate the different penetration levels of PEVs. The incremental investments include reinforcements in the low voltage (LV) network, medium/low voltage transformers and medium voltage (MV) network. In relative terms, the most important cost is that associated with the incremental investment in the LV network. The investments and maintenance costs increase 19% for the highest PEV penetration level. With these results the equivalent investment cost per PEV connected is 6,310 €/PEV.

5.2.2 Investment Needs at Peak Hours with PEV Smart Charging

A control system that avoids that PEVs charge simultaneously during peak hours may reduce the new infrastructure needs. For that purpose, the incremental investment costs have been studied for two different PEV charging strategies (see Figure 17): dumb charging (simultaneity factor equals 1) and smart charging (simultaneity factor lower than 1). Simultaneity factors are calculated as the probability of coincidence of charging PEVs, and take the following values: 0.8 for 0.2C charge rate, 0.15 for 1C charge, and 0.07 for the fastest rate 2C.

As expected, simultaneity factors lower than one highly decreases the need for network reinforcements. With the proposed PEV smart charging strategy investment needs can be reduced by three for the different penetration levels analyzed.

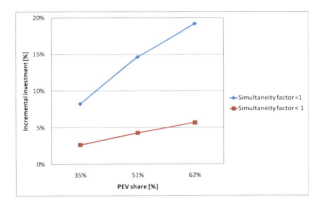

Fig. 17. Incremental investment in peak hours versus charging simultaneity factors

5.2.3 Investment Needs If PEV Charging Is Shifted

When the different PEV penetration scenarios for off-peak hours are analyzed, it is observed that there is no need to reinforce the network designed with the peak demand requirements (as computed in 5.2.1). Therefore, moving PEV charging from the peak demand period to the off-peak period, may reduce the investment requirements obtained for networks designed with the peak load. However, increasing PEV charging at off-peak hours may also increase investments needed at off-peak hours. For instance, in the scenario of PEV 62%, in order to reduce the investment needs by moving battery charging from peak to off-peak hours, up to 636 kW (around 187 PEVs) can be moved to off-peak hours in the studied area.

Figure 18 compares the initial investment needs and the investments required if PEV charging is shifted from peak to off-peak hours. The savings for the different penetration levels indicate that shifting decreases the investment needs, from 6% to 10.5%.

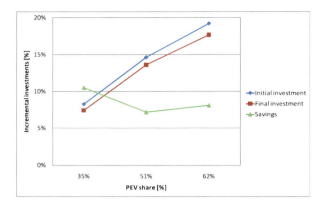

Fig. 18. Savings in network investment by shifting PEV charging from peak to off-peak hours

The results of this analysis show that, with the estimated levels of driving and charging patterns in peak hours, the required network reinforcements can reach up to 19% of total actual network costs in a situation without PEVs. Most of these reinforcements required are associated with the low voltage network. If PEV smart charging strategies are implemented, hence decreasing charging simultaneity factors, the required incremental investments are reduced to 6%. In addition, if strategies are defined so that some of the PEVs that were charged at peak hours are charged at off-peak hours instead, up to 8% of the required investment can be also avoided. A similar analysis for other distribution areas can be found in [Pieltain et al 2010].

Smart charging strategies for PEVs reduce the needs of new infrastructures in the distribution network that should be carried out by DSOs. It is important to note that the previous analysis does not include the costs of charging points or stations themselves, as they would be covered by the promoters of these infrastructures.

6 Concluding Remarks

This chapter has presented the main features of two large scale distribution planning models known as Reference Network Models. Those models present high computational performance being able to design whole distribution grids from scratch, green-field RNM, or incrementally from an existing network, expansion-planning RNM, in regions with millions of electricity consumers.

The selected data models and data architecture based on a layer kernel structure, together with a variety of algorithms from graph theory to electrical grids including reliability assessment, make this development a computationally efficient and robust tool.

Application cases regarding new challenges faced by distribution grids in order to accommodate increasing penetration levels of distributed generation or plug in electric vehicles with smart charging strategies have been presented. The results provided by these planning models convert them in a useful tool to assist utilities, regulators, and policy makers in decision making processes.

References

Al-muhaini, M., Arabia, S.: A New Particle-Swarm-Based Algorithm for Distribution System Expansion Planning Including Distributed Generation. Energy & Environment (2007)

Axsen, J., Burke, A., Kurani, K.: Batteries for Plug-in Hybrid Electric Vehicles (PHEVs): Goals and the State of Technology circa, Institute of transportation Studies, University of California (May 2008) http://pubs.its.ucdavis.edu/

Binato, S., Pereira, M.V.F., Granville, S.: A new benders decomposition approach to solve power transmission network design problems. IEEE Transactions On Power Systems 16, 235–240 (2001)

Carrano, E.G., Cardoso, R.T.N., Takahashi, R.H.C., Fonseca, C.M., Neto, O.M.: Power distribution network expansion scheduling using dynamic programming genetic algorithm. IET Generation, Transmission & Distribution 2, 444 (2008)

Chowdhury, A.A., Kumar Agarwal, S., Koval, D.O.: Reliability Modeling of Distributed Generation in Conventional Distribution Systems Planning and Analysis. IEEE Transactions on Industry Applications 39, 1493–1498 (2003)

Cossent, R., Olmos, L., Gómez, T., Mateo, C., Frías, P.: Mitigating the impact of distributed generation on distribution network costs through advanced response options. In: Proceedings of the 7th Conference on the European Energy Market, Madrid, Spain, June 23-25 (2010a)

Cossent, R., Olmos, L., Gómez, T., Mateo, C., Frías, P.: Distribution network costs under different penetration levels of distributed generation. European Transaction on Electrical Power (2010b) (accepted for publication)

Cossent, R., Olmos, L., Gómez, T., Mateo, C., de Joode, J., Scheepers, M., Nieuwenhout, F., Poot, J., Bongaerts, M., Trebolle, D., Doersam, B., Bofinger, S., Cali, U., Gerhardt, N.: Case studies of system costs of distribution areas. D5 IMPROGRES Project (2009), http://www.improgres.org

Diaz-Dorado, E., Pidre, J., Miguez Garcia, E.: Planning of large rural low-voltage networks using evolution strategies. IEEE Transactions on Power Systems 18, 1594–1600 (2003)

El-Khattam, W., Bhattacharya, K., Hegazy, Y., Salama, M.M.A.: Optimal In-vestment Planning for Distributed Generation in a Competitive Electricity Market. IEEE Transactions on Power Systems 19, 1674–1684 (2004)

EPRI & NRDC, Environmental Assessment of Plug-In Hybrid Electric Vehicles. Nationwide Greenhouse Gas Emissions. 1015325. Final Report, vol. 1, pp. 1–56 (July 2007)

European Smart Grids Technology Platform, Vision and strategy for Europe's electricity networks of the future (2006), http://www.smartgrids.eu/

Gallego, R.A., Alves, A.B., Monticelli, A., Romero, R.: Parallel simulated annealing applied to long term transmission network expansion planning. IEEE Transactions on Power Systems 12, 181 (1997)

Garver, L.L.: Transmission network estimation using linear programming. IEEE Transactions on Power Apparatus and Systems Pas-89, 1688 (1970)

Goswami, S.: Distribution system planning using branch exchange technique. IEEE Transactions on Power Systems 12, 718–723 (1997)

IEEE, IEEE guide for electric power distribution reliability indices, IEEE. Std 1366, 2001 Edition (2001)
Jamasb, J., Pollitt, M.: International benchmarking and regulation: an application to European electricity distribution utilities. Energy Policy 31, 1609–1622 (2003)
Khator, S.K., Leung, L.C.: Power distribution planning: a review of models and issues. IEEE Transactions on Power Systems 12, 1151 (1997)
Kromer, M.A.: Electric Power trains: Opportunities and Challenges in the US Light-Duty Vehicle Fleet. Massachusetts Institute of Technology (MIT), Massachusetts (2007)
Latorre, G., Cruz, R.D., Areiza, J.M., Villegas, A.: Classification of publications and models on transmission expansion planning. IEEE Transactions on Power Systems 18, 938 (2003)
Mateo, C., Gomez, T., Sánchez, A., Peco, J., Candela, A.: A reference network model for large-scale distribution planning with automatic street map generation. IEEE Transactions on Power Systems (2010) (accepted for publication)
Míguez, E., Cidrás, J., Díaz-Dorado, E., García-Dornelas, J.L.: An Improved Branch-Exchange Algorithm for Large-Scale Distribution Network Planning. IEEE Transactions on Power Systems 17, 931–936 (2002)
Ministry of Industry, Tourism and Trade, Royal Decree 222/2008, that establishes the remuneration regime for the electricity distribution activity (February 15, 2008) (in Spanish)
Mori, H., Iimura, Y.: Application of parallel tabu search to distribution network expansion planning with distributed generation. In: Power Tech Conference Proceedings, 2003 IEEE Bologna, June 23-26, vol. 1, p. 6 (2003)
Nara, K., Satoh, T., Aoki, K., Kitagawa, M., Ishihara, T.: Distribution systems expansion planning by multi-stage branch exchange. IEEE Transactions on Power Systems 7, 208–214 (1992)
Navarro, A., Rudnick, H.: Large-Scale Distribution Planning - Part I: Simultaneous Network and Transformer Optimization. IEEE Transactions on Power Systems 24, 744–751 (2009a)
Navarro, A., Rudnick, H.: Large-Scale Distribution Planning - Part II: Macro-Optimization With Voronoi's Diagram and Tabu Search. IEEE Transactions on Power Systems 24, 752–758 (2009b)
Peco, J.: Model for designing electrical distribution grids with geographical constraints (In Spanish). PhD dissertation. Universidad Pontificia Comillas. Madrid (2001)
Peponis, G., Papadopoulos, M.P.: New dynamic, branch exchange method for optimal distribution system planning. IEE Proceedings on Generation, Transmission and Distribution 144, 333–339 (1997)
Pieltain, L., Gómez, T., Cossent, R., Mateo, C., Frías, P.: Assessment of the impact of plug-in electric vehicles on distribution networks. IEEE Transactions on Power Systems (2010) (accepted for publication)
Ping, R., Li-Qun, G., Nan, L., Yang, L., Zhi-Ling, L.: Transmission network optimal planning using the particle swarm optimization method. In: Proc. 2005 International Conference on Machine Learning and Cybernetics, Guangzhou, China (2009)
Ramirez-Rosado, I.J., Bernal-Agustin, J.L.: Genetic algorithms applied to the design of large power distribution systems. IEEE Transactions on Power Systems 13, 696 (1998)
Ramirez-Rosado, I.J., Dominguez-Navarro, J.A.: New multiobjective tabu search algorithm for fuzzy optimal planning of power distribution systems. IEEE Transactions on Power Systems 21, 224 (2006)
Romero, R., Monticelli, A.: A Hierarchical Decomposition Approach For Transmission Network Expansion Planning. IEEE Transactions on Power Systems 9, 373–379 (1994)

Rothwell, G., Gómez, T.: Electricity Economics: Regulation and Deregulation. IEEE/Wiley, Piscataway (2003)

Rudnick, H., Donoso, J.A.: Integration of price cap and yardstick competition schemes in electrical distribution regulation. IEEE Transactions on Power Systems 4, 1428–1433 (2000)

Valentine-Urbschat, M., Bernhart, M.: Powertrain 2020 –The Future Drives Electric. Roland Berger Strategic Consultants Report, pp. 1–95 (September 2009), http://www.rolandberger.com

Electrical Load Modeling and Simulation

David P. Chassin

Pacific Northwest National Laboratory
Richland, Washington USA
David.Chassin@pnnl.gov

Abstract. Electricity consumer demand response and load control are playing an increasingly important role in the development of a smart grid. Smart grid load management technologies such as Grid Friendly™ controls and real-time pricing are making their way into the conventional model of grid planning and operations. However, the behavior of load both affects, and is affected by load control strategies that are designed to support electric grid planning and operations. This chapter discussed the natural behavior of electric loads, how it interacts with various load control and demand response strategies, what the consequences are for new grid operation concepts and the computing issues these new technologies raise.

1 Background

The National Academy of Engineering regards electrification as the single most important technological achievement of the 20th Century. Electricity brought about important social changes and influenced the course of industrialization by allowing factories to be built farther from sources of power than previously possible and making large-scale manufacturing possible. Cities changed and population centers grew at previously impossible rates in places that were previously considered uninhabitable. Driving by new machinery produced in these new factories, agricultural productivity increased dramatically, and a more mobile and educated middle class liberated from manual labor came into being.

The beginning of the 20th Century saw the rapid growth of commercial alternating current (AC) electricity production and distribution pioneered by brilliant engineers like Nikola Tesla and visionary businessmen like George Westinghouse. At the heart of this industry were the prime movers that provided the motive force for the electric generating units. The very first hydroelectric generating units evolved from the water wheels of the past and units like the Niagara station provided a reliable source of power. By 1925 high pressure steam turbines became an efficient alternative to hydroelectric power and facilities like the Edgar Station in Boston were a model for high-pressure steam power generation that used fossil fuels.

The next two major steps in the evolution of the electricity industry were rural electricity and regulation. These led to the rapid expansion of widespread, affordable, and standardized electricity infrastructure, giving access to electricity to a vast majority of Americans. During the 1930s, major infrastructure investments made by the United States government during the Great Depression resulted in the construction of dams such as Hoover and Grand Coulee, as well as the creation of important power authorities such as Bonneville Power Administration (BPA) and the Tennessee Value Authority (TVA). By 1942 these were connected to over 800 rural electric cooperatives through over 350,000 miles of long distance electricity transmission lines. Today, the North American electric system is dominated in most regions by fossil generation, and it touches nearly every aspect of our lives.

The technical challenge of maintaining the continuous operation of these interconnected systems has only increased with the size and complexity of our economy. Because electricity is not stored in the system, the loss of even a single major system element can cause adverse fluctuations in voltage and frequency, making it necessary for system operators to maintain what is now commonly referred to as security for the N-minus one contingency. Compliance with the N-minus one security criterion means that common contingencies such as the interruption of flow over an important transmission line or the disconnection of a major generating unit lead only to system disturbances, rather than service interruptions. Sometimes though, these disturbances can lead to cascading failures that do result in significant numbers of customers losing power. The most important of these include the November 1965 blackout in the Northeast, which led the creation of the North America Electricitiy Reliability Council (NERC) and the N-minus one criterion, and the August 1996 outage in the Western system, which led the widespread of deployment of Phasor Measurement Units (PMUs) to improve system observability. There have been many other significant events, and each provided important lessons to system planners and operators about how transmission systems can fail and led to improved grid planning and operations policies that reduce the incidence of outages.

The organizations responsible for operating the transmission system are a heterogeneous collection of governmental, quasi-governmental, and non-governmental entities known as power marketing authorities (PMAs), regional transmission organization (RTOs), and independent system operators (ISOs). Each performs the same basic function in different ways: they ensure that supply matches demand exactly within the major interconnections they belong to, as shown in Figure 1.

The task of matching supply with demand is complicated by the uncertainty of future demand and the availability of generation resources. In almost all transmission operations, the task begins years in advance with resource planning. This activity is based on forecasts of demand, which is influenced primarily by weather, demographic, economic and regulatory factors. These estimates are used by each regional authority to produce mid- to long-term generation and transmission investment plans. They are also shared with the North American Electricity Council (NERC), which is responsible to producing inter-regional and national planning projections.

Electrical Load Modeling and Simulation

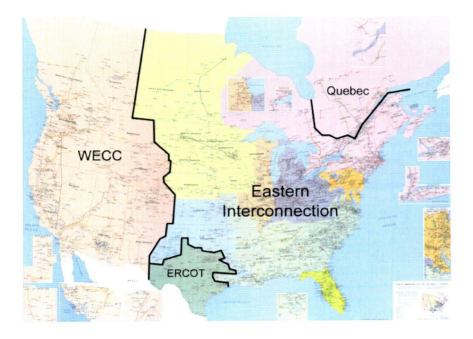

Fig. 1. Major interconnected transmission systems in North America

The next major element in the task of system operators is scheduling of existing generating resources, which is done at least one or two days in advance. This day ahead plan is used to schedule which plants are expected to be operating and at what level. As the time horizons to real-time operation shorten, price forecasts converge on the final price, and the dispatch levels resolve to the actual output of the plant. At each stage, the primary consideration is the balance of supply and demand. Economic and environmental considerations, while always present in the minds of system operators, are often a distant second to consideration of security to the N-minus one criterion.

Recently however, concerns about the impact of carbon emissions from fossil fuel plants have given rise to an effort to increase the amount of low- or no-carbon resources that generate electricity. These concerns have led to public policy measures, such as renewable portfolio standards and economic policies establishing and meeting carbon emission limits. Unfortunately, the idea of increasing renewable resources on the electric system is limited by a number of technical barriers that system planners and operators must overcome.

The first of these is that the number of remaining sites for large scale hydroelectric facilities is very limited. Most of the important river systems capable of supporting major hydroelectric projects have already been fully exploited, and those few that remain are often home to large populations who oppose the proposed projects. It seems highly unlikely that growing demand for electricity can be satisfied by much more hydroelectric power.

Conservation has been traditionally regarded as another resource, but it is in fact neither renewable, nor is it limitless. One can increase the efficiency of air-conditioners only so much before physical limits of Carnot systems are reached. Furthermore, demand reductions through denied or foregone service are only counted as a resource because of the regulatory principle of "obligation to serve." This results in an economic distortion that creates a confusing situation and obfuscates the planning problem. Were it not for this obligation of the utility to produce and the corresponding obligation of the customer to consume, conservation and other demand reduction measures could not be counted at all. (Demand reductions or negawatts only have value because the right to be served with exactly the right amount of energy at the right time was previously acquired either by regulatory fiat or by economic means such as in a capacity or energy auction.)

Nuclear energy has a significant potential to produce very large amounts of power with almost no carbon impact beyond the construction phase. However, the obvious political hurdles remain: it seems unlikely that the United States would countenance the construction of more than 400 new nuclear power plants needed to replace all the fossil based facilities in existence today.

Thus, system planners are left to consider the only significant remaining non-fossil resource that shows any substantial promise: wind. The most significant problem with this resource is the unpredictability of the prime mover itself. The wind intermittency problem imposes added resource requirements on the system so that N-minus one security criteria can be satisfied. There is an ongoing concern that with high levels of penetration of wind resources, the intermittency problem might become too severe to manage. Nonetheless, wind has shown itself to be competitive when installed with sufficient capacity, and countries such as Denmark, Germany, and Spain have achieved significant levels of generation with only modest increases in system cost.

Renewable energy resources are generally regarded as the most important near- to mid-term solution to the carbon-emission problem of electricity production worldwide. The most common of these are hydroelectric, wind, and solar. The installed capacity of hydroelectric power in the United States is about 80 GW with a capacity factor of about 40%. By the beginning of 2010 the installed capacity of wind in the US was about 35 GW with a capacity factor of less than 25%. (Solar is not yet a significant contributor to the total system production capacity, although it deserves mention in this context because it suffers from some of the same intermittency problems similar as wind power.)

Hydroelectric generators have a longstanding history as the basis for power generation. An important advantage of hydroelectric generators over all other standard types of bulk power generation is their ability to ramp output power up and down quickly in response to changing balance between supply and demand. In contrast, other generating facilities have more limited ramping capabilities, particularly so-called base load plants like thermal (e.g., coal, gas, nuclear) steam plants.

Another important factor affecting the behavior of generation resources is the availability of the fuel for the motive force. In the case of thermal plants this is either fossil or nuclear fuel which are more or less readily available and whose energy delivery is readily controlled. In contrast renewable energy resources rely on natural phenomena as the motive force, i.e., rain and snowmelt, air pressure gradients, or sunlight. Generally, hydropower systems are either substantial storage reservoirs or large "run of the river systems" in which there is little to no reservoir capacity but high (and potentially seasonal) sustained volumetric flow rates.

Wind and solar have the added disadvantage that both depend on highly variable prime movers, particularly over very short time spans. Both wind and sunlight can rise and fall significantly in a given geographic area over a matter of minutes. This uncertainty in the power generating capacity of wind and solar plants is a significant challenge for system planners and operators who must ensure that supply always matches demand exactly in a manner the conforms to the N-minus one criterion. Even in regions where wind resources are outstanding, as shown in Figure 2, the problem of high uncertainty and remoteness from load centers often accompanies high wind potential.

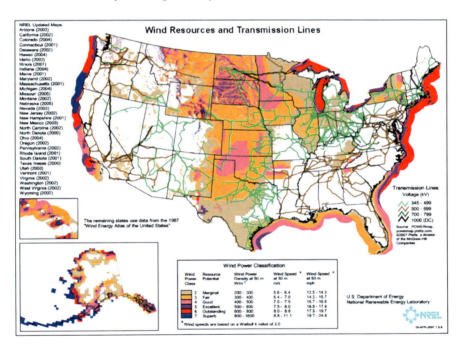

Fig. 2. Wind resources in the United States

The standard approach to addressing uncertainty in the security of the system is to increase the requirement for reserves in the form of thermal and hydroelectric resources that are online but unused. This reserve capacity requirement is typically about 15% of the maximum load and requires that sufficient additional capacity always be retained in case they are needed. Increasing the uncertainty of any generation resource imposes additional constraints on other generation resources to deal with contingencies related to the intermittency of prime mover.

One proposed approach to address the problem of renewable intermittency is to increase the energy storage capacity of the system as a whole. At the current time, the bulk electric system has almost no energy storage capacity at all. That is not to say that the technology does not exist: a back of the envelope calculation that can easily done is to estimate how long the US could be powered by all the automotive batteries in the country. (There are about 250 million vehicles, a 12 V lead-acid battery has about 70 Ah of storage, and the peak load in the US is about 850 GW.) Worst case scenario if all the generating resource in the US stopped at once on peak, the existing lead-acid batteries in our cars could keep the lights on for about 15 minutes. However, electricity storage is relatively expensive, it is costly to connect, and it's not clear that consumers are willing to subject their plug-hybrid electric vehicles to the demands of supporting renewable intermittency.

More recently, the development of the Smart Grid concept has given rise to the notion that instead of trying to regulate the system by controlling generation, we should try to control the demand to the extent possible. Controlling loads is usually called demand response, and there are several flavors of demand response. For example, a utility can simply cut power to a neighborhood or it can give customers rate rebates to allow direct control of loads like air conditioners and electric water heaters. However, utilities are well aware that these strategies have adverse effects on customers and only use them *in extremis*.

Another approach that has been tested and shows promise is the use of real-time prices combined with enabling technology to signal loads when supply is scarce (prices are high) or supply is plentiful (prices are low). When prices are high, loads can adjust in a non-intrusive way to reduce power draw, and conversely increase power draw when prices are low. If prices are updated frequently and load controls are highly interactive, then loads can become as effective at following intermittent generation as standard generation is at following changing load, without causing harm to consumers. One condition for success though is that such demand response be done with minimal or ideally no impact on the consumer. Questions remain as to how exactly this should be achieved and whether it could be accomplished without using prices *per se*, which obviously present some social and regulatory obstacles. While there are certainly numerous intriguing technical and theoretical problems to address, this chapter only considers technical problem of how much demand response might be found, modeled and used.

The standard method of control for interconnected electric power systems involves continuous control of generation to follow load. These controls are usually augmented with fast acting devices such as shunt and series devices to maintain system voltage and frequency at desired operating points. Even in extreme situations, load control and ultimately load shedding is retained only as measures of last resort, the obligation to serve being the paramount duty of the system operators. Whether system control is based on generator speed governor feedback or area generation control, the conventional focus of control is on the supply side of the system and not the demand side.

There has nonetheless been a longstanding interest in using load to assist in managing the balance and supply and demand in power systems. Time-of-use rates have been used for many decades to help reduce load during peak periods, although such economic measures are not generally regarded as load control *per se*, rather a mechanism that encourages consumer behavior through indirect economic signals that lead to load reduction at propitious times. Load control through broadcast of real-time prices has been described and demonstrated but the stability of the open-loop control it gives rise to has been a source of concern. (Interestingly, demonstrations of real-time price systems using bidding strategies for feedback have showed that such concerns may not extend to closed-loop price-based control systems.)

Direct load control can provide a measurable impact on load behavior. Thermostatically controlled loads in particular are more amenable to load curtailment strategies because of the inherent cyclic behavior. However, these strategies have potentially adverse delayed effects such as demand rebound, consumer fatigue and free-rider behavior. Direct control of small numbers of large on/off loads has long been known to not achieve the same results as proportional control of large numbers of small cyclic loads. This has necessitated careful design of direct load control systems such that sufficient numbers of participants provide an adequate equivalent capacity and that control actuation doesn't disrupt load diversity patterns or result in curtailment demands that last too long. Smart grid operations concepts like these have led to a growing interest in using higher-resolution real-time data from loads to establish continuous control of load that follows generation, in spite of persistent concerns about stability, controllability and observability and the lack of suitable load models, control designs, and end-use level metering systems.

The demand for load control to provide services other than peak load reduction has created a growing need for high fidelity models of loads. Indeed, there are already a wide range of potential applications of load control models, including forecasting, billing, load synthesis, control design, program cost/benefit analysis, measurement and verification, optimal control, load relief, load transfer, contingency analysis, customer classification, and unit-commitment and economic dispatch.

Recently load control has been proposed to mitigate wind and solar intermittency. However, this application, more than any other requires very advanced models of load behavior and demand control to address the stability, resource adequacy and economic considerations that inevitable arise when the power system controls are modified. This problem is especially challenging because conventional load control dispatch is essentially one-time curtailment only. Real-time price control systems use continuous signals and there is every reason to believe that only closed-loop control systems will be safe to operate on a large scale.

A more formal and integrated approach to modeling demand response behavior in the frequency domain was recently introduced that is useful for the study of economic power systems control, because it represents the aggregate behavior of loads in a manner that is highly consistent with tools used to consider the stability, controllability and observability of systems. Aggregate load models have been historically challenging and they typically focus on the composition of load by end-use class and rarely recognize the flexibility inherent in their cyclic nature.

2 Smart Grid Impacts

A smart grid uses digital technology to improve the reliability, security, and efficiency of the electricity system. Many stakeholders have brought various perspectives to a lively debate regarding the definition, benefits and purpose of a smart grid. Most place special emphasis on the interests they bring to the anticipated transformation of the utility system.

A significant number of technical areas are included in any reasonable taxonomy of the electricity system that covers the areas of concern to smart grid technology advocates. Progress made in moving toward a smart grid is often described in terms that focus on the interfaces between elements of the system and the systemic issues that reach beyond any single area. The areas of the electricity system that cover the scope of a smart grid include the following:

Area, Regional, and National Coordination Regimes: A series of interrelated, hierarchical coordination functions exists for an economically efficient and reliably operated electricity system. These include balancing areas (BAs), independent system operators (ISOs), regional transmission operators (RTOs), electricity market operations, and government emergency-operation centers. Smart grid elements in this area include collecting measurements from across the system to determine system state and health, and coordinating actions to enhance economic efficiency, reliability, environmental compliance, and response to disturbances.

Distributed-Energy Resource Technology: The most compelling frontier for smart grid advancements, this area includes the integration of distributed generation (DG), storage, and demand-side resources in electricity system operation. Smart appliances and electric vehicles may be an important components of this area, as well as renewable-generation components such as those derived from solar and local wind sources. Aggregate control of distributed energy resources (DER) are also considered.

Delivery Transmission and Distribution (T&D) Infrastructure: This represents the delivery part of the electricity system. Smart grid items at the transmission level include substation automation, dynamic limits, relay coordination and the associated sensing, communication, and coordinated action. Distribution-level items include distribution automation such as feeder-load balancing, capacitor switching, and restoration; and advanced metering such as meter reading, remote-service enabling and disabling, and demand-response gateways.

Central Generation: Generation plants already contain sophisticated plant automation systems because the production cost savings provide clear signals for investment. While technological progress in automation continues, the change is expected to be incremental rather than transformational, and therefore, this area is not usually emphasized in the context of loads in smart grids.

Information Networks and Finance: Information technology and pervasive communications are cornerstones of a smart grid. Though the information network requirements (capabilities and performance) will be different in different areas, their attributes tend to transcend application areas. Examples include interoperability and the ease of integration of automation components, as well as cyber security concerns, such as privacy, integrity, and availability. Information-technology-related standards, methodologies, and tools also fall into this area. In addition, the economic and investment environment for procuring smart-grid-related technology is an important part of the discussion concerning implementation progress.

Prominently figured in any discussion of the smart grid is the role of the consumer. Part of compelling vision of a smart grid is the ability to enable informed participation by customers, making them an integral part of the electric power system. With bi-directional flows of energy and coordination through communication mechanisms, a smart grid should help balance supply and demand and enhance reliability by modifying the manner in which customers use and purchase electricity. These modifications can be the result of consumer choices that motivate shifting patterns of behavior and consumption. These choices involve new technologies, new information regarding electricity use, and new pricing and incentive programs.

A smart grid adds consumer demand as another manageable resource, joining power generation, grid capacity, and energy storage. From the standpoint of the consumer, energy management in a smart grid environment involves making economic choices based on the variable cost of electricity, the ability to shift load, and the ability to store or sell energy.

Consumers who are presented with a variety of options when it comes to energy purchases and consumption are able to:

1) respond to price signals and other economic incentives to make better-informed decisions regarding when to purchase electricity, when to generate energy using distributed generation, and whether to store and reuse it later with distributed storage and
2) make informed investment decisions regarding more efficient and smarter appliances, equipment, and control systems.

In the final analysis, system engineers must be able to understand and incorporate models of consumer demand and the devices they use in the design and operation of a smart grid. The models must include all the salient features of the devices that support the smart grid to quantify the benefits and impacts of smart grid technology on the overall electric system. This section focuses on how those features are modeled and the computational methods used when these models are implemented in simulation and engineering analysis tools.

3 Load Model Fundamentals

Load modeling is the process of describing how electricity is consumed by customers of different classes. The principal classes of customers are residential, commercial, industrial and agricultural electricity customers. In this section we will discuss the models of different classes of loads, how they are represented computationally, considerations in data provisioning of load models, and how they are numerically solved.

Aside from the main distinction of whether the solution is needed in the time domain or in the frequency domain, the approach to modeling loads differs depending on the time scale being considered and the accuracy of the solution sought. Electromechanical dynamics should be considered for time scale of a second or less. Between a second and an hour, state dynamics should be considered. Load shapes should be considered between an hour and several weeks. Problems ranging from several weeks to about a year should consider seasonal load behavior, and beyond a year, load growth should be considered, as shown in Figure 3.

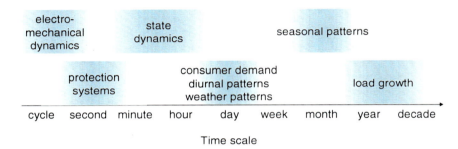

Fig. 3. Electrical load models consider different phenomena according to the time scale

While it is not generally an issue, it is important to recall that there is always an intrinsic limit to the precision with which a solution in either frequency or time domain can be found. This limit exists because the more precision with which we define the time of an event, the less we can know about its frequency and vice versa. In its simplest form, this is expressed as an uncertainty principle

$$\sigma(S)\sigma(\hat{S}) \geq \frac{1}{4\pi} \qquad (1)$$

where S is the normalized power in the time domain, $\hat{S}$ is its Fourier transform (i.e., the normalized power in the frequency domain), and $\sigma(y)$ is the standard deviation of $y(x)$ about $x=0$. This sampling limit can be a problem when considering load behavior that is limited to a one-sided base bandwidth B. The band limited signal's Fourier transform gives us the amplitudes

$$\hat{X}(f) = \int_{-\infty}^{\infty} X(t) e^{2\pi i f t} dt \qquad (2)$$

at nearly the same frequency as it is being sampled, which is particularly a concern for thermostatic or intermittent loads. For example, the measurement of the duty cycle of an air-conditioner relies on observations of at least three state transitions (*on-off-on*, or *off-on-off*), which can take up to several hours to occur. Therefore, a single observation of an air-conditioner's power consumption at an instant in time provides almost no information about its cycling behavior (or whether it is even cycling at all!) The observation of a single cycle provides only enough information to describe that cycle but may not provide very much information about the mean duty cycle, the period or their variances.

3.1 Static Load Models

Depending on the time horizon, load models are used to compute different properties of the electric network's behavior. For the shortest time scales instantaneous voltage and current, or complex power are the quantities sought and the load models must incorporate both the voltage and frequency elements necessary to computing them. For steady state power flow solutions, the frequency dependent elements of the model are omitted so that only RMS voltage and current, or RMS power can be determined. For any time scale beyond the duration of a steady state solution the RMS power changes, so it is typical for the models to compute the energy consumption per unit time instead of RMS power.

ZIP Load Models

For the purpose of computing the steady-state behavior of an electrical system, electrical loads typically include a representation of the three principle *load components*—the constant impedance (Z), current (I) and power (P) load response with respect to voltage, as shown in Figure 4.

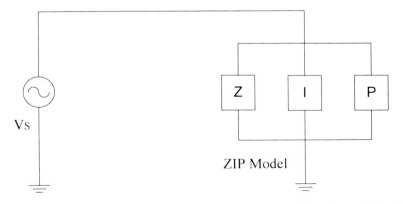

Fig. 4. Static (Z, I, P) load models are the most common representation of electrical loads

The total real power consumed by a ZIP load is

$$P_{ZIP} = L(V^2 Z|F_Z| + VI|F_I| + P|F_P|), \qquad (3)$$

the total reactive power is

$$Q_{ZIP} = L(V^2 Z \sin\cos^{-1} F_Z + VI \sin\cos^{-1} F_I + P \sin\cos^{-1} F_P) \qquad (4)$$

under the constraint

$$Z + I + P = 1 \qquad (5)$$

where
- V is the voltage at the terminals of the load;
- L is the total apparent power consumption of the load;
- Z, I, and P are the constant impedance, current and power component fractions; and
- F_Z, F_I, and F_P are the impedance, current, and power component power factors.

When multiple devices are considered, their individual total real and reactive power contributions to the system load must be computed separately and summed before they can be connected to a bus of a given voltage. Furthermore, if two loads having different power factors are located at two buses of different voltages, their real and reactive power consumptions may not be combined algebraically when solving for the electrical system voltages and currents.

Every type of electrical device has distinct values for the static component fractions and angles. Table 1 provides a summary of common residential plug-in devices component fractions and angles.

For motors, we must compute the motor's RMS power requirements for static load components as shown in Table 2.

For load components that involve heating and cooling applications, the power is a function of the thermal loading and efficiency of the system, as shown in Table 3.

The duty-cycle of air-conditioning and heating systems can be estimated by interpolating between the balance outdoor temperature (at which the duty-cycle is zero) and the design outdoor temperature (at which the duty-cycle is unity). The balance outdoor temperature is found by solving the ETP equation (see below) for outdoor temperature with no heating or cooling. Typical values for balance outdoor temperatures in residential buildings are between 50 and 60°F for heating and 65 and 75°F for cooling, depending on solar gains and natural ventilation. With higher solar gains and/or lower natural ventilation, the balance temperature tends to be lower. In commercial buildings the outdoor balance temperature tends to be lower than residential buildings in the same area by 10°F or more because of higher internal heat gains.

Table 1. ZIP load component fractions and power factors of common residential devices

	ZIP value (per unit)			Power factor		
	Z	I	P	F_Z	F_I	F_P
Light (70W filament)	0.571	0.436	0.003	1.00	-1.00	1.00
Light (13W CFL)	0.409	0.670	0.585	-0.88	0.42	-0.78
Light (20W CFL)	-0.011	1.0	0.011	0.00	-0.81	0.90
Light (42W CFL)	0.487	-0.375	0.888	-0.97	-0.70	-0.79
Television (CRT)	0.002	0.827	0.172	-0.99	1.00	-0.92
Television (Plasma)	-0.321	0.484	0.837	0.85	0.91	-0.99
Television (LCD)	-0.038	0.040	0.999	0.61	-0.54	-1.00
Computer display (LCD)	-0.407	0.463	0.944	-0.97	-0.98	-0.97
Desktop fan	0.733	0.253	0.135	0.97	0.95	-1.00

Table 2. Estimating motor RMS power from motor application

Type of load	Motor RMS power (kW)
Rotating motion	1.42×10^{-4} x *Torque* (ft.lb) x *Speed* (rpm)
Linear motion	2.26×10^{-5} x *Force* (lbs) x *Speed* (ft/min)
Water pumps	1.88×10^{-2} x *Flow* (gpm) x *Head* (ft) / *Efficiency* (%) 4.35×10^{-2} x *Flow* (gpm) x *Pressure* (psi) / *Efficiency* (%)
Fans/blowers	1.17×10^{-2} x Flow (cfm) x Pressure (psi) / Efficiency (%) 0.326 x Flow (cfm) x Pressure (water inches) / Efficiency (%)

Table 3. Estimating residential and commercial heating/cooling load

System type	Power (kW)
Residential	
Baseboard	2.93×10^{-4} × Heating Capacity (Btu/h) × Duty cycle
Heat pump	2.93×10^{-4} × Heating Capacity (Btu/h) × Duty cycle × COP
Air-conditioner	8.33×10^{-5} × Cooling Capacity (tonnes) × Duty cycle / EER
Refrigerators	1.82×10^{-3} × Volume (cf) + 0.022
Freezers	2.93×10^{-3} × Volume (cf) + 0.015
Commercial	
Heating	2.93×10^{-4} × Heating Capacity (Btu/h) × Duty cycle
Air-conditioning	8.33×10^{-5} × Cooling Capacity (tonnes) × Duty cycle / EER
Refrigeration	Walk-in: 0.031 × Volume (cf) Reach-in(15-30 cf): 2.93×10^{-3} × Volume (cf) + 0.058
Freezing	Walk-in: 0.041 × Volume (cf) Reach-in(15-30 cf): 1.37×10^{-3} × Volume (cf) + 0.824

The heating and cooling design capacities are generally calculated based on the heat loss and gain at the heating and cooling design temperatures for the climate, respectively. The heating design temperature is usually the coldest night-time temperature expected, and the cooling design temperature is usually for the sunniest and hottest afternoon expected. When the space conditioning system is a heat-pump, the largest of the two capacities is used. In most climates, the cooling design temperature dominates the system design capacity.

The EER values of air-conditioning units can be estimated from the SEER rating of the unit as follows

$$EER = -0.02\,SEER^2 + 1.12\,SEER \qquad (6)$$

where the value of the SEER rating depends on the make, model and age of the air-conditioning unit. The minimum rating depends on the prevailing building energy codes and equipment standards and the best performing units no more than double the minimum performance.

Sometimes it is necessary to apply a frequency correction to static loads. The most common correction is the application of the factor

$$F = 1 + A\Delta f \qquad (7)$$

where A is the sensitivity of the load to the change in frequency Δf.

Variable Load Models

At the sub-hourly time scale, many loads exhibit state dynamics that give rise to load state synchronization phenomena such as *cold load pickup* and *demand response rebound*. These phenomena are some of the most difficult load behaviors to model. They are typically modeled by estimating changes in the load components using reduced-order models or agent-based models.

Similarly, at the hourly time scales, the load components are not usually constant. Loads are often driven by hour-to-hour changes in consumer behavior, diurnal effects such as sunlight, and weather effects such as outdoor temperature and humidity changes. These effects are often represented by *load shapes*. Most often, load shapes represent the daily pattern of energy demand in loads, which can be converted into hour-by-hour changes in the value of the ZIP loads. This holds true for seasonal and multi-year time scales, even through the driving functions differ—weather patterns and daylight hours for seasonal effects and load growth trends for multi-year time scales.

Load Shapes

When the variation of the static load needs to be considered over time, load shapes are frequently used. Load shapes describe the changes in load on a daily to seasonal timescale. Most load shapes are represented as a time-average of energy, e.g., kWh/h. However seasonal load shapes are sometimes represented as peak value, e.g., MW, even though sometimes that peak value is obtained from the maximum of an average diurnal load shape, e.g., MWh/h. In either case, the load is effectively a power value, as shown in Figure 5.

There are some important characteristics of load shape data that must always be considered when they are used. The significance and impact of these characteristics will depend on the model being developed and its application. Certain end-uses vary widely depending on climate, geographic location, local energy codes, and demographics. The following are some important factors to consider.

- The difference between summer and winter lighting load shapes is greater the further the load is from the tropics.
- For the same outdoor temperature, air-conditioning loads are typically higher in humid climates than dry climates.
- Affluent regions typically have higher loads per capita than less affluent areas.

- Commercial loads are much less dependent on climate and weather than are residential loads.
- Industrial loads are highly sensitive to economic conditions.
- Agricultural loads are highly seasonal and sensitive to seasonal weather patterns.
- Load shape data can change significantly over time because of evolving energy efficiency standards and consumer purchasing habits. A lot of load shape data from the 1980s is still being used, but the penetration of consumer electronics in the residential market has increased significantly since then even though the efficiency of appliances has improved at the same time.
- The composition of the load has changed over the years. Many loads that used to be fundamentally ZIP have slowly become electronic when considered over an entire population of end-uses. This is particularly true in the food preparation, consumer electronics, lighting and motor loads. Constant impedance loads are increasingly the minority of the load, and most cyclic end-use loads are closer to constant power when taken over time intervals exceeding a single duty cycle.

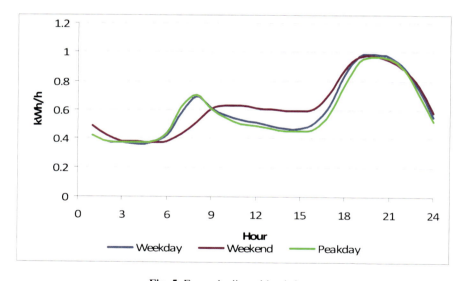

Fig. 5. Example diurnal load shape

Load shape normalization is usually performed over a specific time interval, such as a day or week. Normalized load shapes are required to perform load shape scaling based on daily or weekly energy usage. The N-hour load shape normalization factor E_N is

$$E_N = \sum_{i=1}^{N} L_i \qquad (8)$$

where L_i is the load at the ith hour. The normalized load $l_i = L_i / E_N$ is the fraction of the energy used during the ith hour, and can be used to compute rescaled load L'_i of the ith hour of an arbitrary period with energy use E'_N:

$$L'_i = E'_N l_i. \tag{9}$$

Discrete Load Models

Discrete load models are a class of models that describe the individual components in a population of loads. These models typically have greater fidelity but can require considerably more computational resources. In general, two types of discrete load models are considered, demand models and physics-based models.

Demand models are typically based on load-shape-like data where the shape represents the consumer or end-use demand profile. These profiles represent a rate of activity in a process, such as the gallons-per-minute of hot water consumption or the hamburger grilled in a fast-food restaurant. A single load is represented as a function of the demand, and the aggregate population is modeled as the sum of many single loads.

Physics-based models are typically developed based on an engineering model of the load, such as an air-conditioning system or a hot-water heating system. The derivation and data provision of physics-based models of the system can be challenging, but usually reasonable approximations can be made that will retain fidelity on the time-scale at which the model is being used. The main advantage of physics-based models is they can provide excellent sensitivity estimates even when the absolute values are not very accurate. The most common agent-based physical model that is found is the Equivalent Thermal Parameters (ETP) model, which is based on a simplified thermal network model of the load that is solved as a function of time.

For example, the thermal behavior of a house and its heating/cooling system can be described by the differential equations

$$\begin{aligned} \dot{T}_a &= \frac{T_m U_m - T_a(U_a + U_m) + T_o U_a + Q_a}{C_a} \\ \dot{T}_m &= \frac{U_m(T_i - T_m) + Q_m}{C_m} \end{aligned} \tag{10}$$

where
- U_a is the thermal conductance between indoor and outdoor air;
- C_a is the heat capacity of the indoor air;
- C_m is the heat capacity of the indoor mass;
- U_m is the thermal conductance between the indoor mass and the indoor air;
- T_i is the indoor air temperature;
- T_m is the indoor mass temperature;
- T_o is the outdoor air temperature;
- Q_a is the heat flux to the indoor air; and
- Q_m is the heat flux to the indoor mass.

The solution to the ETP equations is found from the general first-order ODEs that correspond to the ETP equations:

$$\dot{T}_i = c_1 T_i + c_2 T_m + c_3$$
$$\dot{T}_m = c_4 T_i + c_5 T_m + c_6 \quad (11)$$

where the constants c_1 through c_6 are defined as

- $c_1 = -(U_a + U_m) / C_a$;
- $c_2 = U_m / C_a$;
- $c_3 = (Q_a + U_a T_o) / C_a$;
- $c_4 = U_m / C_m$;
- $c_5 = -U_m / C_m$;
- $c_6 = Q_m / C_m$.

The general form of the second-order ODE is

$$p_1 \ddot{T}_i + p_2 \dot{T}_i + p_3 T_i + p_4 = 0 \quad (12)$$

where

- $p_1 = 1 / c_2$;
- $p_2 = -(c_1 + c_5) / c_2$;
- $p_3 = c_1 c_5 / c_2 - c_4$; and
- $p_4 = c_3 c_5 / c_2 + c_6$.

The solution is found using the roots r_1 and r_2 of $p_1 r^2 + p_2 r + p_3 = 0$ and the parameters

$$k_1 = \frac{r_2 \left(T_i(0) + \frac{p_4}{p_3} \right) - \dot{T}_i(0)}{r_2 - r_1}$$
$$k_2 = \frac{T_i(0) - r_1 k_1}{r_2} \quad (13)$$

where the initial change in indoor air temperature $\dot{T}_i(0) = c_1 T_i(0) + c_2 T_m(0) - (c_1 + c_2) T_o + Q_a / C_a$. The time-solutions for the air and mass temperatures of the house are

$$T_i(t) = k_1 e^{r_1 t} + k_2 e^{r_2 t} - \frac{p_4}{p_3}$$
$$T_m(t) = k_1 \frac{r_1 - c_1}{c_2} e^{r_1 t} + k_2 \frac{r_2 - c_1}{c_2} e^{r_2 t} - \frac{p_4}{p_3} + \frac{c_6}{c_2} \quad (14)$$

The internal gains $Q_a + Q_m$ must include the internal heat gains from people, appliances and other sources (or sinks) of heat in the home. For all end-uses, a fraction (if not all) the power provided ends up as heat in the air or the mass of the house, as described in Table 4.

Table 4. Residential end-use heat gain fractions by season

End-use	Heat gain fraction
Lights	0.9
Plugs	1.0
Washer	1.0
Hotwater	0.5
Refrigerator	1.0
Dryer	0.15 (see Note 1)
Freezer	1.0
Dishwasher	1.0 (see Note 2)
Range	0.8 (see Note 3)
Microwave	1.0 (see Note 3)

Notes:
1. The heat gain ratio can be higher is summer and lower in winter.
2. The heat gain can be lower if water heating is enabled in the dishwasher.
3. The heat gain can be lower if the ventilation fan is on.

Given knowledge of the indoor air temperature, a thermostat control strategy must be provided to determine when the heating/cooling system is running. The simplest control strategies employ a set point T_{set} and a hysteresis D to control the state of the heating/cooling system. Often there are two set points, one for cooling and one for heating, as well as an auxiliary heating strategy for heat-pump systems. To compute the time to the next set point change, equation (14) must be solved for the time t as a function of $T_i \pm \frac{1}{2}D$. (Note that there is no closed-form solution to this equation and it must be solved using iterative methods, such a Newton's method.) When the heating system is *on*, the value of Q_a includes the heat flow between the heating/cooling system and the air. Simultaneously, the house electrical load includes the heating/cooling system loads. It is common to avoid this problem by employing solution methods with a fixed time-step. However doing to so will require choosing a time-step short enough to a) not exceed the shortest device cycle time, e.g., about 2 minutes, and b) not give rise to simulation synchronization artifacts that do not exist in reality.

Aggregate Load Models

Under many circumstances, load models can be summed algebraically or scaled by factors such as the number of customers, energy consumption or installed capacity to create aggregate load models. There is one important exception that should be noted: discrete load models where the diversity of load states changes over time cannot always be summed algebraically or scaled. In such cases, a numerical or analytic approach to load aggregation must be considered.

These approaches rely on discrete state-space models of the load state dynamics, such as is shown in Figure 6. The fraction $n(v,t)$ of loads in each state v describes the state x of the aggregate population of loads at the time t. The total load is simply the sum of the loads that are in an *on* state, multiplied by the power of each load.

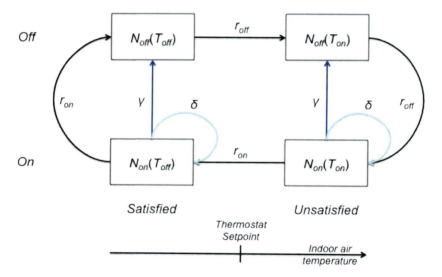

Fig. 6. State space model of a thermostatic load where consumers are *satisfied* or *unsatisfied* and devices of *on* or *off*

When the system of loads is quiescent, the loads follow the natural control regime cycle (shown in blue), turning on and off with regularity in a highly diversified way. However, both consumer behavior and utility load control signals can disturb the natural cycle of the loads, resulting in fluctuations in the aggregate load. State space models can also be used to develop the canonical control model for both consumer behavior and utility load control signals, as in

$$\dot{x} = Ax + Bu$$
$$y = Cx + Du \tag{15}$$

where u is the input signal (i.e., load control signal), y is the output (i.e., load), A is the load response model, B is the control system model, C is the load observation model, and D is the feed-forward control model (if any). Based on this control model, the canonical transfer function for the response of the aggregate load can be easily developed. The discrete transfer function is and can be used to perform analysis such as that shown in Figure 7.

$$H(z) = C(zI - A)^{-1} B + D. \tag{16}$$

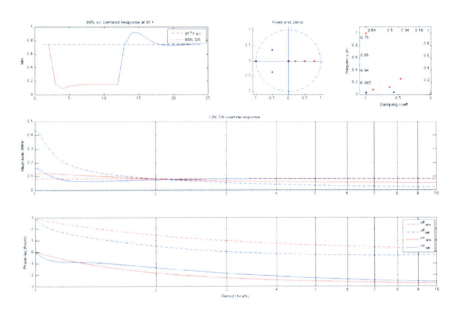

Fig. 7. Time domain (top left), stability analysis (top right), and response (bottom) of an air-conditioning load control program

Load Control

The controllability of loads under load control programs can be tested using aggregate load models. Controllability is determined by considering whether

$$rank[B \quad AB \quad AB^2 \quad \mathsf{L} \quad AB^{n-1}] = n \tag{17}$$

which will generally be true provided that every cycle in the state diagram is influenced by the control signal u. In such cases, it is expected that the load control system will be able to reach every possible state x given an allowable sequence of inputs u. This will be true for both open-loop and closed-loop load control strategies.

The observability of load control programs can also be tested using aggregate load models. Observability is determined by considering whether

$$rank \begin{bmatrix} C \\ CA \\ CA^2 \\ \vdots \\ CA^{n-1} \end{bmatrix} = n \qquad (18)$$

which may *not* be true for certain load control systems because as a general rule, load control systems rely on metering infrastructures that only measures the total load that is on and may not observe the distribution of the internal states of those loads that are not on. They do not typically measure the internal states of individual loads. The implication is that closed-loop control of aggregate loads is not possible unless the internal states of the loads are somehow observed. Such is the case in well-designed price-based load control systems where load bidding is performed using the internal states of the loads (e.g., temperature, status) and the observability criteria is satisfied.

3.2 Dynamic Load Models

For any non-steady condition, the static ZIP model is not sufficient because the current may depend on voltage in non-trivial ways. Electro-mechanical transient phenomena with sub-second time scales require that voltage and frequency responses be included in the load models. Motors and electronic loads are considered separately because the interaction between the mechanical load and the electromotive force driving it cannot be properly described using the static ZIP model, as shown in Figure 8.

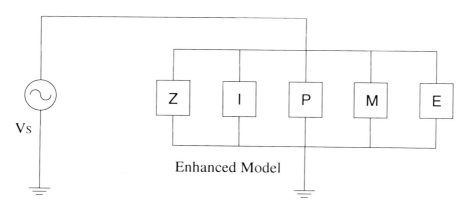

Fig. 8. Enhanced models use additional load components, such as motors (M) and electronics (E)

Induction motors draw current based on the mechanical torque and the load on the motor is

$$I = \frac{sV^2}{\sqrt{R^2 + s^2 L^2 \omega^2}} \quad (19)$$

where
- ω is the angular frequency of the electric power ($2\pi f$);
- R is the rotor resistance (in Ohms);
- L is the rotor inductance (in Henries); and
- s is the rotor slip (in fraction of angular frequency);

Motor stalling and protection behavior must be considered for time scales up to a few minutes. As the terminal voltage on the motor is reduced, the probability of motor stalling rises.

Synchronous motors run at a constant speed that is a function of the electrical frequency f and the number n of magnetic poles in the motor

$$v = \frac{120 f}{n}. \quad (20)$$

Synchronous motors show some interesting properties, which find applications in power factor correction. The synchronous motor can be run at lagging, unity or leading power factor. The control is done using the field excitation, as described below:

- Under-excitation occurs when the field excitation voltage is decreased and the motor runs in lagging power factor. The power factor by which the motor lags varies directly with the drop in excitation voltage.
- Unity power factor occurs when the field excitation voltage is made equal to the rated voltage, the motor runs at unity power factor.
- Over-excitation occurs when the field excitation voltage is increased above the rated voltage, the motor runs at leading power factor. The power factor by which the motor leads varies directly with the increase in field excitation voltage.

The most basic property of synchronous motor is that it can be used both as a capacitor or inductor. The leading application of power factor operation of synchronous motors is power factor correction. Normally, all the loads connected to the power supply grid run in lagging power factor, which increases reactive power consumption in the grid, thus contributing to additional losses. In such cases, a synchronous motor with no load is connected to the grid and is run over-excited, so that the leading power factor created by synchronous motor compensates the existing lagging power factor in the grid and the overall power factor is brought close to 1 (unity power factor). If unity power factor is maintained in a grid, reactive power losses diminish to zero, increasing the efficiency of the grid. This operation of synchronous motor in over-excited mode to correct the power factor is sometimes called a synchronous condenser.

Note that variable frequency drives typically use synchronous motors, but should be considered electronic loads for the purposes of electrical modeling. The mechanical load should be modeled as any other motor load.

Electronic loads exhibit dynamic behavior that is highly dependent on the design the power supply and must be either measured or derived based on an engineering analysis of the combined electrical and mechanical system. This applies to devices that use switching power supplies, power inverters, and variable frequency drives. The following general principles should be considered when modeling electronic loads:

- **Switched power supplies** draw current in the "rabbit ear" waveform, i.e., on the rising and falling edges of the voltage waveform. These types of loads are common in almost all consumer electronic devices, such as computers, compact-fluorescent lights, home entertainment systems, and many appliances.
- **Variable frequency drives** can come in two types. The simpler designs have only a single inverter and do not draw current when the primary voltage falls below the DC bus voltage. More advanced designs include a second inverter and backfeed power when the primary voltage falls below the DC bus voltage. Most drive controllers include setup features that control this type of behavior, although it is not unusual to find that equipment installers leave these features set to the factory defaults.

Load Components

While maintaining the standard constant impedance, current and power models (ZIP), composite load models allow for more accurate representation of induction motors with various inertial characteristics, as well as electronic loads, which decouple from the bulk system under certain conditions. The composite model shown in Figure 9, has two additional critical features: 1) the model includes the electrical "distance" between the transmission bus and the end-use loads, and 2) the model represents the diversity in the composition and dynamic characteristics of the various end-use themselves. These features are required to correctly represent the impact of loads on the voltage during dynamic transient grid events.

Ordinarily, modelers want the minimum number of load model components necessary to maintain fidelity of the model and capture the diversity in the dynamic characteristics of the relevant end-uses. The various metrics for classifying direct-drive motor performance are a) number of phases, b) electrical characteristics, c) load characteristics, d) inertial behavior of combined motor and load, and e) protection and controls. (Note that variable frequency drive motors are classified as electronic loads). It has been shown that the most important characteristic that distinguishes motors is the driven load, i.e., constant torque versus speed-dependent torque. Motors that drive constant torque loads such as compressors have a much higher susceptibility to stalling. These motors consequently have a more significant adverse impact on voltage stability and damping of inter-area power oscillations. Furthermore, the behavior of three

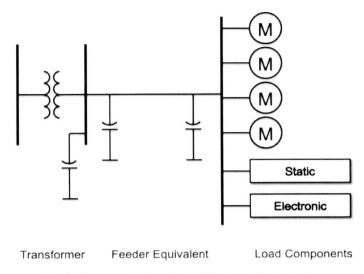

Fig. 9. The composite load model for dynamic simulation

phase motors is quite distinct from single phase motors, which have a higher propensity for stalling closer to normal operating voltages. Finally, the combined motor/load inertia is found to be a significant differentiating factor, particularly in situation where transmission faults are being considered.

Based on an analysis by the Western Electric Coordinating Council (WECC) Load Modeling Task Force (LMTF), the following load components were identified for motors:

1. Motor A represents three-phase motors driving constant torque loads, such a commercial and industrial air-conditioning and refrigeration compressors.
2. Motor B represents all so-called "speed squared" motors with large inertia, such as fan motors.
3. Motor C represents all so-called "speed squared" motors with low inertia, such as water pumps.
4. Motor D represents single-phase motors driving constant torque loads, such as residential air-conditioner, refrigerator and heat pump compressors.

Motor D is of special interest because it can accurately represent behavior of residential air conditioner compressors following a transmission fault. Representing this behavior is critical to emulate the system behavior for so-called fault induced delayed voltage recovery events.

Static and electronic loads continue to be represented with polynomial models. However, some three-phase electronic loads may decouple briefly during voltage sags depending on the design and control of the inverter circuit.

Load Composition

Load composition refers to process of determining the fraction of the total customer electric load that compose the different load components. The methodology used to compute the fraction of load attributed to the installed capacity of each load component was developed initially for WECC LMTF. The approach uses component-based load models that are validated and tuned using dynamic event measurements. This is done in three major steps: 1) identify the structure of the composite load model, 2) collect initial load model data, and 3) calibrate the load model through system performance studies.

The load model includes distribution feeder equivalent data, customer compositions, end-use composition, and end-use load characteristics. The objective is to minimize the number of load components modeled while retaining fidelity to the diversity of dynamic characteristics found among the known end-uses. Electric motors in particular require several distinct characteristics, such as single or three-phase supply, electrical characteristics, nature of the driven load, combined inertia of the motor and load, and motor protection and controls. Electronic load models must also retain elements that identify the response of the power conversion system to voltage fluctuations, particularly with respect to decoupling of the primary and secondary sides.

Load composition data is the most difficult part of this process. Load composition is affected by prevailing ambient conditions (temperature, humidity, and solar gains), the physical and thermal characteristics of buildings, the installed capacities of equipment, and human activities. In some areas, there are also pronounced micro-climate and latitude affects that must also be considered. The WECC LMTF uses a "bottom-up" approach to develop the load composition. For commercial buildings, the California End-Use Survey (CEUS) is used as the primary source for commercial load composition. For residential loads, estimates based on End-use Load Characterization and Assessment Program (ELCAP) data were used with a heating and cooling system response model to allow for regionalization of ELCAP load shapes.

Using Automated Metering Infrastructure (AMI) disaggregation techniques to reconstruct end-use load profiles is not recommended. It has been suggested that the bottom-up approach could be augmented by using AMI data. The proposed enhancement relies on the fact that AMI data typically collects real or total power observations at 15 minute or hourly samples. However, the data is not suitable for a direct observation of the individual end-use loads or the composition, because the additional time-resolution of AMI does not provide any additional observability of the states of individual loads "behind" the meter. To date, the only known reliable methods to measure the states of end-use load are direct end-use metering and real-time pricing systems that employ end-use bidding strategies.

The overall methodology of the component-based load modeling is shown in Figure 10. The load at a distribution feeder can be categorized into different load classes and each load class can be represented by different load components. The existing practice is to collect information from survey or billing data for each load class and identify the load components percentages. There is a potential to use Automated Meter Reading (AMR) data to get better estimates of load classes as opposed to using customer survey information only to estimate the load components. The added benefit of using AMR data is that it will capture seasonal variations in the loads. It may also capture some of the subhourly transient phenomena if they occur within the Nyquist limit of the measurement system.

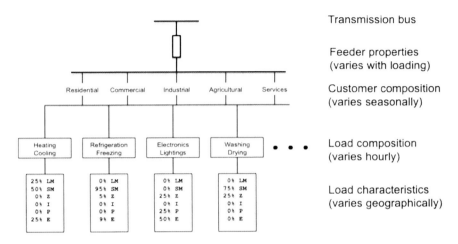

Fig. 10. Electric load composition hierarchy

The component-based load modeling approach is a "bottom-up" technique that consists of aggregating distribution loads according to standard load classes, specifying the types of devices comprised by those load classes, and then assigning the appropriate composition of the aggregated load to the various components of a suitable load model structure. In this approach, a common load model structure and associated set of parameter values are used throughout the system model. What is changed throughout the system model is the percentage composition (e.g. percentage motor load versus static load) from bus to bus. In most cases, even this percentage composition may be kept constant throughout the system model due to lack of better knowledge.

The bottom-up method for load composition estimation uses end-use analysis results from building population surveys, building simulations, and end-use metering programs to develop estimates for the fractions of loads that are operating at any given time. Typically, the data collected is separated hierarchically by customer class, building type, and end-use type: residential, commercial industrial and agricultural loads are broken down in facility types, as illustrated in Table 5.

Table 5. Facility Types by Customer Class

Residential	Commercial	Industrial	Agricultural
Single family	Small/Large office	Idiosyncratic	Idiosyncratic
Multi family	Retail/Lodging Grocery/Restaurant School/Health		

Each facility type has a fraction of end-use equipment associated with it, which is then associated with each load component, as illustrated in Table 6.

Table 6. Example end-use component associations (single-family residences in Portland, Oregon)

| | Electronic | Motor | | | | Z | | I | | P | |
		A	B	C	D	Real	React.	Real	React.	Real	React.
Cooking	30%				10%	60%					
Hotwater						100%					
Lighting						90%		10%	-2%		
Plugs	90%					90%					
Washing					70%	30%					
Heating					20%	80%					
Cooling					100%						
Refrigeration	10%				90%						

Given a known set of conditions, the bottom-up method can identify which end-uses are operating and through a series of successive products, one can compute the final load component fractions. The challenge is always how one determines the prevailing usage of each load at the time being considered. There are several factors that affect which loads are operating at any given time and the amount of power they consume when operating including

Temperature – outdoor air temperature principally affects the duty cycle and the efficiency of heating, ventilating and air-conditioning (HVAC) equipment. When the outdoor air temperature is near a building's balance point (the temperature at which all heat gains and heat losses balance), the duty cycle of the HVAC approaches zero. As the outdoor air approaches the heating or cooling design temperature, the duty cycle approaches unity. Furthermore, air-conditioning system and heat-pump heating system efficiency degrades as the temperature approaches the design temperature. Many heat-pump heating systems have auxiliary heating modes that do not utilize a compressor, but instead use a resistance heating element. The auxiliary heating mode is engaged when outdoor conditions go beyond the capabilities of the heat-pump, the thermostat deviates significantly from the set-point temperature, or the heat-pump runtime is excessive.

Humidity – relative humidity principally affects the latent heat load on air-conditioning systems. When the indoor air humidity is above about 50%, the latent load on the air-conditioner begins to rise. Near saturation, the latent heat load can be as high as 30% of the total load on the air-conditioning system.

Solar – heat gain from insulation can significantly increase the cooling loads and significantly decrease the heating loads. This is especially true when the sun is close to the horizon and solar radiation is normal to the glazing surfaces of buildings, such as in the early morning and late afternoon, or in northern latitudes during the winter.

Occupancy – building occupancy can dramatically affect loads through the behavior of the people. Residential loads often peak once in the morning and once in the evening. This is normally determined by day-of-week and hour-of-day.

Operations – commercial buildings undergo a morning setup load as the building is readied for weekday occupancy. Many residential thermostats have temperature setback schedules that can result in periods when the heating or cooling equipment is running 100% of the time, or not at all, depending on whether the temperature is floating to a new set point, or being driven to the new set point.

Design – the design, construction and type of equipment use has a significant impact on the net electric load observed during various times of the year. In regions where natural gas is readily available, the majority of homes use gas water heaters, cooking, and heating systems. In regions of the country where electricity was historically inexpensive or the natural gas infrastructure is less accessible, electric water heating, cooking, and space heating is more common.

The most commonly used approach to determine load composition is to examine end-use survey data, such the End-use Load and Customer Assessment Program (ELCAP) conducted by the Bonneville Power Administration (BPA) between 1984 and 1994 in the Pacific Northwest region of the United States. However, the highly regional nature of this kind of data and the absence of up-to-date data makes it necessary to update existing data sets, by making adjustments to account for climate variations, changes in construction quality and methods, changes in equipment performance and efficiency, and to account for changes in consumption, purchasing and evolving use habits of building occupants. A variety of approaches can be used to perform this task and results are generally mixed. One way of addressing the issue related to changing load devices is to actually test the behavior of specific load devices in a controlled laboratory environment. In fact, the Electric Power Research Institute (EPRI) and a number of larger utilities have engaged in work of this nature to test commercial air conditioning units in the laboratory. One of the benefits of this type of "component" testing is a better understanding of the discontinuous behavior of load components that can lead to high fidelity in load models, e.g. actions such as motor contactor drop out and air conditioner thermal overload relays.

Overall, bottom-up load composition methods remain difficult to implement with sufficient generality and are only used when sufficient time and effort can be expended on modeling the loads in detail. The advantages of this approach include the following:

- Promotes correlation to the physical characteristic of end-use devices; and
- Utilizes load class data from individual substations, which is generally available.

3.3 Load Response Properties

Aside from the electrical response of ZIP and induction loads, there are a wide range of other responses that must be considered when modeling loads over various time spans, as shown in Table 7.

Table 7. Load responses over various time spans

Time span	Response
Seconds	Equipment protection (e.g., thermal, undervoltage, overvoltage)
Minutes	Protection reset, load curtailment, loss of diversity
Hours	Curtailment rebound, cold-load pickup, diversification, weather
Days	Weekend/weekday cycle, holidays, weather changes
Weeks	Weather patterns
Months	Seasonal effects
Years	Load growth/loss, economic changes, climate changes

Load response can be separated into two distinct behaviors, one which is a one-time irreversible behavior, e.g., a person turns off the lights when leaving home, and one which is reversed later, e.g., a person defers doing a load of laundry until the next day. Some load responses have a little of both. For example, lowering a thermostat by 2°F for 8 hours every day will reduce the heating energy consumption, but the strategy will also introduce a recovery period during which some of the savings received are returned when the thermostat is raised back. A comparison of the reversibility of different residential load responses is shown in Table 8.

Electrical Load Modeling and Simulation

Table 8. Residential load response reversibility

Irreversible responses	Reversible responses
Lighting controls Cooking/heating fuel switch Heating & cooling thermostat setback Energy efficiency retrofits	Heating & Cooling thermostat schedules Washing & dryer deferral Refrigeration & freezer defrost deferral

Irreversible load responses reduce overall energy consumption by the amount of load that responds. The response R is typically estimated in terms of the fraction of the original load $L_{original}$ that is reduced, and then converted to a total load reduction

$$L_{reduced} = R \cdot L_{original} \tag{21}$$

Irreversible load response is simply a change in the load shape results in a net reduction of both energy and maximum power, as illustrated Figure 11 (a). Most irreversible load response requires a one-time investment and the benefit is typically enduring. However, some load responses such as consumer awareness programs may appear to be irreversible over the short term, but do in fact decay in the long term.

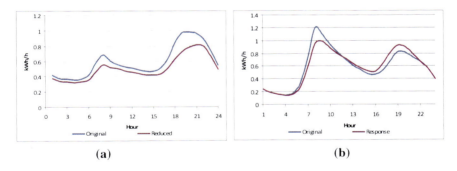

(a)   (b)

Fig. 11. Lighting reduction (a) is an irreversible response where as hotwater curtailment (b) is a reversible response

Reversible load response is a change in the load shape that results in a change in maximum power, but relatively little net change in energy consumption, as shown in Figure 11 (b). This behavior is typical for thermostatic loads, such as heating and cooling systems. These responses are often called load shifting or deferral. Pre-heating and pre-cooling are also reversible load responses, but with the opposite sign (i.e., load increase precedes load decrease). Some reversible load responses can in fact result in a net increase in energy consumption, such a pre-heating or pre-cooling homes in anticipation of a high-price period.

Load Diversity

IEEE defines *diversity factor* as "The ratio of the sum of the individual non-coincident maximum demands of various subdivisions of the system to the maximum demand of the complete system. The diversity factor is always greater than 1. The (unofficial) term *diversity*, as distinguished from *diversity factor* refers to the percent of time available that a machine, piece of equipment, or facility has its maximum or nominal load or demand (a 70% diversity means that the device in question operates at its nominal or maximum load level 70% of the time that it is connected and turned on)."

Unfortunately, this definition is flawed because the reference to diversity as the fraction of a single machine's runtime (i.e., its duty cycle) is inapt. Most diversity phenomena arise from the aggregate fractional load and not just a single device's duty cycle. The aggregate fractional load also includes phenomena such as load state synchronization, load coincidence and other load behaviors, which are properties applicable only to populations of loads and not individual loads.

Consequently, it seems necessary that we refrain from using *diversity factor* as defined by IEEE in favor of a more general definition of *load diversity*, i.e., the fraction of total load that is on the system at a given time. The load diversity is consequently a dimensionless quantity between zero and unity, inclusive. The load diversity for a population of N periodic loads is

$$D = \frac{\sum_{i=1}^{N} P_i}{\sum_{i=1}^{N} L_i} = \frac{\sum_{i=1}^{N} p_i L_i}{\sum_{i=1}^{N} L_i} \qquad (22)$$

where L_i is the maximum power of the ith load, P_i is the actual power of the ith load, and p_i is the fractional power or duty-cycle of the ith load or if the load is bimodal (*on* vs. *off*) the probability that the ith load is on at its maximum power.

Load Duration

A load duration curve is a load shape that is sorted by descending power. Generally, a load duration curve is used to estimate the number of hours that the load exceeds 800 W. For example, the load shape in Figure 12 can be used to determine that the load exceeds 80% of the peak lighting load for 5 hours a day, and exceeds 50% of the peak load for 16 hours a day.

The most common use of a load duration curve is for planning studies, when planners estimate the number of hours per year for which a system resource must be allocated. The load duration curve can also be used to estimate the maximum amount of load that can be curtailed for a certain number of hours.

The annual load duration curve is typically developed for the 8760 hours of the year from the annual load shape, as shown in Figure 13 (a). Typically the annual load duration exhibits a very strong peak for a very small number of hours. Because it is the details of these few hours that tend to determine the resource requirements for the system, it is sometimes helpful to present the load duration curve on a logarithmic scale of hours, as shown in Figure 13 (b).

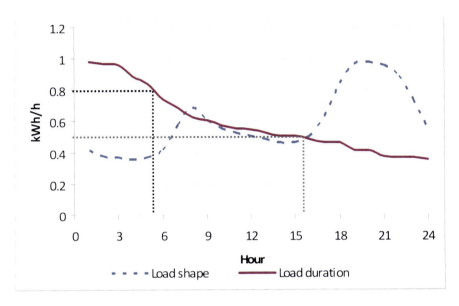

Fig. 12. Load shapes can be used to determine the amount of time the load exceeds a given power

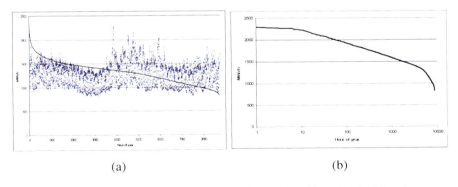

Fig. 13. Annual load duration curve on a linear (a) and logarithmic (b) scale

Load Periodicity

The load duty-cycles and periods can be measured directly in many systems using end-use metering technology. However, end-use metering is quite expensive, and the data collected about any given device is often the superposition of the device's natural behavior and other driving functions such as consumer behavior. So it is often very difficult to clearly identify the fundamental properties of loads. Furthermore, for any time-domain models where the load aggregate is a consideration, it is typically necessary that we observe not only the duty cycles or *on*-probabilities of devices, but also their periods and state phases, as shown in Figure 14.

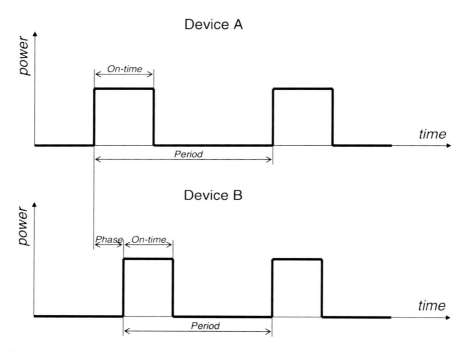

Fig. 14. Device state aggregation considers period, duty-cycle (on-time/period) and state phase

In such situations the state of a single device is the position of the device given as a fraction of the whole cycle. Hence, only devices with the same period can be aggregated and the distribution of their phases will determine the shape of the time-series they produce when aggregated. This is the true sense in which the diversity of the load should be considered.

The load diversity of a load shape L is at its maximum when the load shape is flat over a given time interval T, i.e., the load is constant. When a load is perfectly constant, it has a diversity of 1.0, and a change of the timing of any single device can only result in a decrease in diversity associated with the deviation of L from the flat load. If the load is already fully diversified, then there are no possible changes to any of the devices composing the load that could further increase diversity, a condition referred to as Pareto optimality.

When a load is completely flat, it not only has a diversity of 1.0, but also for all times t the continuous Fourier transform of L

$$\hat{L}(s) = \int_{-\infty}^{+\infty} L(t) e^{st} dt \tag{23}$$

is zero of all values of s except $s \to \pm\infty$. Of course, observations of feeder and substation load data show that this hardly ever occurs, even for relatively short durations. There are always some components of load behavior that are periodic and not so synchronized at a particular frequency that they completely cancel each other out. This is clearly visible in Figure 15, particularly with respect to diurnal cycles, but indeed to some extent at every observable frequency.

The curve $\hat{L}$ is called the *load periodicity* curve or *load profile*. The values of the load periodicity curve describe amplitude of periodic load behavior, and measure how much power use is subject to periodic variations at a given frequency s. The periodicity of infinite duration is the mean load, which is why when the mean load equals the observed load there cannot be periodic variations of finite frequency and the load is therefore flat.

Using this description of periodic load behavior we can determine the energy periodicity of load. The periodic behavior of energy, which occurs when load acts like storage, is found by taking the time integral of the load shape. In the frequency domain the *energy profile*

$$\hat{E}(s) = \frac{1}{s} \hat{L}(s) \tag{24}$$

measures the pseudo-storage potential of load, which generally increases as frequency decreases.

Similarly, we can find the ramping profile with the time differential of the load shape in the frequency domain

$$\hat{R}(s) = s\hat{L}(s) \tag{25}$$

The ramping profile generally increases as frequency increases. These results are illustrated by the 30 second sampling of a 2-month feeder load shape in Figure 15.

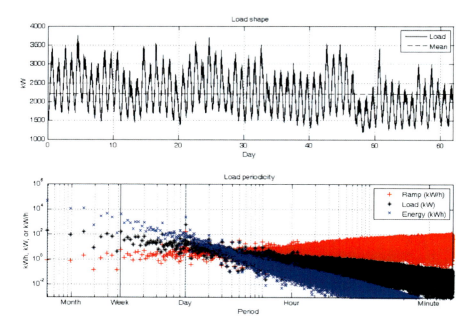

Fig. 15. Load (black), energy (blue), and ramping (red) periodicity of a load shape (top)

It is immediately apparent that a direct comparison of the benefit of any load shifting potential due to the ramp, load, and energy behavior is not directly possible, in spite of the fact that they can so easily be presented in the same graphic, as is done in Figure 15. The reason is that the units of ramping (kW/h), load (kW) and energy (kWh) are not directly comparable, owing to different time factor of each. This problem can be addressed by considering the relative economic potential if each aspect of load behavior is flattened and presenting the three results as costs, as shown in Figure 16.

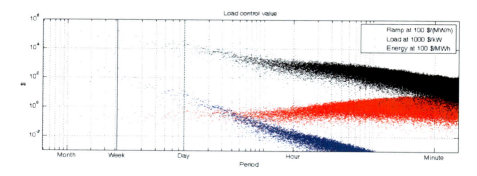

Fig. 16. Cost comparison of ramp, load, and energy behavior of loads

Obviously, there is more potential available if the mean value is changed either through energy efficiency measures or load shifting occurs beyond the time horizon of the periodicity analysis. However, if we assume that neither of these changes is considered, then the maximum "free" potential of load shifting is indeed only that which arises from flattening the load. Any additional shift would tend to invert the load shape, and would simply express itself as an increase in the amplitude of the periodicity with which one would associate a new potential value. This is often discussed in the context of load shifting programs that cause "new peaks"—if such load peaks are non-coincident with system peaks (i.e., out of phase), they may be desirable to system operators. That simply means that the economic value of the potential is made apparent to the consumers through time varying prices. A shift in load would have a relatively high economic value when the systems operators need for load shifting is unsatisfied and relatively low economic value when satisfied.

3.4 Load Response Behaviors

There are several important behaviors of loads that must be considered as a part of any comprehensive load model. These behaviors are typically triggered by initiating events and are generally not part of the linear load model that is usually considered by power system modelers. Nonetheless they can dramatically influence the outcome of system studies under critical conditions, such as under or over-voltage and under or over-frequency events.

Equipment Protection

Electric equipment is typically protected from three principle causes of damage: overcurrent, overvoltage, and undervoltage. Overcurrent and overvoltage protection are usually implemented using direct electrical protection systems that sense the condition. However, undervoltage protection is often implemented using indirect and non-electrical systems, such a thermal protection, and can be more difficult to model as a result.

The undervoltage tolerance of computer equipment is approximately 50 to 80% of nominal voltage for no more than 100 ms. Such equipment is not generally provided with undervoltage protection, but has certain tolerances built in. The exact tolerances of specific equipment will depend highly on a) the power drawn by the load on the secondary supply relative to the design capacity of the power supply and b) on the pre-sag conditions of the primary power supply. In contrast, programmable logic controllers that support high-risk applications such an industrial control systems are provided with undervoltage protection that generally trip at about 75% voltage after just 1 cycle. Some controllers come equipped with DC bus capacitors that will allow for more significant sags of up to 50% for ½ second or more.

Undervoltage protection is generally used for motor protection because of the following three phenomena:

1. Motors can automatically restart when the voltage returns following an interruption of power even though the application of the motor does not require restart or is potentially unsafe to restart;
2. Excessive current draw following an interruption of power can cause overheating and damage motors; and
3. Motors can reverse direction instead of stopping.

Automatic restart protection is usually determined by control requirements and typically is only implemented when the requirement is present. Most residential and small commercial motor loads such as residential air-conditioner compressors and fans have no such protection. The only notable exception is dryers and washing machines, which have a safety requirement to not restart after their cycle is interrupted without an explicit command from a human. Such loads should be removed from load models after an undervoltage event sufficiently large enough to interrupt the operation of the load and should not be returned to service without consideration of how human agency is applied to the restart command.

Three phase motors used in large commercial and industrial applications are usually protected with a magnetic circuit breaker for overcurrent protection and three thermal overload relays for sustained overcurrent protection. All large motor loads (i.e., above 600V) used in large commercial and industrial applications should be presumed to use undervoltage protection. There are many large motors extant that are inductive and require accessories to provide undervoltage protection. For electronically controlled motors protection is usually provided using programmable controls. There is only anecdotal evidence regarding the extent to which the default setup is found in operation, but it suggests that this is the predominant configuration for medium sized commercial loads and only on larger motor loads are the controls programmed based on the design specification of the control system when installed.

For the protection against thermal damage, the vast majority of motors should be assumed to trip between 2 and 15 seconds after rotor lock, with a long tail distribution of motors that can remain stalled for up to several minutes without tripping.

Motor reversal is a possible outcome for single-phase motors when the undervoltage is within a critical range around the stall threshold. Although this is obviously a highly undesirable situation, there appears to be little concern among manufacturers about undervoltage protection systems to prevent this occurrence. This should be regarded as possible for low inertia applications (such as in compressors). In such cases the total load drops significantly when the reverse direction is unloaded and does not return to original load until prevailing conditions stop the motors with unsatisfied loads (e.g., air-conditioners stop because of ambient cooling rather than active cooling).

Generally undervoltage motor protection for all the above mitigation objectives are designed to wait until fault detecting relays have had a chance to clear faults. However, the time delay for undervoltage protection may not be satisfactory because contactors may drop out before protection engages. In addition, fast automatic transfer switches may not be coordinated with undervoltage protection schemes.

Finally, problems do arise because undervoltage protection schemes allow excessive inrush currents after interruption before thermal protection is engaged.

Thermal protection is provided in loads that have the potential to be damaged when operating in certain states. For example, because overheating may cause damage to stator insulation or allow bending forces to prevail, most motors have an over-temperature cut-off switch that will disconnect power from the motor when the internal temperature of the motor gets too high. NEMA Motor Insulation Classes limit the temperature rise because even a 10 C over-temperature can reduce the life of a stator by 50%. Some motors, such as squirrel cage motors have much higher limits, but are still vulnerable to deformation or melting, which are particularly relevant during off-normal operation such as stalling or fast acceleration.

Table 9. General motor undervoltage response parameters for composition load models

Motor type	Application	Stall V	Stall P/Q	Stall T	Trip T	Reset T
Motor A	Large compressors	~0.7 – 0.8	I2R	< 100 ms	≤ 2 s	> 1 m
Motor B	Large fans	~0.6 – 0.7	I2R	> 1 s	≤ 2 s	> 1 m
Motor C	Large pumps	~0.7 – 0.8	I2R	< 100 ms	≤ 2 s	> 1 m
Motor D	Small compressors	~0.6 – 0.7	I2R	< 100 ms	2 – 15 s	> 1 m
Electronic	Active VFD	~0.5 – 0.8	Regen	< 200 ms	< 1 s	varies
	Passive VFD	~0.5 – 0.8	0	varies	< 1 s	varies

Undervoltage protection is often closely related to thermal protection. This is typically used for larger loads and loads where the thermal protection would be too slow to protect equipment from damage. The time and voltage settings of undervoltage protection vary widely, but they are almost always designed to protect the equipment, and not the power system itself. The causes of over-temperature can be varied, but perhaps the most serious is motor stalling, which causes motors to become essentially purely resistive loads—none of the energy put into the motor can go to mechanical work and it all goes to heat instead, as shown in Figure 17.

The time for thermal protection to take effect varies according to the load types and properties of the specific devices, but it can be estimated based on the thermal mass of the material surrounding the temperature sensor, the thermal conductivity of the materials, the ambient temperature, and net energy flow to or from the material. The formal model for the time t to trip at the current I (normalized per unit of rated current) is given by

$$t(I) = \tau \ln \frac{I^2}{I^2 - 1} \tag{26}$$

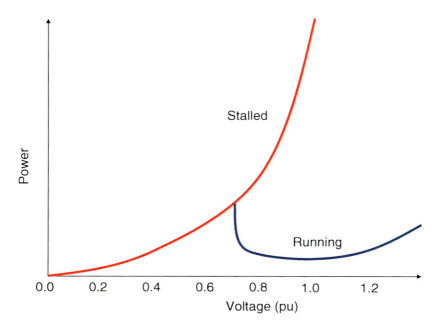

Fig. 17. Motor stalling behavior: running power (blue) increases at the critical stall voltage and once stalled unprotected motors act like constant impedance loads (red) until thermal protection is applied or sufficient voltage is returned to restart them (not depicted)

where τ is the time-constant C/H, C is the specific heat capacity of the motor and H is the running heat dissipation rate. For values of current much larger than the rated current, $I \gg 1$, and we can use the simpler relation

$$t(I) = \frac{\tau}{I^2} \qquad (27)$$

as a satisfactory approximation for the time to trip a motor on thermal overload during undervoltage/overcurrent conditions.

Contactors and solenoids have a very wide range of undervoltage tolerances, ranging from holding at 35% voltage indefinitely to dropping at 75% voltage in 10 ms. The drop-out probability of a contactor can also be very dependent on the initial conditions prior to the voltage sag. Relays, such as "ice-cube" relays can also influence the behavior load significantly because they have high drop-out probabilities, i.e., 70% voltage after as little as 1 cycle and 30% voltage in ¼ cycle. These relays are ubiquitous in emergency stop equipment, door interlocks, compressor starter controls, chiller controls, conveyer controls, over controls, motor controls, variable frequency drive circuits and vending machines.

The great diversity of off-normal voltage and current behavior is a significant challenge to load modelers in power systems. However, the difficulty of acquiring load model data and the obstacles to accurate detailed simulation should never be an excuse for erroneous modeling of loads. The computational capabilities exist to properly model loads, particularly when high-performance computing systems are available. The current versions of PSLF, PSS/E and PowerWorld the composition load models described are employed with no significant adverse impact on performance.

Load Curtailment Controls

Load curtailment is a strategy used by utilities to reduce load during peak hours. Most load curtailment programs use so-called "direct load control" programs. Utilities install service interruption switches on customer end-use loads, such as air-conditioners or electric water-heaters and dispatch load shedding at the onset of a peak load condition. Such programs have been in use for many years and they are found wherever load relief is sought as an alternative to capacity expansion.

Other methods of load curtailment have been tested, including pricing programs and thermostatic control programs. These programs are not as widely used, and modeling these programs remains an idiosyncratic problem. There is little uniformity in the design and deployment of such programs and the models must be developed on a case-by-case basis.

Great care should be taken when the load curtailment programs are in fact load control programs. Such is the case with real-time price based programs where consumer devices "bid" for power use. The bid process communicates information back from the load to the market-based demand response dispatch system. This creates a powerful feedback loop that can be very challenging to model correctly. Standard top-down or bottom-up solvers do not generally properly model such applications and hybrid solvers such as GridLAB-D™ must be used to determine the combined power flow, load behavior and market solutions.

Load Diversity

The distribution of the properties and states of different loads have a strong impact on their diversity. Consider the simple example of a population of thermostatic loads, each of which has a natural duty cycle ρ, and natural period ϕ, and is currently a fraction θ through that period. If all the devices have identical duty cycles and periods, then as time progresses, any non-uniformity in the distribution of θ will be preserved and appear as a regular oscillation or beat in the aggregate load of these devices. Normally the other properties of the load are randomly distributed in some way. As a result, any ordering in the distribution of θ is not preserved as time progresses, and any regular oscillations in the aggregate load of such loads is eventually damped.

This phenomenon is most often seen in models that fail to properly consider the distribution of the properties ρ and ϕ over the population of devices. Models that simply use the mean values cannot consider the evolution of load diversity over time. Such models will always be deficient when load diversity is a factor that must be considered, as shown in Figure 18.

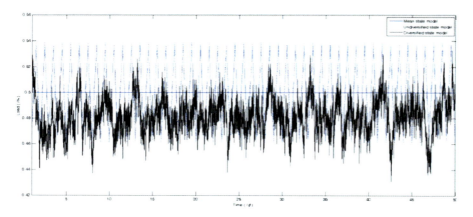

Fig. 18. Comparison of load models that consider diversity (black), mean state models (blue) and undiversified state models (dotted)

Fault-Induced Load Behavior

Incidents where the electric transmission system voltage has failed to recover after a transmission fault have been recorded more frequently in recent decades. Events reported in heavily loaded regions of systems where air-conditioners are prevalent have led system engineers to conclude that the delayed voltage recovery is related to stalling of residential single-phase air-conditioners in the area close to the fault.

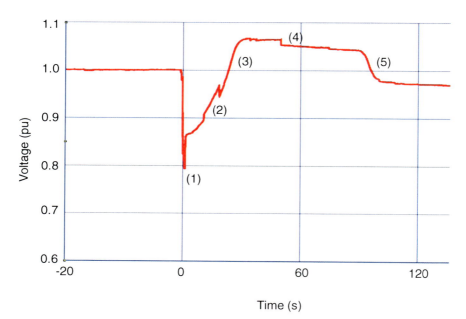

Fig. 19. Typical fault-induced delayed voltage recovery

Figure 19 shows a typical delayed voltage recovery profile on a high-voltage transmission circuit. Right after the fault the voltage decreased to 79% of nominal voltage (point 1) before the fault is cleared by the grid protection scheme. This dip in voltage caused air conditioning compressor motors to stall and the voltage on the distribution circuits drops even lower. The stalled air conditioning units kept the voltage from recovering to nominal level (point 2). When the air conditioning units' thermal overload protection tripped, the voltage gradually recovered but overshot the nominal voltage (in this case 6% above) because the capacitor banks used to raise depressed voltages remained connected to the circuit after they were no longer needed (point 3). The over-voltage caused the capacitor bank to then trip out (point 4). With the capacitors tripped off and the load (including air conditioners load) returning to normal, the voltage then went below the nominal voltage (point 5). Modeling this kind of sequence of events remains an elusive challenge for power system simulators, yet this phenomenon has been observed by multiple utilities over the last ten years.

4 Computational Load Modeling

At the center of all power flow calculations are the power flow calculations, which perform a number of important computational steps. Depending on the method use to solve the non-linear set of simultaneous equations, some of these steps may not lend themselves to exploitation of high-performance computing architectures.

For example, consider the difference between a Gauss-Seidel solver and a Newton-Raphson solver. The Newton-Raphson solver is usually preferred because it converges quadratically on the solution whereas the Gauss-Seidel solver only converges linearly. For large systems, this difference can be very significant when evaluating how much time it takes to find the solution—a Newton-Raphson solution might take on 5 or 10 iterations when a Gauss-Seidel solution for the same problem might take thousands. The reason is that the Newton-Raphson solver uses a Jacobian matrix, which allows each iteration to evaluate the error correction as a function of the entire system error when improving the solution. In contrast, the Gauss-Seidel method uses the errors from its nearest topological neighbors in improving the solution which results in long error propagation delays for large systems.

However, the flip-side of the advantage Newton-Raphson enjoys is that the calculation of the Jacobian is not an intrinsically parallel process, whereas the error correction method of a Gauss-Seidel solution is intrinsically parallel. This suggests that when many processors are available, the Gauss-Seidel implementation might actually complete sooner than Newton-Raphson implementation. It is interesting to note that attempts to exploit this phenomenon have largely not been successful for several reasons:

1. The number of processors needed should be similar to the number of nodes in the system model, which in today's systems often exceeds 10,000 and can approache 100,000.

2. The communication between the processors needs to be very fast relative to the computation time of each processor and the amount of information communicated must be kept to a minimum.
3. The number of iterations required remains large relative to the Newton-Raphson method, thus the system model itself must be sufficiently large to justify the parallel implementation of Gauss-Seidel but not so large that the number of iterations overwhelms the benefits of parallelism.

4.1 Quasi-Steady Power Flow

In general, quasi-steady state flow solvers are in fact solvers that determine the flows in a time series of states that are each assumed to be steady. The assumption is made that the solution of the next state can be reached from the previous state using a trivial transition that need not be simulated using a dynamic solution. Indeed, the vast majority of state changes in the system are trivial from the transient perspective, so the assumption is most often correct. Furthermore, for the types of analyses where transient considerations are not the primary concern, the assumption is quite safe to make. Primary among these types of analyses are retail electricity pricing, rate design, load control/demand response technology evaluation, and business case analysis.

The overall process for quasi-steady power flow analysis is shown in Figure 20. The method allows calculation of longer time-series than obtained using dynamic power flow calculations while preserving the modeling simplicity and performance efficiency of steady state power flow calculation.

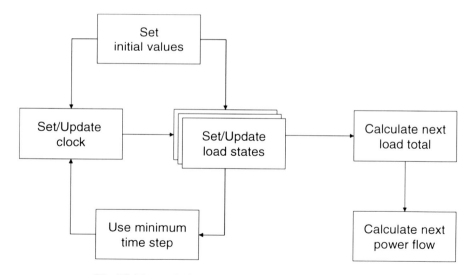

Fig. 20. Time solution for quasi-steady power flow solvers

4.2 Event-Driven Models

Event driven models are a class of discrete time models that can be used to solve quasi-steady simulations. As a rule, event-driven models do not advance time in uniform time-steps. Instead, the clock is updated only when "something interesting happens." An event is a modeling abstraction that is used to representing a discrete action in one part of the physical system at a specified moment in time that results in changes to other parts of the system. Typically, events result in a change to one or more of the state variables defined by the simulation. Consequently, the simulation only advanced the clock from the time of one event to the time of the next event.

From the perspective of the modeling problem, event-driven models must be able to compute both the state of the system components as a function of time, and the time until each component's next event. This requirement often results in the need for an inverse solution to the equations of state, such that given the function F that finds the state change

$$\dot{x} = F(x,t) \tag{28}$$

we also can define the function F^{-1} that determines the elapsed time

$$\Delta t = F^{-1}(\dot{x}',t) \tag{29}$$

until a new state x' along the path of the current system is reached.

Event driven simulation allow for a useful enhancement called *speculative modeling* in which parallel threads can advance farther than synchronized time would allow provided they can "reverse" outcome should it be determined later than an event in another part of the system would yield a different outcome.

4.3 Agent-Based Models

Agent-based modeling is a class of computational modeling used primarily for simulating the interaction of autonomous agents. Agents are typically individual or small collections of entities that behave in a coherent (hence computable) way in response to various external stimuli given their internal states. The purpose of agent-based simulation is to assess the collective effect of the agents interacting with each other on the system as a whole. Often one finds discussions of agent-based models intermixed with discussions of game theory, complex systems, emergence, computational sociology, evolutionary programming and genetic algorithms.

Agent-based models attempt to recreate and predict the emergence of complex phenomena that cannot be reduced to a solvable set of simultaneous equations. Simple microscopic (i.e., individual-based) behaviors frequently combine in highly non-linear ways to give rise to very rich and difficult to understand macroscopic behaviors: the whole is indeed greater than the sum of the parts. To accomplish the agents typically include the following capabilities.

1. Agents are described at multiple scales (e.g., multiple types and scopes).
2. Agents have the ability to change their behavior in response to information from other agents (e.g., internal controls with sensors).
3. Agents can alter their fundamental properties slowly over time (e.g., learning, adaptation).
4. Agents can relate selectively to a limited set of other agents (e.g., topology).
5. Agents are assembled in an environment (e.g., power grid, market, weather system) that include a sense of time.

When an agent-based simulation is run, each agent is given the opportunity to respond to the behavior of other agents, and to the situation in the environment. Sometimes, the environment itself can provide "summary" or system-level behaviors to which agents can respond, such as when an electric power grid computes the grid frequency and agents change their behavior when the frequency changes.

An important example of such a simulation is GridLAB-D™, an open-source smart grid simulation developed under funding from the US Department of Energy Office of Electricity. This tool is an agent based simulator and its solution method is illustrated in Figure 21.

Agent-based models are in many ways complementary to the analysis models historically used. Analytic models permit engineers to characterize power systems in such a way that insight is gained into the various equilibrium phenomena that are of interest. But the mathematics of computing equilibria is often challenging and limit the ability of analytic models to satisfactorily describe the equilibrium processes of large complex systems. Agent-based models can generate equilibria in such cases, but often the deep insights that accompany analytic models are not evident and sometimes altogether absent.

Agent-based models usually mimic reliably the emergence of higher-order behavior in complex systems, but also fail to provide direct explanations of exactly why they arise. In the end, agent-based models usually require strong inductive reasoning to give same insights that analytic models yield deductively. Nonetheless, agent-based models can help analysts identify critical phenomena, limit cycles, and metastable states that would be at best very difficult to find using analytic methods.

The validation of agent-based models can also be very challenging and is an area of ongoing research. The reason is that many agent-based system can be highly sensitive to initial conditions (chaotic), express divergent behaviors (non-linear path dependence) and are highly sensitive to random fluctuations

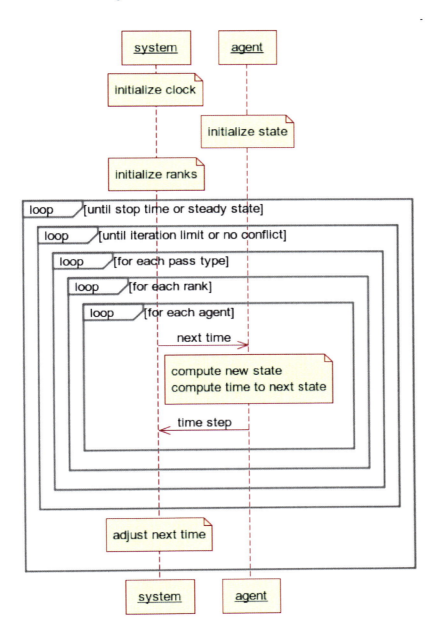

Fig. 21. The solution method of GridLAB-D uses ranks to organize agents into groups that can be solved in parallel, with information generally flowing up and down the ranks

(stochastic). Comparison of agent-based simulation results with data collected from the real world depends on satisfying a number of critical assumptions:

1. The model must accurately describe the salient details of the agents in the real system. Unfortunately, many details are unknowable and even when they are knowable, they vary over time and space with correlations that can be difficult to model accurately.
2. The model and the real world must evolve in a similar manner for enough time that observable properties emerge that are comparable. The chaotic, path-dependent and stochastic behavior of agent-based models makes assuring convergence on the same fixed point a fundamentally under-specified fitting problem.
3. The environments must be fundamentally the same. For some systems (e.g., power grids) this is usually achievable with only moderately more effort than convention simulations. But for many systems (e.g., markets) even the environment changes over time in ways that can be difficult to observe in the real world and can be difficult to model with any reasonable fidelity.

In the end, a significant amount of engineering judgment is required to ensure that agent-based models are accurate enough to draw useful conclusions. In general, modelers use the following steps to assure the correctness of the agents, their topology, and the environment.

1. Agents are tested individually for the full range of behaviors they are required to exhibit. This requires both a well-described specification for the behavior of the agents as well as a close understanding of its implementation. *Units tests* typically require setting an initial condition, establishing a boundary condition and advancing the clock until a steady condition is reached. The steady condition is then compared to the expected condition and any deviation is addressed either through a change to the implementation of the agent, a change in initial conditions, a change in the boundary condition, and a change in the test.
2. Unit tests are followed by sensitivity tests where changes in initial conditions are used to characterize the chaotic behavior of agents, changes in boundary conditions are used to test trajectory sensitivity, and changes in time-evolution are used to test the time solution of state changes. Deviation from the steady state are then observed and an engineering assessment of the sensitivity of the system is made. Again, any unexpected deviation is investigated and corrected either by changing the test and/or the model.
3. Sensitivity tests should then be followed by stochastic sensitivity tests. In this type of testing, a large collection of agents are given properties drawn from both synthetic and realistic distributions of properties. The distribution of the steady states is then examined and again unexpected and inexplainable deviations are investigated and corrected either by improving the tests or the model.

Electrical Load Modeling and Simulation

4. Environment tests are required to assure that the environment in which the agents operate is functioning properly. At the very least, the environment's time behavior must be tested. This is done by varying the time-step, using different starting times, and running the simulation for different durations. Some agent-based systems have time systems that are so realistic that leap-years, time-zones, daylight-savings time rules, and even leap-seconds must be tested. Some environments provide additional elements such as weather, economic or regulatory frameworks that must be testing individually and then collectively.
5. Simple topological tests are used to examine the emergent behavior of the agents in cases that are sufficiently trivial to be solved analytically. While these tests are by their very nature oversimplified with respect to real world cases, they offer a degree of assurance that the agent-based system as a whole, including multiple agents and the environment are behaving correctly when coupled.
6. Finally, comprehensive emergence testing can be completed. Most often, the exact outcome of such tests are not known with sufficient precision to be deemed objective, but as a general rule, engineering judgment can applied to assess the validity of results.

Some of the best work on the topic of agent-based model validation has been done in the area of computational economics. These models have been used frequently in developing pricing strategies. It is assumed that the agents are more likely to use strategies that have been recently successful, and because the success of any given strategy is dependent on prevailing market conditions as well as the strategies used by other agents, the models are quite good at capturing the long-term cycles observed in markets. As a result, agents are forced to respond to changes in both short term and long term conditions and the aggregate results tend to more realistically represent the range of possible outcomes observed in real world systems.

5 Model Validation

Model validation is an essential step in the development of numerical methods used in utility planning and operations. Even when the numerical codes can be shown to properly represent the behavior of system in general, the high sensitivity of models to the large amount of input required can present an overwhelming challenge to those charged with tool validation. The following recommendations are based on the WECC criteria for planning model validation and are a good template for meeting the expectations of utilities and reliability organizations. The WECC model validation requirements establish minimum thresholds for results from dynamic simulations that include load models. These can be considered illustrative but not comprehensive for all power system model validation approaches. Many of the criteria described relate only to dynamic models and other types of models may require different ones.

5.1 Model Documentation

Generally, utilities require a concise description of why a new model is required with particular emphasis on how the existing models inadequately represent the dynamic behavior. Confidential and proprietary information about generating units (and by implication customers) are generally not acceptable as a part of an assessment.

The documentation of the new model includes the following:

a) A description of the equipment that is modeled at a level that reveals the aspects of the equipment that are and are not described by the model.

b) A description on how the model reasonably represents the behavior of the equipment over the frequency range from DC to 3 Hz including voltage and frequency oscillations.

c) A description of the model in mathematical and functional detail including, as appropriate, including information such as Laplace transfer functions, block diagrams, description of physical and logical limits, control logic, interlock, supervisory and permissive actions.

d) A description of the relationship of all parameters to the physical and logical characteristics of the equipment. The documentation of the model must be sufficient to permit the implementation of the model in all of the simulation programs used. This may require that part of the documentation of the model be in the form of code snippets. However it is not anticipated that a complete code of a model is needed in the documentation.

e) Description of any behavior that is not represented by the model.

5.2 Model Validation

The model is validated using strong evidence that it effectively reproduces the behavior of the equipment being modeled. It is generally recognized that dynamic modeling falls into two distinct categories:

a) Models of equipment that can be described explicitly and tested directly either in laboratory conditions or in specially controlled operating conditions. Generators, generator controls, electrical protection elements, and most transmission system elements are in this category.

b) Models of the power system behavior that cannot be described in explicit detail and cannot be tested directly. Most aspects of the modeling of load behavior are in this category.

Validation of models in the first category should include comparisons of simulations made with the model with test results or responses produced by other authoritative sources, such as results from manufacturers' detailed physical design simulations, results from factory acceptance tests, on-line recorder response of the equipment to system disturbances, or performance guarantee documents.

Where possible validation of models in the second category should include comparisons of simulations made using the model with records of events that have

Electrical Load Modeling and Simulation

occurred on the transmission system. Where comparison with actual grid behavior is not practical the characteristics of models in the second category should be demonstrated by simulations of small scale operational situations, probing test and disturbances chosen so that the proposed model is the predominant factor in the response. Validation must cover the behavior of the model in the broad range of operational situations that the equipment is expected to encounter. Validation must cover both the steady state behavior of the equipment over its full operational range and the dynamic behavior in response to events such as:

a) sudden step changes of voltage and frequency at pertinent point of interconnection with the grid;
b) undervoltage fast transients typical to fault clearing and delay fault clearing times;
c) step changes of control references and set-points;
d) ramps of voltage, frequency, references and set-points; and
e) oscillatory behavior in the frequency range from 0.1 to 3 Hz.

Changes of voltage and frequency considered in validation should cover the range of amplitude that will be produced by transmission disturbances from faults to persistent small oscillations.

5.3 Test Independence

For the purpose of model validation, the equipment model should be validated through comparison of simulated and expected behavior (whether based on tests or other means as discussed above) focusing on the behavior of the modeled equipment. The intent is to emulate the test conditions (injected signals, etc.) and compare the simulated behavior of the developed model with the recorded field, factory or other test/results for validation. Full scale power system simulations with proposed models in the simulation set-up should not be used for model validation, since the purpose is to focus on dynamics of the equipment independent of its impact on the system study results.

5.4 Numeric Stability

Evidence should be provided demonstrating that a model is numerically stable for simulations of faults with normal clearing in marginal conditions, e.g., weak systems (short circuit ratio less than 5). Numerical stability can be difficult to verify, but generally for load flow calculations on initial conditions, it is apparent when a model causes the simulation to fail a numerical stability test—convergence will not occur when the non-linear solver iterates. Depending on the specific implementation and data provided it is also possible that specific component models become numerical unstable and cause general failure of the dynamic time-domain simulation.

6 Conclusions

We have shown how electricity consumer demand response and load control are playing an increasingly important role in the smart grid. The advent of smarter grid operations using load management technologies is challenging conventional models of load and now more than ever the behavior of load both affects and is affected by load control strategies that are designed to support the electric grid. The natural behavior of electric loads, how it is affected by various load control and demand response strategies, and the consequences of new concepts such as using load to mitigate the integration of renewable energy resources, are challenging our existing computational systems. With the introduction of new smart grid technologies, new computing challenges arise. New models are being incorporated into existing simulation and new simulation paradigms are being adopted to satisfy the demand for accurate models of increasingly complex power systems.

There remains a great deal of work to do, particularly with respect to models of demand response control systems. Ongoing work in academia, government laboratories and industry is certain to continue supporting the trend toward a more active role for load in the planning and operation of large interconnect electric grids.

Acknowledgments. The following individuals are gratefully recognized (alphabetically) for their contribution to the work described in this chapter through many years of collaboration: Patrick Balducci (PNNL), Donald Davies (WECC), Jason Fuller (PNNL), Anish Gaikwad (EPRI), Harold Kirkham (PNNL), Dmitry Kosterev (BPA), Karanjit Kalsi (PNNL), John Kueck (ORNL), Bernie Lesieutre (UWisc), Jeffry Mallow (Loyola Chicago), Rob Pratt (PNNL), Todd Taylor (PNNL), Leigh Tesfatsion (IA State), and John Undrill.

References

Almon, C.P., Donelson, J.: Performance tests of high-speed load and frequency control equipment. In: Procs. of the AIEE Winter General Meeting (January 1954)

Cardell, J., Anderson, L., Chin Yen Tee, T.: The Effect of Wind and Demand Uncertainty on Electricity Prices and System Performance. In: IEEE Power Engineering Society Transmission and Distribution Conference (April 2010)

Chassin, D.P., Fuller, J.C.: On the equilibrium dynamics of demand response in thermostatic loads. In: Procs. of 44th Hawaii International Conference on Systems Sciences (January 2011)

Chassin, D.P.: Load control analysis for intermittent generation mitigation. Submitted to IEEE Trans. on Smart Grid (November 15, 2011)

Chassin, F.S., Mayhorn, E.T., Elizondo, M.A., Lu, S.: Load modeling and calibration techniques for power systems studies. In: Procs. of 2011 North American Power Symposium (August 2011)

Gilleskie, R.J., Brown, E.E.: Determination of capacity equivalent load factor for air-conditioner cycling at San Diego Gas and Electric. IEEE Trans. on Power Systems PWRS-1(3) (August 1986)

GridLAB-D™, http://www.gridlabd.org

Guojun, J., Guangyong, Y., Liu, T.: Effects of rival whistle-blowing on green product innovation. In: Procs. of 8th International Conference on Service Systems and Service Management, ICSSSM (June 2011)

Hammerstrom, D., et al.: Pacific Northwest GridWise™ Testbed Demonstation Projects: Part I. Olympic Peninsula Project, PNNL-17167, Pacific Northwest National Laboratory, Richland Washington,
http://www.pnl.gov/main/publications/external/technical_reports/PNNL-17167.pdf

Heffner, G., Goldman, C.A., Moezzi, M.M.: Innovative approaches to verifying demand response of water heater load control. IEEE Transactions on Power Delivery 21(1) (January 2006)

Huang, P., Kalagnanam, J., Natarajan, R., Sharma, M., Ambrosio, R., Hammerstrom, D., Melton, R.: Analytics and Transactive Control Design for the Pacific Northwest Smart Grid Demonstration Project. In: Procs. of IEEE International Conference on Smart Grid Communications, pp. 449–454 (October 2010)

Kalsi, K., Chassin, F.S., Chassin, D.P.: Aggregated modeling of thermostatic loads in demand response: A systems and controls perspective. In: Procs. of 50th IEEE Conference on Decision and Control and European Control Conferences (December 2011)

Karki, R., Po, H., Billington, R.: Reliability evaluation considering wind and hydro power coordination. IEEE Trans. on Power Systems 25(2), 685–693 (2010)

Kashyap, A., Callaway, D.: Controlling distributed energy constrained resources for power systems ancillary services. In: Procs. of 11th IEEE Conference on Probabilistic Methods Applied to Power Systems, PMAPS (June 2010)

Kundur, P.: Power System Stability and Control. McGraw Hill, New York (1994)

Laurent, J.C., Malhame, R.P.: A physically-based computer model of aggregate electric water-heating loads. IEEE Trans. on Power Systems 9(3), 1209–1217 (1994)

Lee, S.H., Wilkins, C.L.: A practical approach to load control: a water heater case study. IEEE Transactions on Power Apparatus and System PAS-102(4) (April 1983)

Lively, M.B.: Short run marginal cost pricing for fast responses on the Smart Grid. In: Procs. of the 2010 Innovative Smart Grid Technologies, ISGT (January 2010)

Meliopoulos, A.P.S., Kovacs, R.R., Reppen, N.D., Contaxis, G., Balu, N.: Power system remedial action methodology. IEEE Trans. on Power Systems 3(2), 500–509 (1988)

Novosel, D., Khoi, T.V., Hart, D., Udren, E.: Practical protection and control strategies during large power-system disturbances. In: Procs. of IEEE Transmission and Distribution Conference, pp. 560–565 (1999)

Papaexopoulos, A.D., Hao, S., Peng, T.M.: An implementation of neural network-based load forecasting model for the EMS. IEEE Trans. on Power Systems 9(4) (November 1994)

Reed, J.H., Thompson, J.C., Broadwater, R.P., Chandrasekaran, A.: Analysis of water heater data from Athens load control experiment. IEEE Trans. on Power Delivery 4(2) (April 1989)

Roozbehani, M., Dahleh, M., Mitter, S.: Dynamic Pricing and Stabilization of Supply and Demand in Modern Electric Power Grids. In: Procs. of IEEE International Conference on Smart Grid Communications, pp. 543–548 (October 2010)

Schneider, K., Fuller, J.C., Chassin, D.P.: Multi-state load models for distribution system analysis. IEEE Trans. on Power Systems 26(4), 2425–2433 (2011)

Schweppe, F.C., Daryanian, B., Tabors, R.D.: Algorithms for a spot price responding residential load controller. IEEE Trans. on Power Systems 4(2) (May 1989)

Song, H., Baik, S.D., Lee, B.: Determination of load shedding for power-flow solvability using outage-continuation power flow (OCPF). In: Procs. of IEEE Generation Transmission and Distribution, vol. 153(3), pp. 321–325 (2006)

Turitsyn, K., Sulc, P., Backhaus, S., Chertkov, M.: Distributed control of reactive power flow in a radial distribution circuit with high photovoltaic penetration. In: Procs. of 2010 IEEE Power and Energy Society (PES) General Meeting (July 2010)

Strickler, G.F., Noell, S.K.: Residential air-conditioning cycling: A case study. IEEE Transactions on Power Systems 3(1) (February 1988)

Verdu, S.V., Garcia, M.O., Franco, F.J.G., Encinas, N., Marin, A.G., Molina, A., Lazaro, E.G.: Characterization and identification of electrical customers through the use of self-organizing maps and daily load parameters. In: Procs. of 2004 IEEE PES Power Systems Conference and Exposition., vol. 2, pp. 899–906 (2004)

Vittal, E., O'Malley, M., Keane, A.: A steady-state voltage stability analysis of power systems with high penetrations of wind. IEEE Trans. on Power Systems 25(1), 433–442 (2010)

Widergren, S., Subbarao, K., Chassin, D., Fuller, J., Pratt, R.: Residential real-time price response simulation. In: Procs. of 2011 IEEE Power and Energy Society (PES) General Meeting (June 2011)

On-Line Transient Stability Screening of a Practical 14,500-Bus Power System: Methodology and Evaluations

Hsiao-Dong Chiang[1], Hua Li[2], Jianzhong Tong[3], and Yasuyuki Tada[4]

[1] School of Electrical and Computer Engineering,
  Cornell University, Ithaca, NY, USA
  chiang@ece.cornell.edu
[2] Bigwood Systems, Inc., Ithaca, New York, USA
  hua@bigwood-systems.com
[3] PJM Interconnection, LLC, Norristown, Philadelphia, USA
  tongji@pjm.com
[4] Tokyo Electric Power Company, R&D Center, Yokohama, Tokyo, Japan
  Tada.Yasuyuki@tepco.cojp

Abstract. This paper describes an effective methodology for on-line screening and ranking of a large set of contingencies. An evaluation study of the on-line methodology in a real-time environment as a transient stability analysis (TSA) screening tool is presented. Requirements for an on-line screening and ranking tools are presented. The methodology of BCU classifiers implemented in the TEPCO-BCU package was evaluated on the PJM system as a fast screening tool to improve on-line performance of the PJM TSA system. This evaluation study is the largest in terms of system size, 14,500-bus, 3000 generators, for a practical application of direct methods for on-line TSA. The total number of contingencies involved in this evaluation is about 5.3 million. The evaluation results were very promising and confirm the practicality of the methodology based on direct methods, in particular the BCU method for on-line TSA of large-scale systems with a large set of contingencies.

1 Introduction

For many utilities around the world, there has been considerable pressure to increase power flows over existing transmission corridors, partly due to economic incentives (a trend towards deregulation and competition) and partly due to practical difficulties of obtaining authorization to build power plants and transmission lines (environmental concerns). This consistent pressure has prompted the requirement for extending energy management system (EMS) to take account of transient stability assessment (TSA) and control. Such extension, however, is a rather difficult task and requires several breakthroughs in data management systems, analysis tools, computation methods and control schemes.

From a computational viewpoint, on-line TSA, concerned with power system stability/instability after contingencies, requires the handling of a large set of nonlinear differential equations in addition to the nonlinear algebraic equations involved in the static security assessment (SSA). The computational efforts required in on-line TSA is roughly three magnitudes higher than that for the SSA. This computational challenge may serve to explain why the task of transient stability assessment and control has long remained in off-line activity.

Current power system operating environments have prompted the need to significantly enhance time-domain stability analysis programs to meet new requirements. In addition, it is becoming advantageous to move the task of transient stability analysis from the off-line study area into the on-line operating environment. There are significant operational benefits, engineering, and financial benefits expected from this movement. Indeed, there is always significant incentive to find superior approaches for stability analysis and control [10].

After decades of research and development in energy-function-based direct methods, it has become clear that the capabilities of the controlling Unstable Equilibrium Point (UEP) method and those of the time-domain simulation approach complement each other. The current direction of development is to incorporate the controlling UEP method, the BCU (Boundary of Controlling Unstable Point) method and time-domain simulation programs into the body of overall on-line TSA.

An integrated architecture for on-line TSA and control is presented in Figure 1. In this architecture, there are two major components in the module of on-line TSA: fast dynamic contingency screening and a time-domain stability program for performing detailed stability analysis. When a new cycle of on-line TSA is warranted, a list of credible contingencies, along with information from the state estimator and topological analysis, are applied to the dynamic contingency screening program whose basic function is to screen out contingencies that are definitely stable or potentially unstable. Contingencies classified to be definitely stable are eliminated from further analysis. Contingencies classified to be potentially unstable are sent to fast time-domain simulation for detailed analysis. On-line TSA is feasible due to its ability to perform dynamic contingency screening on a large number of contingencies and to filter out a much smaller number of contingencies requiring further analysis. Contingencies either undecided or identified as unstable are then sent to the time-domain transient stability simulation program for detailed stability analysis.

The block function of control action decisions determines if timely post-fault contingency corrective action, such as automated remedial action, are feasible to steer the system away from unacceptable conditions toward a more acceptable state. If appropriate corrective action is not available, the block function of preventive action determines the required pre-contingency preventive controls should the contingency occur. These controls include real power redispatches, adjustment of phase-shifter or line switching to maintain the system stability. If the system is marginally stable (i.e. critically stable), the block function of enhancement action determines the required pre-contingency enhancement controls to increase the degree of system stability should the contingency occur. In this architecture, a fast yet reliable method for performing dynamic contingency screening plays a vital role in the overall process of on-line TSA.

Suppose a complete on-line TSA assessment cycle will be completed within, say 15 minutes. This cycle starts when all of the necessary data is available to the system and ends when the system is ready for the next cycle. Depending on the size of the underlying power systems, it is estimated that, for a large-size power

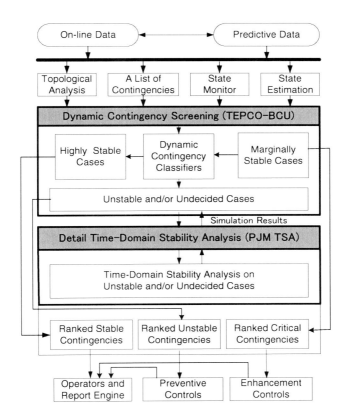

Fig. 1. An Example of Architecture of Transient Stability Assessment (TSA) and Control System

system such as a 15,000-bus power system, the number of contingencies in a contingency list lies between 1,000 to 3,000. The contingency types will include both three-phase faults with primary clearance and single-line-to-ground faults with backup clearance.

The PJM Interconnection has designed and implemented an on-line Transient Stability Analysis (TSA) system based on the time-domain simulation approach within its EMS. The EMS provides a real time snapshot including power flow case data from the state estimator, dynamic data and contingency list to the TSA application. For each real time snapshot, on-line TSA is designed to perform a transient stability assessment against the list of contingencies, calculating transient stability limits on several key transfer interfaces. However, due to the slowness of the time-domain simulation approach, the current PJM on-line TSA system needs a dynamic contingency screening function in order to achieve the requirements of on-line TSA.

The methodology of BCU classifiers implemented in the TEPCO-BCU package was selected as the leading fast screening tool for improving the performance of the PJM TSA system. The TEPCO-BCU software was evaluated for three months in the PJM TSA Test System environment. The three-month evaluation period was chosen to include the scheduled outage season and the summer peak load season. This period historically encountered a wide range of real time scenarios that can happen on the PJM power system. During this evaluation period, both the PJM TSA software and the TEPCO-BCU software were run periodically in parallel, every 15 minutes in the PJM TSA Test System environment. The goal was to evaluate TEPCO-BCU in a real time environment as a transient stability screening tool.

This paper presents a comprehensive evaluation of the dynamic contingency screening function of TEPCO-BCU and an overview of the solution methodologies behind TEPCO-BCU. Requirements for an on-line screening and ranking tool for PJM systems are presented and evaluated. This evaluation study is the largest in terms of system size, 14,500-bus, 3000 generators, for a practical application of direct methods. The total number of contingencies involved in this evaluation is about 5.3 million. The integrated package TEPCO-BCU is based on the controlling UEP method, BCU method, energy function method and the improved BCU classifiers. The controlling UEP method and the boundary of stability-region-based controlling UEP (BCU) method will be described at a level sufficient to comprehend the concept of a transient-stability screening program for TSA systems.

2 On-Line Transient Stability Screening

The strategy of using an effective scheme to screen out a large number of stable contingencies and capture critical contingencies and to apply detailed simulation programs only to potentially unstable contingencies is well recognized. This strategy has been successfully implemented in on-line static security assessment. The ability to screen several hundred contingencies to capture tens of the critical contingencies has made the on-line SSA feasible. This strategy can be applied to on-line TSA. Given a set of credible contingencies, the strategy would break the task of on-line TSA into the following stages:

- screening out a large number of stable contingencies, and capturing critical contingencies,
- ranking the critical contingencies and detectinging potentially unstable contingenices
- applying detailed time-domain simulation programs to only potentially unstable contingencies to identify unstable contingencies

It is the ability to perform transient stability screening on a large number of contingencies and to filter out a much smaller number of contingencies, which are potentially unstable contingencies requiring further analysis that would make on-line TSA feasible. One key objective of the dynamic contingency screening is to identify contingencies which are definitely stable and thereby avoid further stability analysis for these contingencies. It is due to the definite classification of stable contingencies that considerable speed-up can be achieved for on-line transient stability assessment. Contingencies that are identified as critical, or potentially unstable are then sent to the time-domain transient stability simulation program for further stability analysis. The overall computational speed and accuracy of an on-line TSA system depends greatly on the effectiveness of the dynamic contingency screening.

The following requirements are essential for any candidate (classifier) intended to perform on-line transient stability screening for current or near future power systems [15]:

(1) (Reliability measure) absolute capture of unstable contingencies; i.e. no unstable (single-swing or multi-swing) contingencies are missed. In other words, the ratio of the number of captured unstable contingencies to the number of actual unstable contingencies is 1.
(2) (Efficiency measure) high yield of screening out stable contingencies, i.e. the ratio of the number of stable contingencies detected to the number of actual stable contingencies is as close to 1 as possible.
(3) (On-line computation) little need of off-line computations and/or adjustments in order to meet with the constantly changing and uncertain operating conditions.
(4) (Speed measure) high speed, i.e. fast classification for each contingency case.
(5) (Performance measure) robust performance with respect to changes in power system operating conditions.

The requirement of absolute capture of unstable contingencies is a reliability measure for transient stability screening. This requirement is extremely important for on-line TSA. However, because of the nonlinear nature of the transient stability screening problem this requirement can best be met by a method with a strong analytical basis. The third requirement asserts that a desired dynamic contingency screener is the one which relies little on off-line information, computations, fine-tuning and/or adjustments. This requirement arises because under current and near future power system operating environments, the correlation between on-line operational data and presumed off-line analysis data can be minimal or, in extreme cases, the two can be irrelevant to one another. In other words, off-line presumed analysis data may become unrelated to on-line operational data. The first four requirements should not be degraded by different operating conditions as dictated by the requirement for robust performance.

Current power system operating environments call for a great need to develop a transient stability screening scheme which satisfies the above five essential requirements. Several methods developed for on-line transient stability screening have been reported in literature [11,25,26]. These methods can be categorized as follows: the energy-function-based direct method approach [14,15], the time-domain approach and the artificial intelligence (AI) approach [20,21,22]. The time-domain approach involves the step-by-step simulation of each contingency for a few seconds, say 2 or 3 seconds, to filter out the very stable or very unstable contingencies. Experience indicates that this approach may suffer from the accuracy problem in identifying multi-swing stable or unstable contingencies. The artificial intelligence (AI) approaches, such as the pattern recognition technique, the expert system technique, the decision tree technique and the artificial neural network approach, all first perform extensive off-line numerical simulations aiming to capture the essential stability features of the system's dynamic behavior. They then construct a classifier attempting to correctly classify new, unseen on-line contingencies. As such, the AI approach is likely to become ineffective for on-line application to current or near-future power systems if little correlation exists between on-line operational data and presumed off-line analysis data.

The energy-function-based direct methods have evolved in the last several decades 14,15]. The current direct method, controlling UEP method, uses an algorithmic procedure to determine whether the system will remain stable, based on the energy function theory. This controlling UEP method directly assesses the stability property of the post-fault trajectory without integrating the post-fault system. It assesses the stability of the post-fault system whose initial state is the system state when the fault is cleared by comparing the system energy at the initial state of the post-fault trajectory with a critical energy value. The controlling UEP method not only avoids the time-consuming numerical integration of the post-fault system, but also provides a quantitative measure of the degree of system stability. This direct method has a solid theoretical foundation [1,3,27]. The theoretical basis of direct methods for the direct stability assessment of a post-fault power system is the knowledge of the stability region; if the initial condition of the post-fault system lies inside the stability region of a desired post-fault stable equilibrium point (SEP), then one can ensure without performing numerical integrations that the ensuing post-fault trajectory will converge to the desired point. Therefore, knowledge of the stability region plays an important role in the theoretical foundation of direct methods.

In summary, the existing AI-based methods for transient stability screening all rely tremendously on extensive off-line simulation results. These methods unfortunately fail to meet the on-line computation requirement and cannot guarantee the reliability requirement. The time-domain-based approach is heuristic when applied to perform dynamic contingency screening for simulating short-period (say 3 seconds) dynamic behaviors. It has been recognized by many researchers and analysists that the time-domain-based approach is ineffective for dynamic contingency screening. In this regard, the improved BCU classifiers [1], which is acombination of controlling UEP method, BCU method and energy function based method, can meet the five requirements.

3 Controlling UEP Method

Of the several direct methods which have evolved over the last several decades, the controlling UEP method is the most vital one for direct stability analysis of practical power system models. The controlling UEP method assesses the stability property of a post-fault trajectory, whose initial state is the system state when the fault is cleared and all the SPS actions are completed, by comparing the system energy at the initial state of post-fault trajectory with the energy value at the associated CUEP. The controlling UEP method not only avoids the time-consuming numerical integration of the post-fault system, but it also provides a quantitative measure of the degree of system stability. The controlling UEP method has a solid theoretical foundation [3,11].

Before we further explain the controlling UEP method, the following concept of an exit point is useful (see Figure 2). Let X_s^{pre} be a pre-fault SEP and $X_f(t)$ be the corresponding fault-on trajectory. Let $\partial A(X_s)$ denote the stability region of the post-fault SEP X_s. Then, the exit point is a relationship between the fault-on trajectory and the post-fault system.

Definition: (Exit point)
The point at which a sustained fault-on trajectory intersects with the stability boundary of the post-fault SEP is called the exit point of the fault-on trajecto (relative to the post-fault system). In addition, the fault-on trajectory exits the stability region after the exit point.

A function is an energy function for a nonlinear system if the following three conditions are satisfied:

Condition (i): The derivative of the energy function along any system trajectory is nonpositive

Condition (ii): Along any nontrivial trajectory, the set that the derivative of the energy function is zero has measure zero in time.

Condition (iii): That a trajectory has a bounded value of energy function implies that the trajectory is also bounded.

Theorem 1: (Global behavior of trajectories)
If a function satisfying condition 1 and condition 2 of the energy function exists for a nonlinear system, thenevery bounded trajectory of the nonlinear system converges to one of the equilibrium points.

There has been some misunderstanding regarding the energy function method's ability to detect the first swing step-out phenomena. Theorem 1 requires no condition in the way of the trajectory in step-out. If the kinetic energy is larger than that of the height of the rim at the crossing point of the bowl, transient instability is determined, but there is still no information as to whether the contingency's waveform is multi-swing step-out or first swing step-out. When using an energy function in combination with the controlling UEP method, not only can the first swing step-out be directly captured, but the multi-swing instability can be captured as well.

We next discuss a rigorous definition of the controlling UEP: The controlling UEP is the first UEP whose stable manifold is hit by the fault-on trajectory $X_f(t)$ at the exit point. The existence and uniqueness of the controlling UEP defined above are assured. This definition is motivated by the facts that a sustained fault-on trajectory must exit the stability boundary of a post-fault system and that the exit point (i.e. the point from which a given fault-on trajectory exits the stability boundary of a post-fault system) of the fault-on trajectory must lie on the stable manifold of a UEP on the stability boundary of the post-fault system. ThisUEP is the controlling UEP of the fault-on trajectory. Note that the controlling UEP is independent of the energy function used in the direct stability assessment.

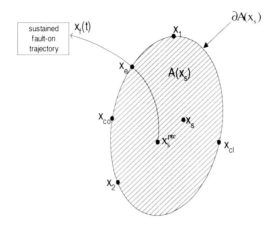

Fig. 2. The sustained fault-on trajectory moves toward the stability boundary $\partial A(x_s)$ and intersects it at the exit point, x_e. The exit point lies on the stable manifold of the controlling UEP.

The controlling UEP method asserts that the energy value at the controlling UEP be used as the critical energy for the fault-on trajectory $X_f(t)$ to directly assess the stability of the post-fault system. Theorem 2 below gives a rigorous theoretical justification for the controlling UEP method for direct stability analysis of post-fault systems by comparing the energy value of the state vector at which the fault is cleared with the energy value at the controlling UEP. We define the following two components. $S(r) \triangleq$ the connected component of the set $\{ X \in R^n : V(X) < r \}$ containing X_s, and $\partial S(r) \triangleq$ the (topological) boundary of the set $S(r)$.

Theorem 2: (Controlling UEP Method) [3,11,27]
Consider a general post-fault system which has an energy function $V(\cdot): R^n \to R$. Let X_{co} be an equilibrium point on the stability boundary $\partial A(X_s)$ of a SEP X_s of this system. Then, the following results hold:

[1]. The connected constant energy surface $\partial S(V(X_{co}))$ intersects with the stable manifold $W^s(X_{co})$ only at point X_{co}; moreover, the set $S(V(X_{co}))$ has an empty intersection with the stable manifold $W^s(X_{co})$.

[2]. Any connected path starting from a point $P \in \{S(V(X_{co})) \cap A(X_s)\}$ with $W^s(X_{co})$ must also intersect with $\partial S(V(X_{co}))$.

Results [1] and [2] of Theorem 2 assert that, for any fault-on trajectory $X_f(t)$ starting from a point $X_s^{pre} \in A(X_s)$ and the energy at the point $V(x_s^{pre}) < V(x_{co})$, if the exit point of this fault-on trajectory $X_f(t)$ lies on the stable manifold of X_{co}, then this fault-on trajectory $X_f(t)$ must pass through the connected constant energy surface $\partial S(V(X_{co}))$ before it passes through the stable manifold of X_{co} (thus exiting the stability boundary $\partial A(X_s)$). Therefore, the connected constant energy surface $\partial S(V(X_{co}))$ is adequate for approximating the relevant part of the stability boundary.

The task of computing the (exact) controlling UEP of a given fault for general power system models is very challenging due to the following complexities:

- The controlling UEP is a particular UEP embedded in a large-degree state-space.
- The controlling UEP is the first UEP whose stable manifold is hit by the fault-on trajectory (at the exit point).
- The task of computing the exit point is very involved; it usually requires a time-domain approach.

The task of computing the controlling UEP is complicated further by the size and the shape of its convergence region. It is known that (1) with respect to a selected numerical method, each equilibrium point has its own convergence region, i.e., the region from which the sequence generated by the numerical method starting from a point in the region will converge to the equilibrium point, (2) from observation and theoretical investigation by several researchers, under the Newton method, the size of the convergence region of the UEP can be much smaller than that of the SEP, and (3) the convergence region of either an SEP or a UEP is a fractal; unfortunately, finding an initial guess sufficiently close to the controlling UEP is a difficult task.

The above complexities also call into doubt the correctness of any attempt to directly compute the controlling UEP of a power system stability model. The only one method that can directly compute the controlling UEP of an (original) power system stability model is the time-domain approach. These complexities serve to explain why many methods proposed in the literature fail to compute the controlling UEP. It is because these methods attempt to directly compute the controlling UEP of the power system stability model, which is difficult if not impossible to compute without using the time-domain approach.

The ability to compute the controlling UEP is vital to direct stability analysis. Since computing a controlling UEP is very challenging, it may prove fruitful to develop a tailored theory-based method for finding the controlling UEP by exploiting special properties as well as some physical and mathematical insights of the underlying power system model. We will discuss such a theory-based method, called the BCU method, along this line for finding the controlling UEP.

4 BCU Method

We next discuss a method which does not attempt to directly compute the controlling UEP of a power system stability model (original model), instead it computes the controlling UEP of a reduced-state model and then it relates the controlling UEP of the reduced-state model to the controlling UEP of the original model.

A systematic method, called boundary of stability region based controlling unstable equilibrium point method (BCU method), to find the controlling UEP was developed [12,13]. The method was also given other names such as the exit point method [25,29] and the hybrid method [4]. The BCU method has been evaluated in a large-scale power system and it has been compared favorably with other methods in terms of its reliability and the required computational efforts [29,30]. BCU method has been studied by several researchers, see for example, [31-36]. Descriptions of the BCU method can be found in books such as [4,19,25,26,27]. The theoretical foundation of BCU method has been established in [12,27,37]. BCU method and BCU classifiers have several practical applications. For example, a demonstration of the capability of the BCU method for on-line transient stability assessments using real-time data was held at two utilities, the Ontario Hydro Company and the Northern States Power Company [38,39]. The BCU method was implemented into a EPRI TSA software which was integrated into an EMS installed at the control center of Northern States Power Company [40]. A TSA system, composed of the BCU classifiers, the BCU method and a time-domain simulation engine, was developed and integrated into the Ranger EMS system [43]. The TSA system has been installed and commissioned, as part of an EMS system at several energy control centers. The BCU method has been applied to fast derivation of power transfer limits [5] and applied to real power rescheduling to increase dynamic security [42]. BCU method has been improved, expanded and extended into the integrated package of TEPCO-BCU [2,6,7,8,9,16].

We next present an overview of the BCU method from two viewpoints: numerical aspects and theoretical aspects. In developing a BCU method for a given power system stability model, the associated artificial, reduced-state model must be defined. The fundamental idea behind the BCU method for computing a controlling UEP can be explained as follows [12,13,27]. This explanation is illustrated on a general power system transient stability model. We review a network-preserving BCU method for direct stability analysis of the following generic network-preserving transient stability model [10]:

$$\begin{aligned}
0 &= -\frac{\partial U}{\partial u}(u,w,x,y) + g_1(u,w,x,y) \\
0 &= -\frac{\partial U}{\partial w}(u,w,x,y) + g_2(u,w,x,y) \\
T\dot{x} &= -\frac{\partial U}{\partial x}(u,w,x,y) + g_3(u,w,x,y) \\
\dot{y} &= z \\
M\dot{z} &= -Dz - \frac{\partial U}{\partial y}(u,w,x,y) + g_4(u,w,x,y)
\end{aligned} \qquad (1)$$

where $u \in R^k$ and $w \in R^l$ are instantaneous variables and $x \in R^m$, $y \in R^n$, and $z \in R^n$ are state variables. T is a positive definite matrix, and M and D are diagonal positive definite matrices. Here the differential equations describe generator and/or load dynamics while the algebraic equations express the power flow equations at each bus. The function U(u,w,x,y) is a scalar function. The physical limitations of variables are not expressed for the sake of convenience. Existing network-preserving transient stability models can be rewritten as a set of the above general differential-algebraic equations (DAEs) [28].

Given a power system transient stability model, called the original model, which admits an energy function, the BCU method first explores the special properties of the original model (1) so as to define an artificial, reduced-state model such that certain static and dynamic properties of the original model are captured by the reduced-state model. We next discuss a reduced-state model associated with the original model (1). Regarding the original model, we choose the following differential-algebraic system as the artificial, reduced-state model.

$$\begin{aligned}
0 &= -\frac{\partial U}{\partial u}(u,w,x,y) + g_1(u,w,x,y) \\
0 &= -\frac{\partial U}{\partial w}(u,w,x,y) + g_2(u,w,x,y) \\
T\dot{x} &= -\frac{\partial U}{\partial x}(u,w,x,y) + g_3(u,w,x,y) \\
\dot{y} &= -\frac{\partial U}{\partial y}(u,w,x,y) + g_4(u,w,x,y)
\end{aligned} \quad (2)$$

It can be shown that the original model (1) and the artificial, reduced-state model (2) satisfy the following static and dynamic properties [12,13,27].

Static Properties
(S1) The locations of equilibrium points of the reduced-state model (2) correspond to the locations of equilibrium points of the original model (1).
(S2) The types of equilibrium points of the reduced-state model (2) are the same as that of the original model (1).

Dynamical properties
(D1): There exists an energy function for the reduced-state model (2).
(D2): An equilibrium point is on the stability boundary of the reduced-state model (2) if and only if the corresponding equilibrium point of the original model (1) is on the stability boundary of the original model (1).
(D3): It is computationally feasible to efficiently detect when the projected fault-on trajectory of the original fault-on trajectory intersects the stability boundary of the post-fault reduced-state model without resorting to detailed time-domain simulation.

The purpose of defining the reduced-state model is to capture all the UEPs lying on the stability boudary of the original model as indicated by the dynamic property (D2). The dynamic property (D3) plays a pivotal role in the development of the BCU method for circumventing the difficulty of applying an iterative time-domain procedure when computing the exit point of the original model [12,13]. The BCU method computes the controlling UEP of the original model by exploring the special structure of the stability boundary and the energy function of the reduced-state model. Given a power system stability model, there exists a corresponding reduced-state model and a corresponding version of the BCU method. Note that the BCU method does not compute the controlling UEP directly on the original model since, as pointed out, the task of computing the exit point of the original model, a key step to computing the controlling UEP, is very difficult and usually requires the time-domain approach.

A conceptual BCU method

Step 1. From the fault-on trajectory of the network-preserving model (1), detect the exit point at which the projected trajectory exits the stability boundary of the post-fault reduced-state model (2).

Step 2. Use the exit point, detected in Step 1, as the initial condition and integrate the post-fault reduced-state model to an equilibrium point. Let the solution be

$$(u_{co}, w_{co}, x_{co}, y_{co})$$

Step 3. The controlling UEP with respect to the fault-on trajectory of the original network-preserving model (1) is

$$(u_{co}, w_{co}, x_{co}, y_{co}, 0)$$

The energy function at the controlling UEP is the critical energy for the fault-on trajectory.

Step 1 and Step 2 of the conceptual BCU method compute the controlling UEP of the reduced-state system. Note that starting from the exit point, Step 2 of the conceptual BCU method, will converge to an equilibrium point. The controlling UEP always exists and is unique, and the stable manifold of controlling UEP of the reduced-state system contains the exit point. Step 3 relates the controlling UEP of the reduced-state system (with respect to the projected fault-on trajectory) to the controlling UEP of the original system with respect to the original fault-on trajectory.

There are several possible ways to numerically implement the conceptual BCU method for network-preserving power system models. A numerical implementation of this method along with several numerical procedures necessary are presented in this section.

A Numerical BCU Method

Step 1: Integrate the fault-on system of the original model to obtain the (sustained) fault-on trajectory until the point at which the projected fault-on trajectory reaches its first local maximum of the numerical potential energy function along the projected trajectory. Let this point be termed as exit point.

Step 2: Integrate the post-fault system of the reduced-state system starting from the exit point until the point at which the (one-dimensional) local minimum of the following norm of the post-fault, reduced-state system is reach;

$$\left\| \frac{\partial U}{\partial u}(u,w,x,y) + g_1(u,w,x,y) \right\|$$
$$+ \left\| \frac{\partial U}{\partial w}(u,w,x,y) + g_2(u,w,x,y) \right\|$$
$$+ \left\| \frac{\partial U}{\partial x}(u,w,x,y) + g_3(u,w,x,y) \right\|$$
$$+ \left\| \frac{\partial U}{\partial y}(u,w,x,y) + g_4(u,w,x,y) \right\|$$

Let the local minimum of the above norm be occurred at the point $(u_0^*, w_0^*, x_0^*, y_0^*)$

Step 3. Use the point $(u_0^*, w_0^*, x_0^*, y_0^*)$ as the initial guess and solve the following set of nonlinear algebraic equations

$$\left\| \frac{\partial U}{\partial u}(u,w,x,y) + g_1(u,w,x,y) \right\|$$
$$+ \left\| \frac{\partial U}{\partial w}(u,w,x,y) + g_2(u,w,x,y) \right\|$$
$$+ \left\| \frac{\partial U}{\partial x}(u,w,x,y) + g_3(u,w,x,y) \right\|$$
$$+ \left\| \frac{\partial U}{\partial y}(u,w,x,y) + g_4(u,w,x,y) \right\| = 0$$

Let the solution be $(u_{co}, w_{co}, x_{co}, y_{co})$

Step 4. The controlling UEP with respect to the fault-on trajectory of the original system is $(u_{co}, w_{co}, x_{co}, y_{co}, 0)$.

Remarks:
[1]. In Step 1, the projected trajectory can be viewed as the projection of the original fault-on trajectory on the state-space of the reduced-state system (2). The first local maximum of the numerical potential energy function along the projected trajectory is an approximated exit point at which the projected trajectory intersects with the stability boundary of the reduced-state system (2).
[2]. In Step 2, a stability-boundary-following procedure, presented below, is developed to guide the search process for CUEP of the reduced-state system starting from the point by moving along the stability boundary of the reduced-state system (2) toward the CUEP. During the search process, the point has the local minimum

of the norm among all the computed points in the search process. The norm is a measure of distance between the current point and an equilibrium point. This point is also termed as the minimum gradient point (MGP).

[3]. The reduced-state system can be numerically stiff and a stiff differential equation solver should be used to implement Step 2 of the numerical network-preserving BCU method.

[4]. Without the stability-boundary-following procedure implemented in Step 2, the search process can move away from CUEP; making the corresponding MGP distant from the CUEP and cause the divergence of Newton method.

[5]. In step 3, the minimum gradient point (MGP) is used as an initial guess for the Newton method to compute the controlling UEP. It is well known that if the MGP is sufficiently close to the controlling UEP, then the sequence generated by the Newton method starting from the MGP will converge to the controlling UEP; otherwise, the sequence may converge to another equilibriumpoint or diverge.

[6]. Note that Steps 1 to 3 of the above numerical network-preserving BCU method compute the controlling UEP of the reduced-state system (2) and Step 4 relates the controlling UEP of the reduced-state system to the controlling UEP of the original system (1).

5 Improved BCU Classifiers

Improved BCU classifiers have been developed for performing transient stability screening. The design objective of improved BCU classifiers is to meet the five essential requirements for transient stability screening. The improved BCU classifiers are based on the three key steps of BCU method and on the properties of energy functions.

When a numerical method is applied to compute the post-fault SEP starting from the pre-fault SEP, the SEP convergence problem may arise due to the following:

1. the numerical divergence problem (i.e. there is a divergence problem in computing the post-fault SEP starting from the pre-fault SEP),
2. the incorrect convergence problem (i.e. it converges to a wrong post-fault UEP, instead of the post-fault SEP).
3. the non-existence of post-fault SEP.

If a SEP convergence problem occurs, then the study contingency is then sent to the time domain program for further analysis. Recall that the following fundamental assumption required in the direct methods:

- Fundamental assumption: the pre-fault stable equilibrium point lies inside the stability region of the post-fault SEP.

This assumption is trivial if the pre-fault system equals the post-fault system. Since the stability region is an open set with certain size, this assumption is plausible for a great majority of contingencies. When this fundamental assumption is not satisfied for certain very extreme contingencies, then the time-domain simulation approach must be used for stability analysis of these contingencies.

The analytical basis of BCU classifiers are based mainly on the three steps of BCU method and the dynamic information derived during the computational procedure of BCU method. A large majority of the computational efforts required in the improved BCU method are involved in computing the three important state points: the exit point (step 1), the minimum gradient point (step 2) and the controlling UEP (step 3). Useful stability information can be derived from these three points for developing effective schemes for dynamic contingency screening. We next present the design of each BCU classifier along with its analytical basis. The improved BCU classifiers are composed of the following 7 classifiers in the specified sequential order:

Classifier I (classifier for network islanding)
BCU classifier I is designed to screen out highly unstable contingencies that result in a separation of the underlying power network.

Classifier II (classifier for the SEP problem)
This classifier is designed to detect potentially unstable contingencies that suffer from the so-called SEP convergence problem, when a numerical method is applied to compute the post-fault SEP starting from the pre-fault SEP. The SEP convergence problem may arise from either the numerical divergence problem (i.e. there is a divergence problem in computing the post-fault SEP starting from the pre-fault SEP), or the incorrect convergence problem (i.e. it converges to a wrong post-fault UEP, instead of the post-fault SEP). If the SEP convergence problem occurs, then the study contingency is then sent to the time domain program for further analysis.

Classifier III-A (for large stability regions)
This classifier is designed to screen out highly stable contingencies that result in sufficiently large size of the post-fault stability region. A study contingency is definitely stable if it is captured by this classifier. Screening out highly stable contingencies can greatly improve the goal of high yield for dynamic contingency screening. If the PEBS crossing can not be found in a specified time interval, and if the potential energy difference is greater than zero but less than a threshold value, and if the maximum angle difference is less than a threshold value, then the contingency case is highly stable and no further analysis is needed.

Classifier III-B (for the exit point problem)
Given a study contingency, if the exit point problem arises (which means the exit point cannot be determined) during the search for the exit point along the fault-on trajectory, then the contingency is classified as potentially unstable and is then sent to the time domain program for further analysis.

Classifier IV (for the ray adjustment problem)
If a numerical problem arises during the stability-boundary-following procedure (i.e. the search for the minimum gradient point (MGP) fails), then the study contingency is classified as potentially unstable and is then sent to the time domain program for further analysis.

Classifier V (for the controlling UEP convergence problem)
This classifier is designed to detect the so-called controlling UEP convergence problem when a numerical method fails to compute the controlling UEP starting from the MGP. In this classifier, a maximum number of iteration in computing the controlling UEP starting from the minimum gradient point is used to detect such a problem. If the required number of iteration is more than a pre-specified number, then the corresponding contingency is viewed as having a numerical divergence problem and is classified as unstable.

The convergence region of the Newton method is known to have a fractal boundary. Using the Newton method, it has been observed that the region of the starting point that converges to a stable equilibrium point is more significant than that of a unstable equilibrium point such as the controlling UEP. Thus, in regular power flow calculation, the initial guess can be chosen near the stable equilibrium point to safely lie within the convergence region so that, after a few iterations, it converges to the desired stable equilibrium point. This explains why the fractal nature of the convergence region of newton method has been unnoticed in power flow study for so long. As power flow study has been expanded to compute unstable equilibrium points for applications such as direct transient stability analysis and low voltage power flow solutions for voltage collapse analysis, the fractal nature and the different size of the convergence region have become more pronounced. Unfortunately, this nature must be taken into account when a unstable equilibrium point is to be sought.

Classifier VI (classifier for incorrect CUEP problem)
The problem is described as following: it converges to a wrong controlling UEP, i.e. the minimum gradient point lies in such a region that another UEP, instead of the controlling UEP, is obtained when a nonlinear algebraic solver is used to solve . In this classifier, both the coordinate of the obtained UEP and the angle difference between the MGP and the obtained UEP are used as indices to detect the problem.

Classifier VII (controlling UEP classifier)
The remaining unclassified contingencies are then sent to BCU classifier VII for final classification. This classifier uses the energy value at the controlling UEP as the critical energy to classify each remaining contingency as (definitely) stable or (potentially) unstable. If the energy value at the fault clearing time is less than the energy value at the controlling UEP, then the corresponding contingency is (definitely) stable; otherwise it is (potentially) unstable. The theoretical basis of this classifier is the CUEP method. The index used in this classifier is the energy value at the fault clearing time while the threshold value for this index is the energy value at the controlling UEP.

In summary, given a list of credible contingencies to the seven BCU classifiers, the first classifier is designed to screen out those contingencies with convergence problems in computing post-fault stable equilibrium points. The second and third classifiers, based on Step 1 of the BCU method, are the fastest ones. They use the energy value at the exit point on the stability boundary of the reduced-state model as an approximation for the critical energy. The second classifier is designed to drop those contingencies which are highly stable, while the third is designed to screen out those contingencies which may cause computational difficulties for the BCU method, and hence may damage the reliability of the following BCU classifiers. The fourth classifier screens out those contingencies which cause failure in finding the MGP. The fifth classifier screens out the contingencies with the problem of converging to the controlling UEP The sixth classifier drops those contingencies with the problem of incorrect CUEP. The seventh classifier uses the energy at the controlling u.e.p as the critical energy to classify every contingency left over from the previous classifiers into two classes: stable contingencies and unstable contingencies. This classifier is based on Step 3 of the BCU method.

Contingencies which are classified as definitely stable at each classifier are eliminated from further analysis. It is due to the definite classification of stable contingencies that considerably increased speed for dynamic security assessment can be achieved. Contingencies which are either undecided or identified as unstable are then sent to the time-domain transient stability simulation program for further stability analysis. Note that the conservative nature of the BCU method guarantees that the results obtained from the seven dynamic contingency classifiers are also conservative (i.e. no unstable cases are mis-classified as stable). Classifying a stable contingency, either first-swing or multi-swing, as unstable is the only scenario in which the BCU classifiers give conservative classifications.

6   TEPCO-BCU Package

TEPCO-BCU is an integrated package developed for fast yet exact transient stability assessments of large-scale power systems for the on-line mode, on-line study mode, or off-line planning mode [16]. The package is reliable and yet fast for calculating the energy margin for every contingency; furthermore the critical energy value computed by the method is compatible with that computed by the controlling UEP method. TEPCO-BCU is composed of the BCU method, improved BCU classifiers, and the BCU guided time-domain method [8]. The architecture of TEPCO-BCU has two major components: a set of improved BCU classifiers for dynamic contingency screening and a fast and reliable time-domain transient stability simulation program and a BCU-guided time-domain method.

The main functions of TEPCO-BCU include the following:

- fast screening of highly stable contingencies
- fast identification of insecure contingencies
- fast identification of critical contingencies
- computation of the energy margin for transient stability assessment of each contingency
- BCU-based fast computation of critical clearing time of each contingency
- contingency screening and ranking for transient stability in terms of energy margin or critical clearing time
- Detailed time-domain analysis of selected contingencies.

The following data are needed in on-line transient stability assessment:

Powerflow. This corresponds to a real time system condition captured by EMS and solved by the state estimator and a power flow solver.

Contingencies. The "raw" contingency information (including only circuit/generator losses) is obtained from EMS/Contingency Analysis. A contingency processing function expands the information for transient stability analysis. This involves:

- Adding fault. Both three-phase and single-line-to-ground faults will be considered.
- Both primary and backup fault clearance times are provided.
- For single-line-to-ground faults, additional circuits will be lost as a result of backup clearance. This is specified for each contingency.
- Parameters used in simulations, such as simulation length, integration step size, etc. are provided.

Dynamic model and data : This refers to dynamic models and data matching the real time powerflow, including load models and the following:

- Generator: classical to two-axis 6th order modals
- Excitation system: all IEEE standard exciter/AVR and PSS models and common extended models
- Load: ZIP model, voltage dependent model, discharge lighting model
- SVC
- User-defined modeling (transfer function)

Monitored quantities: these are the quantities that should be monitored in simulations and will be available to system operators to examine system responses.

Transaction definition: this is required only if transfer limits are to be determined. Two sets of data are required: (i) definition of the source and link in the transaction, (ii) definition of the interface to be monitored.

A practical contingency list may contain the following contingencies

- Faults may be at buses or anywhere on lines
- For unbalanced faults, fault impedances can be specified or computed with sequence network data provided
- Branch (single or three phase) tripping and reconnection, shunt switching, adding or modifying branch
- Generator tripping, exciter or governor reference set point changes
- Load shedding

To meet the on-line dynamic contingency screening for large power systems, TEPCO-BCU is implemented on a parallel processing architecture. Parallel processing is the simultaneous execution of the same task (split up and specially adapted) on multiple processors in order to obtain faster speed. The parallel nature can come from a single machine with multiple processors or multiple machines connected together to form a cluster. It is well recognized that every application function benefit from a parallel processing with a wide-range of efficiency. Some application functions are just unsuitable for parallel processing.

7 Evaluation of TEPCO-BCU

TEPCO-BCU package is installed on the PJM TSA Test System, which receives the following input data including real time data from the PJM EMS system:

- Real time network model
- State Estimation Solution
- A list of Contingencies
- Generator dynamic data for each generator
- Stability limit calculations data

TEPCO-BCU package runs automatically on the PJM TSA Test System every 15 minutes in order to synchronize with the PJM TSA software execution cycle. The stability assessments by TEPCO-BCU software include the following:

(i) stability assessment of each contingency
(ii) energy margin of each contingency
(iii) estimated CCTs of each contingency
(iv) ranking of each contingency in terms of energy margins or in terms of estimated CCTs
(v) computation time consumed.

PJM time-domain simulation (TSA) software was also installed on the PJM TSA Test System and executed periodically. Basically, the PJM TSA is a time-domain simulation program. Both PJM TSA and TEPCO-BCU package received the same input data and the same real time snapshot data from the PJM EMS. The time-domain simulation results from the PJM TSA were used as benchmarks to examine the stability assessment results of TEPCO-BCU.

The three-month evaluation period was chosen to include the scheduled outage season and the summer peak load season. This period historically encountered a wide range of real time scenarios that can happen on the PJM power system. During this evaluation period, both the PJM TSA software and the TEPCO-BCU software were run periodically in parallel, every 15 minutes in the PJM TSA Test System environment. Both packages read the same set of time-domain dynamic data, same list of contingencies, and same real-time input data from the EMS for each execution. The contingency list contains about 3000 contingencies with a three-phase fault/single phase with CB stuck. The outputs of both packages (i.e. evaluation results) were written into two separate text files. In each day, there were about 96 pairs of evaluation results stored in each text file. Another set of software was developed to compare the evaluation results as follows:

- Stability status (i.e. stable/unstable) of each contingency
- The required CPU time of each contingency by TEPCO-BCU software

The time-domain simulation approach numerically integrates both the fault-on system and the post-fault system. The stability of the post-fault system is assessed based on the simulated post-fault trajectory. The typical simulation period for the post-fault system is 10 seconds and can go beyond 15 seconds if multi-swing instability is of concern; making this conventional approach can be time-consuming in assessing stability.

The evaluation of TEPCO-BCU method was focused on the following performance measures:

(1). Reliability measure: as a screening tool, TEPCO-BCU is designed to give conservative stability assessment results, but not to give any over-estimated results in stability assessment.

(2). Speed measure: as a screening tool, speed is one of the key requirements. The speed of TEPCO-BCU in assessing the stability properties of each contingency on a 14,000-bus PJM system should be on the average **1.0 second** or less.

(3). Screening measure (i.e. Percentage of stable contingencies screened out): For a list of credible contingencies, the maximum number of unstable/critical contingencies in each assessment cycle is assumed to be 15 % of the total number of contingencies. TEPCO-BCU should screen out more than 85% of stable contingencies.

(4). On-line measure: No on-line adjustment of parameters is needed for the screening method.

(5). Robustness measure: the perofrmances of screening measure, reliability measure and speed measure are insensitive to the operating conditions and network topologies.

Evaluation Results

(1) **Reliability measure:** TEPCO-BCU consistently gave conservative stability assessments for each contingency during the three-month evaluation time. TEPCO-BCU did not give over-estimated stability assessment for any contingency. For a total of 5.29 million contingencies, TEPCO-BCU captures all the unstable contingencies.

Table 1. Reliability Measure

Total number of contingencies	Percentage of capturing unstable contingencies
5293691	100%

(2) **Speed:** The speed assessment is shown in Table 2. For a total of 5.29 million contingencies, TEPCO-BCU consumes a total of 717575 CPU seconds. Hence, on average, TEPCO-BCU consumes about 1.3556 second/node for the stability assessment of each contingency.

On-Line Transient Stability Screening of a Practical 14,500-Bus Power System   355

Table 2. Speed Assessment

Total number of contingencies	Computation Time	Time/per contingency/per node
5293691	717575 seconds	1.3556 second

(3) **Screening measure:** The percentage of screening for stable contingencies is shown in the Table 2. Depending on the loading conditions and network topologies, the screening rate ranges from 92% to 99.5%; meaning the TEPCO-BCU screens out, for every 100 stable contingencies, between 92 contingencies and 99.5 contingencies which are definitely stable.

Table 3. Screening Assessment

Total number of contingencies	Percentage of screening out stable contingencies
5293691	92% to 99.5 %

A summary of overall performance of TEPCO-BCU is shown in Table 4 which indicates that TEPCO-BCU is an excellent screening tool for on-line transient stability analysis of large-scale power systems. Of the 5.3 million contingencies, all the unstable contingencies verified by detailed time domain simulation were classified by TEPCO-BCU as unstable. These unstable contingencies exhibit first-swing instability as well as multi-swing instability.

Table 4. Overall performance of TEPCO-BCU for on-line transient stability screening

Reliability measure	Screening measurement	Computation speed	on-line computation	Robust performance
100%	92% to 99.5%	1.3 second/per node/per contingency	Yes	Yes

8 Concluding Remarks

This paper has presented a fairly detailed description of the BCU method from both theoretical and numerical viewpoints. It also has presented a comprehensive evaluation study of the TEPCO-BCU package in a real time environment as a screening tool for on-line transient stability assessment. This evaluation study was focused on the reliability measurement, speed, on-line computation, screening performance and robustness measure. In order to evaluate the robustness of the TEPCO-BCU, the three-month evaluation period was chosen to include the scheduled outage season and the summer peak load season. This period historically encountered a wide range of real time scenarios that can happen on the PJM power system.

Based on the evaluation results obtained in this study, it is believed that the TEPCO-BCU package is an excellent transient stability screening tool for on-line transient stability analysis of large-scale power systems. This evaluation study represents the largest practical application of the stability region theory and its estimation of relevant stability region behind the BCU methodology in terms of the size of the study system which is a 14,000-bus power system dynamic model with a total of 5.3 million contingencies.

This paper has also illustrated one practical application of the controlling UEP method and the theory-based BCU method on large-scale power systems with practical data in an on-line environment. This confirms the authors' belief that theory-based solution methods can lead to practical applications in large-scale nonlinear systems. In particular, theory-based methods are suitable for power system on-line applications which require both speed and accuracy requirements.

References

1. Chiang, H.D., Tada, Y., Li, H.: Power System On-line Transient Stability Assessment. In: Wiley Encyclopedia of Electrical and Electronics Engineering. John Wiley & Sons, Inc. (2007)
2. Tada, Y., Kurita, A., Zhou, Y.C., Koyanagi, K., Chiang, H.D., Zheng, Y.: BCU-guided time-domain method for energy margin calculation to improve BCU-DSA system. In: IEEE/PES Transmission and Distribution Conference and Exhibition, Yokohama, Japan, October 6-10 (2002)
3. Chiang, H.D., Wu, F.F., Varaiya, P.P.: Foundations of direct methods for power system transient stability analysis. IEEE Trans. on Circuits and Systems CAS-34, 160–173 (1987)
4. Pai, M.A.: Energy Function Analysis for Power System Stability. Kluwer Academic Publishers, Boston (1989)
5. Tong, J., Chiang, H.D., Conneen, T.P.: A sensitivity-based BCU method for fast derivation of stability limits in electric power systems. IEEE Trans. on Power Systems PWRS-8, 1418–1437 (1993)
6. Tada, Y., Takazawa, T., Chiang, H.D., Li, H., Tong, J.: Transient stability evaluation of a 12,000-bus power system data using TEPCO-BCU. In: 15th Power system Computation Conference (PSCC), Belgium, August 24-28 (2005)
7. Tada, Y., Chiang, H.D.: Design and Implementation of On-Line Dynamic Security Assessment. IEEJ Trans. on Electrical and Electronic Engineering 4(3), 313–321 (2008)
8. Tada, Y., Kurita, A., Zhou, Y.C., Koyanagi, K., Chiang, H.D., Zheng, Y.: BCU-guided time-domain method for energy margin calculation to improve BCU-DSA system. In: IEEE/PES Transmission and Distribution Conference and Exhibition, Yokohama, Japan, October 6-10 (2002)
9. Chiang, H.D., Zheng, Y., Tada, Y., Okamoto, H., Koyanagi, K., Zhou, Y.C.: Development of an on-line BCU dynamic contingency classifiers for practical power systems. In: 14th Power System Computation Conference (PSCC), Spain, June 24-28 (2002)
10. Chiang, H.D., Chu, C.C., Cauley, G.: Direct stability analysis of electric power systems using energy functions: Theory, applications and perspective (Invited paper). Proceedings of the IEEE 83(11), 1497–1529 (1995)

11. Chiang, H.D.: Analytical results on the direct methods for power system transient stability analysis. In: Control and Dynamic Systems: Advances in Theory and Application, vol. 43, pp. 275–334. Academic Press, New York (1991)
12. Chiang, H.D.: The BCU method for direct stability analysis of electric power systems: pp. theory and applications. In: Systems Control Theory for Power Systems. IMA Volumes in Mathematics and Its Applications, vol. 64, pp. 39–94. Springer, NewYork (1995)
13. Chiang, H.D., Wu, F.F., Varaiya, P.P.: A BCU method for direct analysis of power system transient stability. IEEE Trans. Power Systems 8(3), 1194–1208 (1994)
14. Chadalavada, V., Vittal, V., Ejebe, G.C., et al.: An On-Line Contingency Filtering Scheme for Dynamic Security Assessment. IEEE Trans. on Power Systems 12(1), 153–161 (1997)
15. Chiang, H.D., Wang, C.S., Li, H.: Development of BCU Classifiers for On-Line Dynamic Contingency Screening of Electric Power Systems. IEEE Trans. on Power Systems 14(2) (May 1999)
16. Kim, S.G., Tada, Y., Hur, S.I.: Transient Contingency Screening of Large Power Systems using Tepco-BCU Classivier. In: The International Conference on Electrical Engineering (ICEE 2009), Shenyang, China, July 5-9 (2009)
17. Chiang, H.D., Tada, Y., Li, H.: Power System On-line Transient Stability Assessment (an invited Chapter). In: Wiley Encyclopedia of Electrical and Electronics Engineering. John Wiley & Sons, New York (2007)
18. Ejebe, G.C., Jing, C., Gao, B., Waight, J.G., Pieper, G., Jamshidian, F., Hirsch, P.: On-Line Implementation of Dynamic Security Assessment at Northern States Power Company. In: IEEE PES Summer Meeting, Edmonton, Alberta, Canada, July 18-22, pp. 270–272 (1999)
19. Kundur, P.: Power System Stability and Control. McGraw Hill, New York (1994)
20. Ernst, D., Ruiz-Vega, D., Pavella, M., Hirsch, P., Sobajic, D.: A Unified Approach to Transient Stability Contingency Filtering, Ranking and Assessment. IEEE Trans. on Power Systems 16(3) (August 2001)
21. Mansour, Y., Vaahedi, E., Chang, A.Y., Corns, B.R., Garrett, B.W., Demaree, K., Athay, T., Cheung, K.: B.C.Hydro's On-line Transient Stability Assessment (TSA): Model Development, Analysis and Post-processing. IEEE Transactions on Power Systems 10(1), 241–253 (1995)
22. Mansour, Y., Vaahedi, E., Chang, A.Y., Corns, B.R., Tamby, J., El-Sharkawi, M.A.: Large scale dynamic security screening and ranking using neural networks. IEEE Transactions on Power Systems 12(2), 954–960 (1997)
23. Morison, K., Wang, L., Kundur, P.: Power System Security Assessment. IEEE Power & Energy Magazine 2(5) (September / October 2004)
24. Wang, L., Morison, K.: Implementation of On-Line Security Assessment. IEEE Power & Energy Magazine 4(5) (September / October 2006)
25. Fouad, A.A., Vittal, V.: Power System Transient Stability Analysis: Using the Transient Energy Function Method. Prentice-Hall, New Jersey (1991)
26. Sauer, P.W., Pai, M.A.: Power System Dynamics and Stability. Prentice-Hall, New Jersey (1998)
27. Chiang, H.D.: Direct Methods for Stability Analysis of Electric Power Systems: Theoretical Foundation, BCU Methodologies, and Applications. Wiley, New York. IEEE Press, New Jercy (2011)

28. Chu, C.C., Chiang, H.D.: Constructing Analytical Energy Functions for Network-Preserving Power System Models. Circuits Systems and Signal Processing 24(4), 363–383 (2005)
29. Rahimi, F.A., Lauby, M.G., Wrubel, J.N., Lee, K.L.: Evaluation of the transient energy function method for on-line dynamic security assessment. IEEE Transaction on Power Systems 8(2), 497–507 (1993)
30. Rahimi, F.A.: Evaluation of transient energy function method software for dynamic security analysis, Final Report RP 4000-18, EPRI, Palo Alto, CA (December 1990)
31. Llamas, A., De La Ree Lopez, J., Mili, L., Phadke, A.G., Thorp, J.S.: Clarifications on the BCU method for transient stability analysis. IEEE Trans. Power Systems 10, 210–219 (1995)
32. Ejebe, G.C., Tong, J.: Discussion of Clarifications on the BCU method for transient stability analysis. IEEE Trans. Power Systems 10, 218–219 (1995)
33. Paganini, F., Lesieutre, B.C.: A critical review of the theoretical foundations of the BCU method. MIT Lab. Electromagnetic Electr. Syst., Tech. Rep. TR97-005 (July 1997)
34. Paganini, F., Lesieutre, B.C.: Generic Properties, One-Parameter Deformations, and the BCU Method. IEEE Trans. on Circuits and Systems, Part-1 46(6), 760–763 (1999)
35. Alberto, L.F.C.: Transient Stability Analysis: Studies of the BCU method; Damping Estimation Approach for Absolute Stability in SMIB Systems, Escola de Eng. de So Carlos - Universidade de So Paulo (1997) (in Portuguese)
36. Yorino, N., Kamei, Y., Zoka, Y.: A New Method for Transient Stability Assessment based on Critical Trajectory. In: 15th Power Systems Computation Conference, Liege, Belgium, August 22-26 (2005)
37. Chiang, H.D., Chu, C.C.: Theoretical Foundation of the BCU method for direct stability analysis of network-reduction power system model with small transfer conductances. IEEE Trans. on Circuits and Systems, Part I: CAS-42(5), 252–265 (1995)
38. Kim, J.: On-line transient stability calculator. Final Report RP2206-1, EPRI, Palo Alto, CA (March 1994)
39. Mokhtari, S., et al.: Analytical methods for contingency selection and ranking for dynamic security assessment. Final Report RP3103-3, EPRI, Palo Alto, CA (May 1994)
40. Electric Power Research Institute, User's Manual for DIRECT 4.0, EPRI TR-105886s, Palo Alto, CA (December 1995)
41. Zou, Y., Yin, M., Chiang, H.-D.: Theoretical Foundation of Controlling UEP Method for Direct Transient Stability Analysis of Network-Preserving Power System Models. IEEE Transactions on Circuits and Systems I: Fundamental Theory and Applications 50, 1324–1356 (2003)
42. Kuo, D.H., Bose, A.: A generation rescheduling method to increase the dynamics security of power systems. IEEE Trans. on Power Systems, PWRS-10(1), 68–76 (1995)
43. Chiang, H.-D., Subramanian, A.K.: BCU Dynamic Security Assessor for Practical Power System Models. In: IEEE PES Summer Meeting, Edmonton, Alberta, Canada, July 18-22, pp. 287–293 (1999)

Application of HPC for Power System Operation and Electricity Market Tools for Power Networks of the Future

Ivana Kockar

Institute for Energy and Environment, Department of Electronic and Electrical Engineering
University of Strathclyde, Glasgow, Scotland, UK
ivana.kockar@eee.strath.ac.uk

Abstract. Development and application of SmartGrids or Intelligrids, including roll-out of smart meters and electrical vehicles, is of a great importance if the UK and other countries are to achieve significant carbon emission reductions and realize sustainable energy systems. These new grids will offer the opportunity to increase the level of renewable energy integrated into the system. They will also allow customers, including small households, to actively participate and adjust their demand depending on energy availability and price. This will further lead towards improved energy efficiency, as well as offer possibilities to reduce overall consumption and reduce or postpone investments into new large generation and infrastructure facilities.

To achieve these goals, a number of technical, economical and policy issues need to be addressed and resolved. The development of new generations of extremely fast software tools that can solve power system problems with large number of nodes will also be important to help resolve these issues. For example, distribution system and network operators, as well as trading entities such as aggregators, will get a better coordination of system operation though the possibility to engage with even smaller generators, and especially smaller customers. This control at lower voltage levels will allow for the aggregation of responses which will then propagate to higher voltage levels.

Currently, the discussion regarding the operation of future power systems is looking into two different options. One is to develop methodologies that will allow decentralization of network operation with the reduced level of coordination at the high level of system operation. However, the new software developed to exploit the benefits of the HPC architecture may open a possibility for businesses and policy makers to investigate and compare operation of centralized vs. decentralized operation over areas with large number of participants.. These new HPC power system analysis tools will enable more frequent price signal calculations and bring the possibility to define policies which will ensure engagement with customers to reduce their energy consumption or shift it towards off-peak periods, as well as allow for the coordination of charging of electric vehicles and their use as storage devices. Such tools will be useful for both decentralized and centralized operation, however they will be crucial for the latter.

This chapter will first give an overview of the changes in future power system operation and then outline power system analysis tools such as power flow, optimal power flow, generation scheduling and the security assessment. It will then discuss current status of the parallel techniques and HPC applications for the power system operation tools. It will also discuss the formulation requirements, achievements and possible obstacles in the application of techniques suitable for HPC and for power system operation problems such as power flow, optimal power flow (OPF), security constrained OPF. Finally, it will look how new developments in the HPC/Numerical Analysis area, and even more powerful Extreme Computing together with new algorithms developed for this next-step class of machines may help improve power system operation and electricity markets tools.

1   Nomenclature

a_i, b_i, c_{oi}	parameters of an offer /cost curve of generator i
$B(P_{di}(t))$	benefit function of load j at time t
$C_i(P_{gi}(t))$	offer curve of generator i at time t
$k_{cong\,k}$	congestion proportionality factor for line k
k_{reg}	regulatory coefficient
$P_{dj}(t)$	active power demand at bus j at time t
$P_d(t)$	total system demand at period t
$P_{gi}(t)$	active generation output at bus i at time t
$P_{gi}^{min}, P_{gi}^{max}$	lower and uppers generation limits of unit i
$STC_i(t)$	start-up cost of generator i at period t
$T_{gi}^{up}, T_{gi}^{down}$	min-up and min-down times of generator i
$\sigma_{gi}^{on}(t), \sigma_{gi}^{off}(t)$	a number of hours generator i has been on or off at period t since it has been turned on or off
Y_{ij}	magnitude of ij element of Y_{bus} matrix
ψ_{ij}	angle of ij element of Y_{bus} matrix
u_i	binary unit commitment variable indicating if unit i is on or off at period t
δ	vector of bus voltage angles
$\mathbf{P}_f^{max}$	vector of actual real line flows limits, submitted as a second part offer of Transmission Owners
$\mathbf{P}_g$	vector of active generation outputs

$\mathbf{P}_g^{min}$, $\mathbf{P}_g^{max}$	vectors of upper and lower active power generation limits
$\mathbf{Q}(\mathbf{V},\delta)$	vector of reactive power injections
$\mathbf{Q}_d$	vector of bus reactive power loads
$\mathbf{Q}_g$	vector of generation reactive power injections
$\mathbf{Q}_g^{min}$, $\mathbf{Q}_g^{max}$	vectors of upper and lower reactive power generation limits
$\mathbf{V}$	vector of bus voltage magnitudes
$\mathbf{V}^{min}$, $\mathbf{V}^{max}$	vectors of upper and lower limits of voltage magnitudes
$\mathbf{Y}_{bus}$	bus admittance matrix

2 Introduction

One of the major challenges in the energy sector is to ensure secure and sustainable energy supply. An increase in the utilization of renewable resources which are usually connected at lower distribution levels, as well as a need for active demand participation will lead towards electricity networks that are more complex. Operation of these new SmartGrids and Intelligrids[1] will require a step-change in capabilities of operational tools due to two factors (i) a significant increase in the number of control variables, and (ii) reduced time intervals between which power system analyses are carried out, including more frequent calculation of generation outputs and prices.

The first factor is influenced by the number of nodes, i.e. injection/consumption points on the network, and its increase will be affected by a number of distributed generators and, even more, by utilization of smart meters and electric vehicles that can act both as consumers and as source/storage devices. For example, the number of nodes for the UK system is currently around 2000, and the roll-out of smart meter and electric vehicles could increase it to millions. Furthermore, active demand-side participation and smart metering will significantly benefit from dynamic accurate pricing signals, which are calculated frequently.

To achieve these goals, two opposing suggestions are usually considered. One is to operate a system in a decentralized way, while the other is based on the assumption that the System and Network Operators should try to retain certain levels of centralized operation as it allows for a better coordination of system operation through the possibility to engage with even smaller generators, and especially smaller customers. Although development of the HPC based tools will be useful for both approaches, it will be more critical for the latter as it will require a new generation of system operation and electricity markets tools that are able to cope with the explosion of decision variables. It will also enable solution of the large scale operation and short-term planning and scheduling problems, and will help

---

[1] Term IntelliGrids is usually used in the US, while SmartGrids is mainly used in other parts of the world, especially in Europe.

devising the decentralized problem formulation and then testing how coordination of a large number of areas performs against the centralized operation.

3 Power System Transformation

The structure of present-day electricity networks has been shaped over several decades to reflect a centralized operation of vertically integrated utilities. This means that generators and loads are interconnected by relatively strong networks that allow for central coordination and control of a power balance and other network operating constraints. Although in the last 10-15 years electricity industry has undergone a significant change through restructuring and instigation of market forces and decentralization on both the supply and load sides, transmission and distribution networks are still mainly regarded as sectors that need to remain centrally operated and under a tight regulation.

The above mentioned liberalization of electricity sector has been carried out in a number of countries, and in most cases it involved introducing competition in the generation sector. This means that generators have to compete in the market to sell their energy by offering it at competitive prices. In order to ensure that all generators have equal chances to reach the customers transmission network has be separated from the generation business and operated by a System Operator (SO) in manner that will ensure open access to all market participants. The structure of this market organization is shown in Fig. 1. Some electricity markets, like the one in the UK, have gone a step further and also introduced a competition at the retail level, as indicated it Fig. 1. This means that even small customers, such as households, can choose their supplier according to the offered tariff and other preferences. The retail competition is enabled by separating supply businesses from the operation of distribution networks which is now in the domain of Distribution Network Operators (DNOs). Similarly to the competition at the generation side, there could be a number of entities, referred to as *suppliers* or *Load Serving Entities* (LSE) that can offer energy to small customers. The roll-out of smart meters will also enable development of dynamic pricing and various different services that these suppliers can offer to sell to, as well as to buy from, small customers.

Thus, the key paradigm shift in the operation of the future power systems is a new level of flexibility that would enable better integration of new technologies and ensure sustainable energy supply. An increase in the utilization of renewable resources, which are usually connected at lower distribution levels, as well as a need for active demand participation, will lead towards electricity networks that are even more complex. For example, to accommodate new, more distributed generation resources and active demand participation, which are crucial for achieving a low carbon economy, current UK electricity infrastructure will need to change so to adapt to new circumstances. As indicated in Fig 1., the transition from the present structure to the future one is motivated by a number of drivers such as the cost of operation and infrastructure reinforcement, technological innovations, achievement of the renewable targets

Application of HPC for Power System Operation and Electricity Market Tools

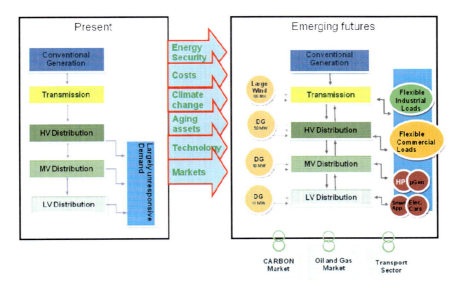

Fig. 1. Power System Transformation [HiDEF 2008]

as well as the reduction of emissions, and ensuring the security of energy supply. This figure also indicates that in the current network design, generation is mainly in the form of large generating installations connected to the transmission network that serves as the backbone. In addition, distribution networks have been mainly designed to transfer power from high voltage transmission networks to medium and small consumers.

The structure of the emerging power systems, as illustrated in Fig 1., is changing so that generation is becoming connected to all voltage levels. For example, while large generating plants and large wind farms are still connected to the High Voltage (HV) transmission lines, a number of distributed generators are connected to the distribution network. In addition, the new emerging system includes flexible demand on all levels (and of all sizes), heat-pumps, microgirds, as well as energy storage and, eventually, electric vehicles.

Thus, connections of distributed generation, flexibility of load, and other above mentioned devices can significantly change power system network operating conditions, so that network reinforcement and utilization of new technological solutions, together with more flexible market rules, may become necessary. In addition, in the UK, US and many other developed countries, electricity networks at both high and low voltage sides are in a need of significant investments to replace ageing infrastructure. On one hand, this opens the opportunity to introduce new technologies and solutions that will improve power system operation and bring new possibilities of generation and load management. Examples include distributed generation, smart metering, or "smart homes" that will have more flexibility in deciding when and what type of services to use.

On the other hand, new equipment and technological solutions may need to be introduced gradually, and have to operate along with existing, and often relatively old, equipment. What makes things even more complex is that investments in electricity infrastructure are expensive and may take a long time to complete. Moreover, a life cycle of a typical power system network and equipment is around 40 years, which may hinder the pace of introducing innovation in technology, and network design.

Finally, accounting mechanisms that are used to evaluate financial aspects and costs associated with traditional plants may not be sufficiently flexible to properly capture the benefits and behavior of new technologies. For example, cost and maintenance expenses of solar and wind farms may follow different rules, since fuel charges for these technologies are non-existent.

The change introduced by these transformations will make operation of the power system more complex and computationally demanding. The first is the result of the above mentioned changes in the utilization of the network that was not planned and build for these new conditions. It will also come from the interactions between various devices and participants, either as market participants that respond to economic incentives or simply as variations in the patterns of generating and demand.

Thus, the increased computational power will be required due to a number of generators, load, storage device, electric vehicle, etc. that will significantly affect the size of the network models. Note that this increase will come from increased granularity of the necessary models rather than from the increase in the geographical size (although interconnections in the EU and US can also impact size of the network model).

In addition, due to the variability of renewable generation output and active demand side participation, frequency of calculations to verify network conditions or clear the market is likely to increase. This will particularly be necessary if dynamic pricing becomes accepted. Furthermore, dynamic pricing may bring additional opportunities to provide various ancillary services and, in turn, lead towards more complex business models that will require mathematical formulation with additional capabilities to help in decision making. All of these will need to be solved for systems with larger number of nodes, far more decision variables and in shorter time scales.

4   Tools for the Network and Market Operation

Typically, a power system is an electric network with a number of nodes, i.e. buses and lines (branches) that connect them. A number of branches and nodes can be very large, for example a typical size of the transmission network model of the Great Britain consists of around 2,000- 2,500 buses, while in the US a Texas model has over a 4,500 buses. Some of the buses have generators, while most of the buses have loads connected to them. Transmission lines are used to transport energy from generators to loads, and other equipment such as transformers, additional sources of reactive power such as Static VAr Compensators (SVCs) etc., are used to support power transfer and maintain system security.

A number of analyses are carried out by system and market operators, as well as network planners in order to ensure secure and economic energy supply. These include power flow, economic dispatch, optimal power flow, security constrained optimal power flow, various market clearing procedures as well as generation scheduling. This section briefly outlines mathematical formulation of these problems while the application of the HPC approaches for their solutions are described in section 5.

4.1 Power Flow

A load-flow analysis determines the power flows along each line in a power system network, and is considered to be one of the most fundamental approaches for investigating power system operation in a steady-state regime. For the given network topology and parameters, a load-flow solution specifies four variables at each bus: active power, P_i, reactive power, Q_i, voltage magnitude, V_i and voltage phase angle, δ_i, which can be then used to calculate the power flows, and fully describe the state of the system. The power injections and the voltage at each bus, as well as the network parameters, are sufficient to calculate the required power flows.

Thus, power flow analyses are important for the management of power systems operating under normal steady-state conditions, as well as for the study of failures, optimal system operation, and in planning studies of power system upgrading. A substantial amount of time and effort has been invested in developing algorithms and improving established methodologies, and therefore it is a well-established problem.

In the general case of a n-bus system, the complex power $\underline{S}_i = P_i + jQ_i$ injected into bus i is, as indicated in Fig.2., defined by,

$$\underline{S}_i = \underline{S}_{g_i} - \underline{S}_{d_i} = P_{g_i} - P_{d_i} + j(Q_{g_i} - Q_{d_i}) \tag{1}$$

where the subscripts g and d denote generation and demand, respectively.

At each bus i, values of active, P_{di} and reactive, Q_{di} demand is know. In addition, for all buses with generators, active generation output, P_{gi}, and voltage magnitude, V_i are also given. The exceptions are the slack bus for which active power generation is free and calculated at the very end, and the reference bus for which the voltage angles is specified. As mentioned above, each bus is described by four variables at each bus, i.e. active power, P_i, reactive power, Q_i, voltage magnitude, V_i and voltage phase angle, δ_i. However, only two of these are

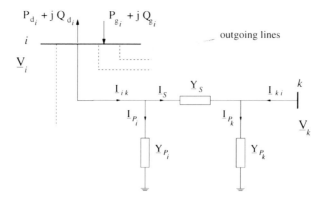

Fig. 2. Interconnections between bus i and other surrounding buses

known for all buses except the slack and the reference. The load-flow problem consists of determining the other two values by solving the following set of non-liner active and reactive power balance equations,

$$P_i = V_i \sum_{k=1}^{n} V_k Y_{ik} \cos(\delta_i - \delta_k - \psi_{ik}) \tag{2}$$

$$Q_i = V_i \sum_{k=1}^{n} V_k Y_{ik} \sin(\delta_i - \delta_k - \psi_{ik}) \tag{3}$$

Known and calculated values of active power, P_i, reactive power, Q_i, voltage magnitude, V_i and voltage phase angle, δ_i, can then be used to calculate the power flows, and fully describe the state of the system.

Note that the bus admittance matrix, $\mathbf{Y}_{bus}$, with elements $\underline{Y}_{ik} = Y_{ik} \angle \psi_{ik}$, models the passive portion of the n-bus network. This matrix can be very sparse because the values of the matrix elements for non-existing lines are equal to zero. Typically, $\mathbf{Y}_{bus}$ can be 95% sparse, or more, and this is often exploited in numerical methods used for the solution of the above system of non-linear equations.

Generally, there are a number of different methods that have been proposed and used for power flow analysis, some of which have become more popular and widespread, and some that remain only of academic interest. Depending on the type of analysis for which the load-flow solution is used (power system planning, operational planning, or operation/control studies), different types of solutions (like accurate or approximate, off-line or on-line, unadjusted or adjusted, and single case or multiple case) and different properties of the load-flow solution method (such as high speed of computation, low computer storage space, reliability of the solution for ill-conditioned systems, versatility and simplicity) are required (Stott [1974]).

4.2 Economic Dispatch

While power flow analysis consider that the active power generation output is known, finding how much each of the units should generate is another important question. The Economic Dispatch (ED) tool seeks to find the allocation of generation such that the cost of generation is minimized, while the system and unit constraints are respected. The mathematical formulation of the ED problem is as follows,

$$\min \sum_{i=1}^{n_g} C(P_{gi}) \qquad (4)$$

subject to

$$\sum_{i=1}^{n_g} P_{gi} = \sum_{j=1}^{n_d} P_{dj} \qquad (5)$$

$$P_{gi}^{\min} \leq P_{gi} \leq P_{gi}^{\max} \quad \forall i = 1 \ldots n_g \qquad (6)$$

In the above formulation constraints (5) is the power balance equation which ensures that supply matches demand at any moment in time, while (6) are upper and lower limits defined for each generator. This formulation does not include network constraints and losses. In some markets this is the formulation that is used to calculate the System Marginal Price that is applied uniformly to all generators and loads. However, the downside of the ED is that it does not reflect network constraints. The Optimal Power Flow (OPF) described in the following subsection overcomes this limitation.

4.3 Optimal Power Flow

Another important tool in the power system analysis is the Optimal Power Flow (OPF), originally defined by Carpentier (1979). In the initial formulation its aim was to allocate power among generators so that the demand was supplied at the least cost subject to system and network constraints. The market clearing procedure proposed by Caramanis et al. (1982) and Schweppe et al. (1988), and used in centralized markets, is based on the OPF method, and can also be extended to include demand elasticity.

In a general form it can be expressed as the following optimization problem:

$$\max \sum_{i=1}^{n} B(P_{di}) - C_i(P_{gi}) \qquad (7)$$

subject to

$$\text{s.t.} \quad (\mathbf{P}_g, \mathbf{Q}_g, \mathbf{V}, \boldsymbol{\delta}) \in S \qquad (8)$$

$$-\mathbf{P}_f^{\lim} \leq \mathbf{P}_f(\mathbf{V}, \boldsymbol{\delta}) \leq \mathbf{P}_f^{\lim} \qquad (9)$$

The set S in (8) denotes the security region of the power system in the space of generation levels and bus voltages, ($\mathbf{P}_g, \mathbf{Q}_g, \mathbf{V}, \boldsymbol{\delta}$). Such a region is defined by:

➤ the load flow equations,

$$\mathbf{P}_g = \mathbf{P}_d + \mathbf{P}(\mathbf{V}, \boldsymbol{\delta}) \quad \text{and} \quad \mathbf{Q}_g = \mathbf{Q}_d + \mathbf{Q}(\mathbf{V}, \boldsymbol{\delta}),$$

➤ the range of real and reactive generation,

$$\mathbf{P}_g^{\min} \leq \mathbf{P}_g \leq \mathbf{P}_g^{\max} \quad \text{and} \quad \mathbf{Q}_g^{\min} \leq \mathbf{Q}_g \leq \mathbf{Q}_g^{\max},$$

➤ the voltage magnitude limits,

$$\mathbf{V}^{\min} \leq \mathbf{V} \leq \mathbf{V}^{\max}$$

Furthermore, equation (9) defines limitations on power flows in transmission lines. Also, in the above formulation, objective function is typically modeled as a quadratic function, with the cost function of the generator defined as,

$$C_{gi} = c_{oi} + a_i P_{gi} + \frac{1}{2} b_i P_{gi}^2 \qquad (10)$$

In a competitive electricity market, generators submit their offers, $\Omega_{dj}(P_{dj})$, that are not necessarily the same as their cost curves, $C_i(P_{gi})$, however they are usually modeled in the same way as the cost function[2]. Similarly, benefit functions of loads, $B_{dj}(P_{dj})$, are also quadratic functions of the consumed power, P_{dj}.

The OPF problem is a non-linear optimization problem, and there were a number of approaches to its solution. A detailed overview of methods based on linear approaches as well as quadratic programming methods based on the Kuhn-Tucker equations is given by Huneault and Galiana [1991]. In addition to those, often referred to as classical method, there was a growing application of the Interior Point Methods, that can be applied to both linear and non-liner problems. General overview and history of the Interior Point Method developments is given by Gondzio [2011], while the application in the area of power systems, including the solution

---

[2] Microeconomics theory shows that in the perfect market suppliers will be best off to submit an offer that is the same as their true costs, however in practice generators are free to decide on their offer curves.

of the OPF problem is discussed by Quintana et al. [2000]. In addition, Mamoh et al. (1999, 1999a) discuss various approaches to the OPF solution. Finally, there were a number of heuristics methods based on the artificial intelligence techniques, such as genetic algorithms, evolutionary programming or other types of biologically inspired techniques that were used to address this problem.

Solution of the OPF yields not only values of the power generation outputs, but also locational prices of electricity that are used in a number of electricity markets. As discussed above application of dynamic prices for smaller customers will require solution of models with a very large number of decision variables, and will need to be carried out frequently (every 30 min or even more often).

4.4 Generation Scheduling

The generation scheduling or Unit Commitment (UC) problem seeks to address the following question: given a load profile (e.g. values of the load for each hour of a day) and a set of units that are available to produce energy, find when should each unit be started, stopped and how much should it generate to meet the load at minimum cost?

$$\min_{\mathbf{P}_g, \mathbf{u}_g} \quad Cost(\mathbf{P}_g, \mathbf{u}_g) = \sum_{t=1}^{T} \sum_{i=1}^{N} u_i(t) \left[C_{gi}(P_{gi}(t)) + STC_i(t) \right] \qquad (11)$$

subject to the following constraints:

> power balance constraint,

$$\sum_{i=1}^{N} u_i(t) P_{gi}(t) = P_d(t) \qquad (12)$$

> upper and lower limits on generation outputs,

$$u_i(t) P_{gi}^{\min} \leq P_{gi}(t) \leq u_i(t) P_{gi}^{\max} \qquad (13)$$

> min-up and–down time constraints,

$$\sigma_{gi}^{on}(t) \geq T_{gi}^{up} \qquad (14)$$

$$\sigma_{gi}^{off}(t) \geq T_{gi}^{down} \qquad (15)$$

Note that in the UC problem, start-up costs are usually modeled in two ways. The first one is to the simple approach in which start-up costs can have only one constant value, β_i,

$$STC_i(t) = \begin{cases} \beta_i & if \quad u(t-1)=0 \quad and \quad u(t)=1 \\ 0 & otherwise \end{cases} \quad (16)$$

The second method is to take into consideration that it is less expensive to start a warm than a cold unit. Thus, depending on the number of hours a unit has been off, its start-up costs can have two values,

$$STC_i(t) = \begin{cases} \alpha_i & if \quad t < \tau_i \\ \beta_i & if \quad t \geq \tau_i \end{cases} \quad (17)$$

From a mathematical viewpoint UC is a mixed-integer combinatorial problem that is, in general, non-linear, however it is often linearized. These type of problems are difficult to solve, and currently there are few techniques that are most often used. They include: (i) Lagrangian relaxation Zhuang and Galiana [1988] and Merlin and Sandrin [1983]; (ii) mixed-integer linear programming Dillon et al. [1978] and Median et al. [1999]; (iii) dynamic programming Hobbs et al. [1988] and (iv) heuristics such as proposed by Lee (1998).

5 Application of High Performance Computing

As mentioned above, it is expected that the new paradigm in power system operation will call for carrying out various power system analyses for systems with large number of variables. Moreover, these larger, and in some cases more complex formulations will need to be solved in shorter time scales, so to allow for more dynamic adjustments of operation that is close to real-time operation. In the past, there were applications of certain improvements, such as utilization of sparse techniques, to accelerate the solution of large systems, however the more widespread availability of powerful multi-core computers brings new possibilities. This subsection explores possible approaches to the utilization of HPC to improve some of the above described power system analysis tools.

5.1 Developments in the Computer Architectures

The application of the High Performance Computing in power systems area is not new. As a review by Falcão [1997] showed, it was regarded as a way forward to improve modeling capabilities of more complex power systems. This was based on the developments in numerical analysis methods that were typically used in power system analysis tools. It was also due to the advances in various processing methods such as parallel and vector techniques that were suitable for the use of the HPC approaches. The main advantage of the HPC was to develop faster and more efficient software. As pointed out by Falcão [1997], a number of computer architectures that emerged enabled these HPC applications. They included:

- Superscalar processors that were single but able to execute concurrently more than one instruction per clock cycle
- Vector processors designed to optimize arithmetic operations of long vectors
- Shared memory multiprocessors where few processors shared and communicated via a global memory
- Single Interaction Stream Multiple Data Stream (SIMD) massively parallel machines that consisted of a large number of simple processors suitable for parallel executions of the same instructions such as data analysis
- Distributed Memory Multicomputer where each processor had its own dedicated memory, and the communication among them was via communication network
- Heterogeneous network of workstations that can be uses as a virtual power machine that use specially developed communication and coordination software

As mentioned above, these different generations of computer architectures correlated with the increase in flops, i.e. computational speed as outlined in DOE [2011]. For example vector machines, such as Cray, led way into gigascale computing which dominated supercomputing in 1980s. Massively parallel processing, which can be compared to an approach of having thousands of computers working together enabled terascale computing. In this case, parallel processing machines break big problems into smaller pieces, each of which is simultaneously solved (in parallel) by a host of processors.

The architecture generation that has followed are current petaflops computers whose architecture includes linking different processors (e.g. enhanced version of the Cell processor used commercially in Sony's Playstation 3 video game console). Today's parallel supercomputers are in this range, like Jaguar which has more than 224,000 processors, and with a theoretical peak speed of 1.75 petaflops is the fastest computer in the United States (as of November 2010). Since October 2010 the fastest computer in the world is the Tianhe-1A (China), with 2.5 petaflops.

The next generation of supercomputers will reach a speed of exaflops, and that will require significant technological breakthroughs due to need to connect a very large number of parallel processors. It is worth noting that the exascale system will not be realized until the 2015–2020 timeframe (DOE [2007]), although, as it will be discussed later, projects looking into the way to tap into this vast computational power have been already initiated ORNL [2007].

Having the computer architecture is only one aspect, with the development of algorithms and software that will harness this power being another important step. The following subsection discusses the current applications of HPC algorithms in power system applications. They rely on the algorithms and approaches that are currently available. However, as it will be discussed in subsection 4.3, researchers in the HPC and numerical algorithm area are looking into the step-change in their approach towards developing more general algorithms and reusable software tools that could be applied to a number of different problems, including those from the power system area.

5.2 Applications of HPC in Power System Analysis and Market Operations

Initial review of applications in power system applications by Falcão [1997] indicated a number of areas where the HPC approach could be potentially useful, and these include:

- *Real-time control* – where a complexity of new models for the energy management, power flow and security assessment are examples of large scale problems whose solutions will be enhanced by the application of HPC methods. Furthermore, security constrained optimal power flow is another example of the problem that is still difficult to solve sufficiently fast so that it can be used for real-time operation. Considering possible new approaches and decentralized operation applied for SmartGrids, parallel algorithms used in HPC seem as good option to address highly intensive computational requirement of this type of the real-time operation problems.
- *Real-time simulation* – often include analysis of system dynamics such as electromechanically and electromagnetic transients. These simulations will not be considered here.
- *Optimization* – under the vertically integrated operation, the problems of economic dispatch, optimal power flow and generation scheduling were solved by the system utility, and tools based on the same underlying power system analysis tools are used by system and market operators to clear the market while ensuring secure operation. As it will be discussed later, this set of problems is not always easy to convert into formulation that is suitable for parallel computers. However, there are techniques that have been developed and are more suitable for the HPC approach than others. Furthermore, as it will be discussed later, optimization, and its application in the area of energy and environment has been recognized as one of the topics that have attracted attention for a future research of new numerical methods and HPC tools.
- *Probabilistic assessment* – this typically included assessment of system reliability. However, with an increase in the application of renewable generation as well as active demand side participation, uncertainties and stochastic behavior will have even larger influence on system operation and decision making of system operators as well as market participants. This, for example, can include more sophisticated models of reserve procurement, or more complex models of generating units, especial those with renewable resources, when deciding on contractual positions or participation in the energy and ancillary services markets. Some of the techniques often used in these analysis are suitable for application of parallel algorithms.

- ***Intelligent Tools for Analysis, Synthesis and Operation*** – these are the tools that seek to include a number of scenarios and evaluate different options. In the case of Agent-Based modeling techniques they also may include a number agents that need to evaluate their own objectives, communicate with the environment (including other agents) and make decisions based on all of the information they have gathered from the dynamic operation of the system. Because of the number of scenarios, or a number of agents, problems that are based on these type of methods are quite suitable for the parallelization.

Power Flow and Contingency Analysis

Although a power flow is one of the most commonly and often used power system analysis tools, currently applied algorithms can solve this highly non-linear problem quite fast for the typical networks of the order of 2,000-3,000 nodes. In addition, parallelization of the power flow problem formulation is difficult. Thus, although there were attempts to apply parallel algorithms suitable for HPC such as proposed by Yang et al. [2005], most attempts seek to simplify the problem by exploiting diagonal form of Jacobian matrix as proposed by Chen et al. [2000] or Flueck et al. [2000]. Additional application of parallelized algorithms is for the simplified linearized formulation of the power flow problem such as Decoupled load flow or DC load flow as investigated in Wu and Bose [1995]. Furthermore, Galiana et al [1994] proposed and analyzed application of preconditioned Conjugate Gradient iterative method for the solution of the linear set of equations. This approach can be used to improve performance of iterative Conjugate Gradient method which can then be considered for parallel implementation. Note that although the solution of the linear equations, which is the main computational task in the power flow problems, is often carried out via direct methods, iterative methods may prove to give better results for the solution of power system problems on parallel processors.

Difficulties mentioned above, as well as the relatively fast solutions of power flow problem without HPC, meant that there were little need and application of the parallel approaches to its solution. However, with an increase in the size of future networks due to an increase in number of nodes, as well as additional complexity brought by utilization of different devices this may change.

The situation with contingency analysis, however, is different as this includes solving a number of power flow problems to evaluate system operation under contingency conditions. Most of the approaches evaluate N-1 security. Some of these distribute different contingencies among different processors (or machines), while others seek to manage allocation of contingencies in different ways. For example, Balduino et al. [2004] evaluate two modes. The first is a synchronous mode which is based on the assumption that the calculation time for all of the contingencies is similar and where tasks are equally distributed among processors so each of them stores cases that it is solving. Balduino et al.[2004] also have evaluated asynchronous mode where each processor is assigned new calculation after it finishes the previous one, while the allocation process is managed centrally. The work also compares two different architectures, (i) parallel virtual machine (PVM) which, as described by Geistet al. (1989), connects different machines or processors to form a virtual machine and (ii) the message passing interface (MPI) which provides a means of communication between different processors (Snir et al. [1996]).

In contrast, work by Huang et al. [2009] seeks to exploit HPC to carry out contingency analysis based on N-x criteria. This significantly increases the number of credible scenarios that need to be evaluated, and the power of the HPC approach enables this type of analysis. Although most systems operate under the assumption of the N-1 contingency, more elaborate N-x criteria makes it possible to account for some cascading failures. This is of a particular interest to analysis related to electricity market trading and calculation of transmission capacity availability which is used for simultaneous feasibility tests associated with auctioning of Financial Transmission Rights. In addition, parallel algorithm for the evaluation of the N-x security dynamic cascading analysis is developed by Khaitan et al [2009]. Their analysis is based on the solution of a large number of differential algebraic equations using algorithms for the parallel architecture.

Optimal Power Flow Analysis

Optimal power flow (OPF) is another important tool used by system and market operators. It is an optimization problem which is not easy to parallelize if classical solution methods are used. However, optimization is at the core of many applications in engineering and science and, therefore there is a great interest in developing good numerical approaches that can exploit HPC. As discussed in the next subsection, this is the reason why HPC and numerical analysis community has set a goal to address this issue.

Notwithstanding the above, several methods to solve the DC OPF problem were proposed. Few utilize the decentralized OPF formulations proposed by Kim et al. [1997] or Conejo [1998] et al. One of the approaches is carried out by Bakirtzis et al. [2003] is based on the decomposition of power systems into regions around tie lines and is based on the KKT conditions of the original large-scale problem in sub-problems that are identical to the KKT conditions of the original problem. The extension of that work carried out in Biskas [2005] reduces the size of the sub-problems and implements it on the network of workstations.

Further HPC approaches that seek to exploit heuristic search optimization techniques such as Genetic Annealing or Simulated Annealing could be well suited for the HPC applications. As discussed above, utilization of techniques proposed by Gondzio [2009] for the Interior Point Methods is also promising.

Finally, Security Constrained OPF (SCOPF), which is the extension of the problem defined in (7) - (9) and includes evaluation of a number of post-contingency configurations, is often addressed via the Benders decomposition techniques. That formulation consists of a master problem representing a base case, and a number of sub-problems that evaluate post-contingency configurations. Due to the separability of the sub-problems, SCOPF is a good candidate for the HPC approach.

Generation Scheduling

As mentioned above, Generation Scheduling, or Unit Commitment (UC) problem is a mixed integer, and thus the most difficult of all here mentioned problems to solve. There were few attempts to apply HPC techniques, although improving solution times of this class of large-scale problems will be very useful. Murillo-Sanchez and Thomas [2000] investigated the application of the Lagrangian Relaxation Algorithm with variable duplication for the thermal unit commitment problem including AC network power flow constraints. Furthermore, heuristic methods often based on the genetic algorithms seem to be good candidates for the HPC applications, and this includes utilization of the parallel Genetic Algorithms investigated by Arroyo and Conejo [2002].

5.3 *Future Developments in Numerical Algorithms for HPC*

Recent developments in this field include coordination of research activities and development of HPC/Numerical Analysis roadmaps. Namely, it was felt by scientist developing new numerical tools that there was a failure of the computational science to develop appropriate software, and that was affecting scientists' ability to tackle large scale and grand challenge applications. The reason behind this worry was that the algorithms underlying current software were not any more able to cope efficiently with the increasingly difficult problems that arise in multiscale, multiphysics applications. To address this problem a project on the Development of the HPC/Numerical Analysis roadmap was instigated in the UK[3], but the workshops included international community as well.

The collaboration and the development of the roadmap was based on the fact that many applications from different areas of research actually share a common numerical algorithmic base (Trefethen [2009]). As discussed in Trefethen [2009, 2008, 2008a and 2009b]. The aim of the roadmap was to capture the elements of that common base and then identify main activities that would help scientific community, as well as industry, recognize and address challenges in the application of the HPC algorithms and software tools in various disciplines. One of the major challenges that the project identified was that there was a need to develop algorithms and software which could be reused by different disciplines. These should be developed in the form of high-quality, high-performance sustained software components, libraries and modules.

Reports from the project also indicate that there are a number of areas where power system problems have a common ground and can rely on universal algorithms and software tools (Trefethen [2008]). These include using direct solvers or optimization packages for scheduling (together with network constraints). In some instances the application may not require any adjustments, while for others it would be worthwhile investigating if some additional considerations need to be taken into account.

---

[3] Details of the project could be found at www.oerc.ox.ac.uk/research/hpc-na

Furthermore, the discussion in subsection 4.2 revealed that there was little coordination in the development of HPC methods for power system applications. This includes approaches towards both the computer architecture as well as numerical analysis algorithms.

Besides the above roadmap project in the UK, researchers in the US were invited to identify computational science requirements for future exascale systems (ORNL [2007]). These are the next generation of computers capable of an exaflop, or 10^{18} floating point operations per second. As mentioned above, the exascale system will not be realized until the 2015–2020 timeframe. However, there are a number of challenges associated with the application of exascale that need to be addressed and the preparations in that directions have started (DOE [2007] and ORNL [2007[). One of the major issues is a major increase – a huge increase – in parallelism. Exascale programs must harness 1-billion-way parallelism, as opposed to current 250,000-way parallelism. The task of coordinating such a large number of processors may present a significant challenge.

The findings in ORNL (2007) were a result of a number of workshops organized by the US government, and, among other areas, it included looking into how this new computational power can be used for solving problems that may be also useful for power system analysis. This includes new algorithms for optimization, implicit non-linear systems, as well as agent based methods (ORNL [2007].

6 Concluding Remarks

The chapter summarized some of the applications and developments in the area of the HPC application for power system analysis tools. It outlined the major changes that were expected in the operation of future networks and the reasons why new, faster power system analysis tools were needed. As indicated, this is driven by the deployment of Smart Grid technology, penetration of renewable resources, especially on the distribution level, active demand participation and the plans for electric vehicles.

This chapter mainly concentrated on power flow, optimal power flow, economic dispatch and generation scheduling problems, and looked how HPC has been applied that are not commonly associated with the solution approaches that are easy to parallelize. Despite these obstacles, there were a number of methods that were developed. However, as power system tools use numerical methods which are common for other disciplines they seem to share the same problem of the relatively low level of the systemized approach when developing new approaches and software tools. Thus, it can be expected that the power system community will benefit from the above discussed recent developments is the HPC/Numerical Analysis which recognized a need to develop a closer collaboration and a research roadmap. One of the major challenges that this roadmap project identified was that there was a need to develop algorithms and software which can be reused by different disciplines. These should be developed in the form of high-quality, high-performance sustained software components, libraries and modules. Utilization of these will help power system community develop next generation of power system analysis tools that can help operation, management and planning of future power systems and electricity markets.

References

Arroyo, J.M., Conejo, A.: A parallel repair genetic algorithm to solve the unit commitment problem. IEEE Transactions on Power Systems 17(4), 1216–1224 (2002)

Bakirtzis, A.G., Biskas, P.N.: A decentralized solution to the DC-OPF of interconnected power systems. IEEE Transactions on Power Systems 18, 1007–1013 (2003)

Biskas, P., Bakirtzis, A., Macheras, N., Pasialis, N.: A decentralized imple-mentation of dc optimal power flow on a network of computers. IEEE Transactions on Power Systems 20(1), 25–33 (2005)

Balduino, L., Alves, A.C.: Parallel processing in a cluster of microcomputers with application in contingency analysis. In: 2004 IEEE/PES Transmission and Distribution Conference and Exposition: Latin America, 2004 IEEE/PES, pp. 285–290 (2004)

Caramanis, M.C., Bohn, R.E., Schweppe, F.C.: Optimal Spot Pricing: Practice and Theory. IEEE Transactions on Power Apparatus and Systems 101(9), 3234–3245 (1982)

Carpentier, J.: Optimal power flow. International Journal of Electrical Power and Energy Systems 1, 3–15 (1979)

Chen, S.-D., Chen, J.-F.: Fast load flow using multiprocessors. International Journal of Electrical Power Energy Systems 22(4), 231–236 (2000)

Conejo, A.J., Aguado, J.A.: Multi-area coordinated decentralized dc optimal power flow. IEEE Transactions on Power Systems 13, 1272–1278 (1998)

Dillon, T.S., Edwin, K.W., Kochs, H.D., Tand, R.J.: Integer programming approach to the problem of optimal unit commitment with probabilistic reserve determina-tion. IEEE Transactions on Power Apparatus and Systems PAS-97(6), 2154–2166 (1978)

DOE Modeling Simulation at the Exascale for Energy and the Environment (Town Hall meeting (2007), http://www.er.doe.gov/ascr/ProgramDocuments/Docs/TownHall.pdf

DOE, Leap to the extreme scale could break science boundaries (2011), http://ascr-discovery.science.doe.gov/feature/exa_ov1.shtml

Dongarra, J.J.: Performance of Various Computers Using Standard Linear Equations Software (Linpack Benchmark Report) (2009)

Falcao, D.M.: High Performance Computing in Power System Applications. Selected papers from the Second International Conference on Vector and Parallel Processing, pp. 1–23 (1997)

Flueck, A.J., Chiang, H.-D.: Solving the nonlinear power flow equations with an inexact Newton method using GMRES. IEEE Transactions on Power Systems 13(2), 267–273 (1998)

Galiana, F.D., Javidi, H., McFee, S.: On the application of a pre-conditioned conjugate gradient algorithm to power network analysis. IEEE Transactions on Power Systems 9(2), 629–636 (1994)

Geist., A., Dongarra., J., Hang, W., Sunderam, V.: PVM: Parallel Virtual Machine, an user's guide tutorial for networked parallel computing. London: Library of congress cataloging-in-publication data (1989)

Gondzio, J., Grothey, A.: Exploiting Structure in Parallel Implementation of Interior Point Methods for Optimization. Computational Management Science 6, 135–160 (2009)

Hobbs, W.J., Hermon, G., Warner, S., Sheblé, G.B.: An enhanced dynamic programming approach for unit commitment. IEEE Transactions on Power Systems 3, 1201–1205 (1998)

Huang, Z., Chen, Y., Nieplocha, J.: Massive contingency analysis with high performance computing. In: IEEE Power & Energy Society General Meeting, pp. 1–8. IEEE (2009)

Huneault, M., Galiana, F.D.: A survey of the optimal power flow literature. IEEE Transactions on Power Systems 6(2), 762–770 (1991)

S.K. Khaitan, Fu, C., McCalley, J.: Fast parallelized algorithms for on-line extended-term dynamic cascading analysis. In: IEEE Power Systems Conference and Exposition, March 15-18 (2009)

Kim, B.H., Baldick, R.: Coarse-grained distributed optimal power flow. IEEE Transactions on Power Systems 12, 932–939 (1997)

Lee, F.N.: Short-term thermal unit commitment—A new method. IEEE Transactions on Power Systems 3, 421–428 (1998)

Momoh, J.A., Adapa, R., El-Hawary, M.E.: A review of selected optimal power flow literature to 1993 — I: Nonlinear and quadratic programming approaches. IEEE Transactions on Power Systems 14(1), 96–104 (1999)

Momoh, J.A., El-Hawary, Adapa, R.: A review of selected optimal power flow literature to 1993 — II: Newton, linear programming and interior point methods. IEEE Transactions on Power Systems 14(1), 105–111 (1999a)

Medina, J., Quintana, V.H., Conejo, A.J.: A clipping-off interior-point technique for medium-term hydro-thermal coordination. IEEE Transactions on Power Systems 14, 266–273 (1999)

Merlin, A., Sandrin, P.: A new method for unit commitment at Electricité de France. IEEE Transactions on Power Apparatus and Systems PAS-102(5), 1218–1225 (1983)

Murillo-Sanchez, C., Thomas, R.: Parallel processing implementation of the unit commitment problem with full ac power flow constraints. In: Proceedings of the Annual Hawaii International Conference on System Sciences, p. 9 (January 2000)

ORNL, Scientific Application Requirements for Leadership Computing at the Exascale, Report ORNL/TM-2007/238 (2007), http://www.nccs.gov/wp-content/media/nccs_reports/Exascale_Reqms.pdf

Schweppe, F.C., Caramanis, M.C., Tabors, R.D., Bohn, R.E.: Spot Pricing of Electricity. Kluwer Academic Publishers (1988)

Snir, S., Otto, S., Huss-ledetman, S., Walker, D., Dongarra, J.: MPI: the complete reference. Library of Congress catalogin-in- publication Data, London (1996)

Quintana, V., Torres, G.L., Medina-Palomo, J.: Interior-Point Methods and Their Applications to Power Systems: A Classification of Publications and Software Codes. IEEE Transactions on Power Systems 15(1), 176–179 (2000)

Stott, B.: Review of Load-Flow Calculation Methods. Proceedings of IEEE 62(7), 916–929 (1974)

Yang, F., He, M., Tang, Y., Rao, M.: Inexact block Newton methods for solving nonlinear equations. Applied Mathematics and Computation 162(3), 1207–1218 (2005)

Wu, J.Q., Bose, A.: Parallel solution of large sparse matrix equations and pa-rallel power flow. IEEE Transactions on Power Systems 10(3), 1343–1349 (1995)

Zhuang, F., Galiana, F.D.: Towards a more rigorous and practical unit commitment by Lagrange relaxation. IEEE Transactions on Power Systems 3, 763–773 (1988)

Author Index

Borges, Carmen L.T.   1

Chassin, David P.   281
Chavarría-Miranda, Daniel   151
Chen, Yousu   151
Chiang, Hsiao-Dong   335
Cossent, Rafael   247

Falcao, Djalma M.   1
Frías, Pablo   247

Gómez, Tomás   247

Huang, Zhenyu   151

Johnson, Jeremy   211

Kamiabad, Amirhassan Asgari   229
Khaitan, Siddhartha Kumar   43, 189
Kockar, Ivana   359

Li, Hua   335

Mateo, Carlos   247
McCalley, James D.   43, 189

Nagvajara, Prawat   211
Nwankpa, Chika   211

Sánchez, Álvaro   247

Tada, Yasuyuki   335
Taranto, Glauco N.   1
Tate, Joseph Euzebe   229
Tong, Jianzhong   335

Wu, Wenchuan   101

Zhang, Boming   101
Zhao, Chuanlin   101
Zhou, Mike   71

Printed by Publishers' Graphics LLC